Hydraulic Engineering: Computation, Analysis and Modeling

Hydraulic Engineering: Computation, Analysis and Modeling

Edited by Roman Morris

SYRAWOOD
PUBLISHING HOUSE
New York

Published by Syrawood Publishing House,
750 Third Avenue, 9th Floor,
New York, NY 10017, USA
www.syrawoodpublishinghouse.com

Hydraulic Engineering: Computation, Analysis and Modeling
Edited by Roman Morris

International Standard Book Number: 978-1-64740-426-0 (Hardback)

Cataloging-in-publication Data

Hydraulic engineering : computation, analysis and modeling / edited by Roman Morris.
p. cm.
Includes bibliographical references and index.
ISBN 978-1-64740-426-0
1. Hydraulic engineering. 2. Hydraulics. I. Morris, Roman.
TC145 .H93 2023
627--dc23

TABLE OF CONTENTS

PREFACE

Hydraulic engineering is a branch of civil engineering that deals with the flow of fluid, typically water and sewage, through conduits and makes use of gravity for the movement of fluid. It involves designing structures with the capacity to remove or divert water from the roadways and pass the collected water from under the roadway. Hydraulic engineering is also concerned with the technical challenges involved in sewerage design and water infrastructure. It is used for creating bridges, sewers, dams, canals, etc. The computational aspects of hydraulics with respect to civil engineering problems are dealt with under the discipline of computational hydraulics, which contains methods and techniques for numerical simulation of water flows in natural or manmade systems with the aid of computers. In such systems, flow and transport is modeled using computer tools such as computer graphics, statistical analysis methods, electronic databases, and spreadsheets. This book contains some path-breaking studies related to computation, analysis, and modeling within hydraulic engineering. It will serve as a valuable source of reference for graduate and postgraduate students.

This book is the end result of constructive efforts and intensive research done by experts in this field. The aim of this book is to enlighten the readers with recent information in this area of research. The information provided in this profound book would serve as a valuable reference to students and researchers in this field.

At the end, I would like to thank all the authors for devoting their precious time and providing their valuable contribution to this book. I would also like to express my gratitude to my fellow colleagues who encouraged me throughout the process.

Editor

1

Hydraulic Engineering at 100 BC-AD 300 Nabataean Petra

Charles R. Ortloff [1,2]

[1] CFD Consultants International, 18310 Southview Avenue, Los Gatos, CA 95033, USA; ortloff5@aol.com
[2] Research Associate in Anthropology, University of Chicago, Chicago, IL 60637, USA

Abstract: The principal water supply and distribution systems of the World Heritage site of Petra in Jordan were analyzed to bring forward water engineering details not previously known in the archaeological literature. The three main water supply pipeline systems sourced by springs and reservoirs (the Siq, Ain Braq, and Wadi Mataha pipeline systems) were analyzed for their different pipeline design philosophies that reflect different geophysical landscape challenges to provide water supplies to different parts of urban Petra. The Siq pipeline system's unique technical design reflects use of partial flow in consecutives sections of the main pipeline to support partial critical flow in each section that reduce pipeline leakage and produce the maximum flow rate the Siq pipeline can transport. An Ain Braq pipeline branch demonstrated a new hydraulic engineering discovery not previously reported in the literature in the form of an offshoot pipeline segment leading to a water collection basin adjacent to and connected to the main water supply line. This design eliminates upstream water surges arising from downstream flow instabilities in the two steep pipelines leading to a residential sector of Petra. The Wadi Mataha pipeline system is constructed at the critical angle to support the maximum flow rate from a reservoir. The analyses presented for these water supply and distribution systems brought forward aspects of the Petra urban water supply system not previously known, revising our understanding of Nabataean water engineers' engineering knowledge.

Keywords: Petra; Nabataean; water systems; hydraulic analysis; CFD; canals; reservoirs; pipelines; flow stability

---

1. Introduction

The history of Petra's monumental architecture and historical development has been described by many authors [1–6]. Some scholars have concentrated on technical and location aspects of water supply and distribution systems within Petra [7–21], while other surveys [16,22–27] have concentrated on the water control and distribution technology available to Nabataean water engineers from Roman and other eastern and western civilizations through trade and information transfer contacts during Petra's expansion period (100 BC–AD 300) period. This paper was designed to add further depth to the hydraulic engineering technology used in the design and operation of Petra's three major pipeline water supply systems serving the urban center of Petra: The Siq system sourced from the Ain Mousa spring, the Ain Braq system, and the Wadi Mousa system.

Petra's openness to foreign influence is demonstrated in the city's monumental architecture that reflect elements of Greek, Persian, Roman, and Egyptian architectural styles integrated into Nabataean monuments [1–6,28,29]. Later Roman occupation of Petra past 106 AD exhibits Roman pipeline technologies employed to expand the marketplace, the Paradeisos Pool Complex [8,9], and city precincts responding to increased water demands for an expanding population as the city's status advanced as a key trade and emporium center. Petra's ability to manage scarce water resources to provide constant potable water supply for its permanent population of ~20,000 citizens with reserves for large caravan arrivals and drought periods was vital to its centuries of prominence as the nexus of a trade network between African, Asian, and European cities for luxury goods. While knowledge of hydraulic technologies from foreign sources was available, the rugged mountainous terrain, distant spring water sources, and brief rainy periods posed unique water supply challenges that required technical innovations consistent with the site's ecological and geophysical constraints. Effective utilization of scarce and seasonally intermittent water supplies required technical expertise to ensure optimum system functionality from the use of distant interconnected spring-supplied pipeline transport and distribution systems (Figure 1) that guided water to the city's densely populated urban core and surrounding agricultural districts.

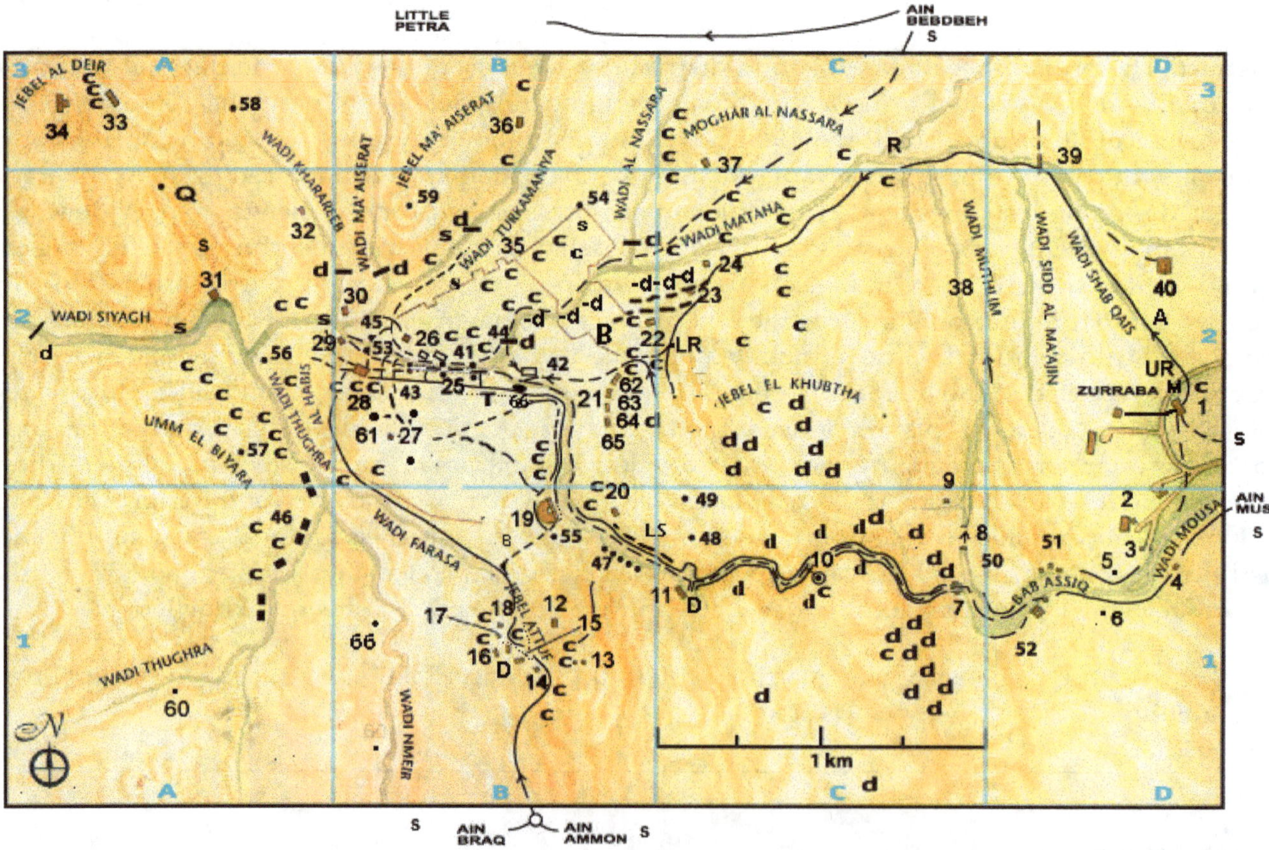

Figure 1. Site map of Petra: Site names corresponding to numbers given in the Appendix A.

Examination of three of Petra's major water conveyance pipeline systems using computational fluid dynamics (CFD) analysis permitted discovery of the rationale behind design selections utilized by Nabataean hydraulic engineers. Solution of fluid dynamics equations using CFD finite-difference methodology together with models of existing pipeline structures [30] within Petra's urban center graphically demonstrate internal water flow patterns within pipelines that revealed damaging hydraulic phenomena that would be observationally familiar to Nabataean engineers from past trial-and-error experience. What the CFD solutions revealed were the problems and their solutions used to guide their pipeline design/selection progress to avoid system failures. CFD results demonstrating flow patterns within pipelines showed what Nabataean hydraulic engineers intended (or avoided) in their hydraulic designs and give insight into their civil engineering knowledge base. Given the low survival rate of documents and descriptions of hydraulic phenomena from ancient authors and from the descriptions of water engineering phenomena from the few surviving ancient documents, the descriptions are given in prescientific terms with little correlation to modern terminology to understand the ancient author's meanings of hydraulic phenomena. Given this lack of descriptive material to interpret the water engineering base underlying their water system designs and functions, the analysis of internal flow patterns within pipelines using CFD methodology presents a viable recourse to calculate, visualize, and analyze hydraulic engineering problems Nabataean hydraulic engineers encountered and solved in their piping network designs. Within the archaeological remains of Petra's pipeline systems lies insight into Nabataean technical processes and the knowledge base available to their water engineers that served as the foundation for their engineering decision making. Use of CFD analysis provided graphic displays of problems encountered and solutions developed by Nabataean water engineers and helped to understand what lay behind field observations of archaeological remains relevant to Petra's water supply and distribution systems.

Additionally, descriptive terms from ancient texts describing hydraulic phenomena given in prescientific terms unfamiliar to western notations can now be associated with actual phenomena through CFD modeling of the flow patterns through water conveyance structures.

2. The Water Infrastructure of Petra

To begin the discussion of Petra's water system development and progress toward utilizing all possible water resources to meet population requirements and the greatly increased water demands from arriving caravans (which according to the ancient author Strabo, consisted of many hundreds of camels and hundreds of accompanying support personnel), the solid and dashed lines of Figure 1 detail the known water supply and distribution pipeline systems leading water into the city's urban core. Numbered locations denote major site features listed in the Appendix A. Shown are major and minor catchment dams and multilevel stepped dams (-d, d, D), cisterns (c), water distribution tanks (T), and springs (s). The (-d) dams located across streambeds are stone barrier structures built to hold large quantities of rainfall runoff water for redistribution to urban structures and local agricultural areas. Dams denoted (d) are minor catchment structures that stored water in mountainous areas to limit descent and flooding of Petra's low altitude urban center and have channels leading water to cisterns. Solid lines represent original pipeline paths originating from distant springs which in many cases now show only barren channels that once contained interlocked terracotta piping elements. The superimposed grid system (A, B, C; 1, 2, 3) defines 1.0-km^2 grid boxes to enable the location of numbered Appendix A features. Figure 1 represents the present state of knowledge of the location of pipeline structures from personal field exploration and survey results from previous site explorers [7–16,18,21]. Due to erosion and soil deposition landscape degradations over ~2000 years, as well as reuse and pillaging of piping elements by later inhabitants of the area, many terracotta pipeline sections are now missing from their original holding channels, obscured by subterranean placement, or are yet to be discovered. Extrapolation of pipeline continuance between available visible pipeline segments then represents the pipeline paths shown in Figure 1 and is, at best, a first approximation to the entirety of the total pipeline system as much of the site remains unexcavated and subject to landscape change from millennia of soil erosion and deposition events caused by flood episodes, landslide events, and frequent earthquakes. Further hidden subterranean pipelines buried for defensive security purposes may exist to supplement the known pipelines shown in Figure 1. Recent on-site discoveries appear to add new information in this regard: For example, Site 66 in Figure 1 represents a large newly discovered (in 2016) platform with a centrally located temple structure yet to be excavated to determine its role in Petra's administrative and ceremonial life [28]. The water supply system serving this newly discovered complex remains to be discovered but surely exists given that all ceremonial structures presently known are accompanied by elaborate water systems for both aesthetic display and supply of potable water.

Additionally, as elements of major pipelines were intentionally built subterranean and as yet undiscovered due to limited excavations performed in the urban center and distant reaches of Petra, Figure 1 can be considered a first approximation of the total piping network. Some major monuments located at high altitude exclude the possibility of canal water supplies from springs located at lower levels; in this case, channels directing rainfall capture to cisterns located nearby provided the water supply to high-level sites.

Petra's urban core lies in a valley surrounded by rugged mountainous terrain. Figure 2 demonstrates a section of the mountain terrain behind the ~10 m-high Q'asar al Bint temple (Feature 29, Appendix A) located at the base of the Jebel al Deir mountain and gives a perspective on the rugged mountainous terrain encompassing the urban core of Petra. Seasonal rainfall runoff passes into the valley through many streambeds and was drained out primarily through the Wadi Siyagh streambed in antiquity (Figure 1, A-2). Although present measurement of the Wadi Siyagh streambed slopes now precludes drainage resulting from water-transferred sediment deposits over millennia, this path was the main drainage channel during the city's existence. Water flow into the city [10] originated from captured rainfall runoff into cisterns from Wadi al Hay to the north, Wadi al Hudayb to the south, and from the watershed area supplying runoff into the Wadi Mousa streambed. Surface flows from these sources passed through the Siq (Figure 1, D-1 to C-1; Appendix A Feature 10) which is the ~2 km long narrow entry passageway into the urban center of Petra (Figures 1 and 2A). The Siq is a natural, narrow passageway through the Jebel el Khubtha mountain range, varying from 5 to 10 m in width; the north and south walls of the Siq are near vertical with maximum heights up to 80 m.

By construction of a diversion dam and tunnel shown in Figure 3 (Figure 1; Feature 8 in Appendix A) in the first century AD, flood water from the Wadi Mousa River was deflected from the Siq entrance and led to Wadi Mataha (Figure 1, C-2, 3) that linked up with the section of the Wadi Mousa streambed within the city center; water then drained into Wadi Siyagh (Figure 1, A-2) through a channel passing through the city center. This dam and tunnel construction served to divert the silt-laden floodwater away from entry into the Siq that would compromise the structural integrity of many of the monuments within Petra's urban core region. At present, the Wadi Siyagh drainage path is partially blocked from silt deposits, causing flooding of Petra's urban center during the rainy season. As the chemical composition of Petra's stone monuments contain different salt varieties [31], exposure to flood water causes destructive flaking from monument walls; this effect is apparent from observation of many of Petra's urban sector monuments.

Runoff water from the southern sector Wadi Thughra, Nmeir, Farasa, and northern sector Wadi Kharareeb, Ma'aiserat, and Turkamanya gullies (Figure 1) additionally drained rainfall runoff into the Wadi Siyagh away from the city center [10]. In sector B-2 of Figure 1, a local depressed low area lies between the Jebel el Khubtha Mountain and sites to the west. Currently, a modern bridge crosses over the southern narrow end of this area. If the southern narrow drainage end of this basin was dammed in antiquity, rain water or a Wadi Mataha diversion channel could produce an internal shallow lake within city confines that would serve local gardening, nearby ceramic and metal working workshop areas as well as city aesthetic beautification purposes consistent with Nabataean plans to utilize all available water before disposal into the Wadi Siyagh drainage. As mentioned in Strabo ([32]), " . . . the inside parts (of Petra) having springs both for domestic purposes and for watering gardens . . . ", which would be consistent with an internal shallow lake within city precincts whose borders and groundwater could serve as gardening areas and for specialty crops. At this stage of research, this feature awaits verification for its existence in ancient times.

(**A**)

(**B**)

Figure 2. (**A**) The Q'asar al Bint Temple at the base of Jebel al Deir Mountain. (**B**) Narrow part of the Siq passageway; note the figure located on the passageway for scale of the height of Siq near vertical walls.

Figure 3. Nabataean drainage tunnel leading water away from the Siq entry to the Wadi Sayagh through channels extending from and parallel to the Wadi Mataha streambed (Figure 1, solid line path through 2-D, 3-C, 2-C sectors).

Spring-fed water supply to the city before the first century BC was through a slab-covered, open channel dug into the floor of the Siq, conducting water from the distant Ain Mousa spring into the urban center of the city. A trace of this system in the vicinity of the south Nymphaeum along Colonnade Street may exist from GPS data [33]. This channel perhaps extended as far as the Temenos Gate (Figure 1, B-2; Feature 43 in Appendix A) later built in early Roman occupation times past AD 106 to honor the visit of Hadrian in 131 AD. First century BC and first century AD Nabataean builders replaced this early Siq channel water supply system with a pipeline system along the north wall of the Siq supplied by a ~14-km pipeline carrying water from the Ain Mousa spring. Details of most of this pipeline system from the Ain Mousa spring are largely unknown due to later urban construction overlay except for a pipeline segment before the Siq entrance. First century BC and early century AD construction revealed a more integrated approach to site water system management as demonstrated by new and novel features. The major Siq northern wall channel supporting a pipeline (Figure 4) was an additional feature eliminating earlier use of the Siq floor channel (Figure 1; B-C-D-1 sectors). New features included surface cisterns to capture rainfall runoff, deep underground cisterns (one located in the eastern part of the site, Appendix A Feature 28, Figure 1), multiple pipeline systems sourced by different springs and storage reservoirs, floodwater control through diversion dams and tunnels, and pipeline supply system redundancy to ensure water delivery from multiple spring and high-level reservoirs—notably the massive Zurraba reservoir that supplied lower level smaller distribution and water storage reservoirs such as the M reservoir (Figure 1, D-2 and Figure 5B. The Figure 5B M reservoir no longer exists due to construction expansion of the nearby town and may have been one of several other distribution reservoirs sourced by networks of canals from the Wadi Mousa spring.

Figure 5A, Appendix A Feature 1, shows the high-level Zurraba reservoir supplied by a branch channel from the Ain Mousa spring that supplied water to high population concentration parts of the city. This reservoir also served as a drought remediation measure and served to provide large water supply storage and delivery for large caravan arrivals into the city. The lower level M reservoir (Figure 5B, Figure 1, 2-D) was connected to the upper level Zurraba reservoir through a pipeline; this lower reservoir provided water for occupation zones close to the Siq entrance. Due to modern road building and urban construction obliterating many of the ancient pipeline connections between reservoirs, the totality of the reservoir network pipeline system sourced by the Wadi Mousa spring remains to be determined.

Major pipelines, primarily on the north wall of the Siq, provided sediment particle filtration and removal basins (Figure 6) and served a sophisticated hydraulic function particular to partial pipeline flow as described in detail in a later section. The Siq south-side water open channel, perhaps sourced by a south-side spring, likely served as the water supply for camels and horses accompanying personnel passing through the Siq. The north-side Siq pipeline had four open basins supplying potable water for personal use (typical of Figure 6); this feature was an integral part of the hydraulic design of the pipeline as discussed in detail in a later section.

Figure 4. Channel cut to support pipeline water supply to the city center along the north wall of the Siq passageway. Pipeline continuation extends to the Outer Siq and further on to the city center.

(A)

Figure 5. *Cont.*

(B)

Figure 5. (**A**) Upper level Zurraba reservoir (Figure 1, 2D), (**B**) Lower-level M reservoir.

Figure 6. Typical open basin between segments of the Siq pipeline.

The new hydraulic features developed at later times reflected the need to bring high-quality potable water into the city center and serve city hillside occupation zones above the valley floor. Late first century BC and early first century AD developments demonstrated continual evolution of the Petra water system and reflected application of acquired technologies from exterior sources integrated with indigenous hydraulic engineering innovations to respond to the water needs of the city as a sustainable population center. Contingency water needs resulting from caravan arrival and water restocking anticipating transit to distant cities in North Africa and Mediterranean cities were vital considerations to maintain Petra's importance as a vital trade emporium. Beyond serving as a transit center for foreign goods to distant areas through known trade routes, the Nabataeans had a monopoly on incense trade originating from Southern Arabia—a vital and profitable product that the ancient world valued for ceremonial functions.

In summary, the means to capture and store a fraction of rainfall runoff through dams and cisterns, to build flood control systems, and build pipelines and channels to deliver water from distant springs to provide a constant water supply to the city was vital to understand Nabataean contributions to hydraulic engineering and water management practice. While water storage was vital to the city's survival, springs internal and external to the city (Ain Mousa, Ain Umm Sar'ab, Ain Braq, Ain Dibdiba, Ain Ammon, Ain Beidha, and Ain Dibidbeh) supplied water channeled and/or piped into the city to provide the main water supply (Figure 1). Several spring-supplied reservoirs exterior to the city (M–Zurraba, Figure 5A,B) provided additional water storage reserves. No accounts of other lower level, near city reservoirs exist at present due to modern urban construction; thus the complete reservoir water storage system remains unknown. Additional on-demand use of reservoir water for industrial ceramic and metal working workshop areas located at the base of the Jebel el Khubtha Mountain and for ceremonial use to supply triclinium rituals for tomb celebratory functions are indicated in [4,10,12]. Of the many springs used in antiquity, the Wadi Siyagh spring near the quarry (Figure 1, Appendix A Feature 31) remains functional for local inhabitants living in remote sections of the ancient city together with the Ain Mousa spring which now provides water for the modern town of Wadi Mousa and the water display structure in the town center. Water taken from the open channel within this structure has ritual use as evidenced by members of holy sects collecting water in containers for ceremonial functions.

3. Details of the Siq Pipeline System

The main city water supply of ancient Petra originated from the Ain Mousa spring about 7 km east of the town of Wadi Mousa (Figure 1, D-1), combined with the waters from the minor Ain Umm Sar'ab spring. The flood bypass tunnel (Figure 3, Appendix A Feature 8) at the Siq entrance, together with an entry dam and elevated paving of the Siq floor, reduced flooding into the urban center from rainy season runoff into the Wadi Mousa River. Part of the early slab-covered channel water supply to the urban core of Petra now lies under hexagon slab pavement in front of the Treasury construction attributed to Aretas IV (9 BC–AD 40). Recent excavations of this area revealed an empty tomb structure below the front part of the Treasury (Figure 7), although the intended use remains conjectural.

Later construction of the early Nabataean Siq water channel was replaced by the Siq pipeline water supply system supplied from the Ain Mousa spring source. The four open surface water basins between Sig piping segments (Bellwald, principal Siq excavator, personal communication) typical of those shown in Figure 6, give indication as to the hydraulic technology available to Nabataean water engineers as further discussion indicates. It is of note that in several locations along the ~2-km length of the Siq pipeline, top portions of the pipeline have been opened up, revealing heavy sinter deposits on pipeline bottoms that taper off about halfway up the pipeline side walls. This observation would indicate that a partial flow existed in at least some portions of the pipeline by observation of the sinter deposit patterns. If a partial flow in Siq pipeline sections was the original design intent of Nabataean water engineers, then opening up the top part of the pipelines made no difference in the pipeline flow rate and may have had the additional benefit of flow observance and access to cleaning sinter from pipeline inner surfaces. Additionally, if partial flow in the Siq pipeline was the design intent, then the Siq pipeline system was not full-flow pressurized as an air space existed over the partial flow, thus reducing the possibility of pressurized water leakage at pipeline joints. Details of the design intent of the Siq pipeline based upon these preliminary observations are next considered to proceed toward hydraulic analysis of the Siq pipeline and its Outer Siq continuation to the city center.

Figure 7. The Treasury, located at the Siq exit.

With increasing water needs for increasing population in early centuries AD, an elevated ~1-km long, north-side pipeline extension (Figure 8; LS, Outer Siq, Figure 1 1-B, C) extended the utility of the Siq pipeline system [14–16,25,27] to supply potable water to further reaches of Petra's urban core. This pipeline extension provided water supply to structures located in the Figure 1 B-2 district, as well as providing pipeline water supply to the Nymphaeum (Appendix A Feature 42, Figure 1). Further pipelines emanating from the Nymphaeum to the Temple of the Winged Lions (Figure 9), major tombs fronting the Wadi Mataha housing structures, and elite palace structures in early stages of excavation [18,19] existed, although details await further excavation. Continuance of water flow through a pipeline to a bridge across the Wadi Mataha led water to the south side of the site as further discussion details.

Figure 8. Elevated Outer Siq pipelines located ~6 m above the current site surface. These pipelines lead to an extension of the Siq pipeline to the city center. Several meters below the current site surface is the original paved floor of the area in front of the Treasury.

Figure 9. Temple of the Winged Lions [4].

To fully analyze the design intent of the Siq pipeline, note first that within pipelines, destructive hydraulic instabilities may exist [34]. These include transient pressure waves, flow intermittency, internal pipeline hydraulic jumps, transient turbulent drag amplification zones, partial vacuum regions, and subcritical to supercritical (and vice versa) flow transitions—these instabilities largely depend upon pipeline slope and internal pipeline wall roughness—all of which affect flow stability and pipeline flow rate. Such hydraulic problems affecting the design of pipeline systems required development of advanced technologies to produce stable pipeline flows that matched a sizeable fraction of the Ain Mousa spring source input. The total pipeline design from the Ain Mouse spring source to the Siq and through the Outer Siq needed to be designed to produce the maximum flow rate the total pipeline system could sustain. Further design considerations involved limiting full flow conditions within pipelines (pipeline cross-section fully occupied by water) to limit hydrostatic pressure within a pipeline. This design consideration limited leakage at the thousands of terracotta pipe-joint-connections along long lengths of pipelines. For a typical pipe element length of ~0.35 m and total pipeline length over ~14 km from the Ain Mousa spring origin location to the end of the Siq and Outer Siq pipeline with a total height drop well over ~50 m (data derived from the contour *Map of Petra*), high hydrostatic pressure in a full flow pipeline would increase the likelihood of pipeline element connection joint leakage at~42,000 pipeline joint connections. Solutions to mitigate many of the design and operational problems provide insight into Nabataean water management expertise.

CFD calculation [30] of full-flow volumetric flow rates in 14 cm diameter(D) piping were made for internal wall roughness $\varepsilon/D > 0.01$ (where ε is the mean-square internal pipeline roughness height); this ε/D value holds for roughness typical of terracotta internal pipeline walls, demonstrating casting corrugations and erosion usage pitting [35] for typical Reynolds number Re ~10^5 values. For pipelines at 2-, 4-, and 6-degree declination slopes for 1.0 and 3.0 m supply head, Figure 10 reveals that past ~400 m pipeline length, an increase of input head does not substantially increase the flow rate [17]. Past this length range, even a 3× head increase does not substantially increase flow rate. The implication for the ~14-km long Ain Mousa spring-supplied pipeline to the Siq and Outer Siq extension is that sophisticated hydraulic engineering had to be in place to match the estimated [12,13] maximum Ain Mousa spring output of~1000–4000 m^3/day; (~0.01–0.04 m^3/s) through the Siq pipeline without invoking any of the flow instability problems previously alluded to. Flow transmitted through the Siq pipeline at the maximum Ain Mousa output flow rate clearly required knowledge of flow conditions in the pipeline before the Siq entrance (which is likely full flow due to internal pipeline wall roughness flow resistance considerations and low pipeline slope) coupled with knowledge of Siq and Outer Siq pipeline designs that mitigate flow resistance and flow instabilities that would propagate upstream into the subcritical supply flow pipeline.

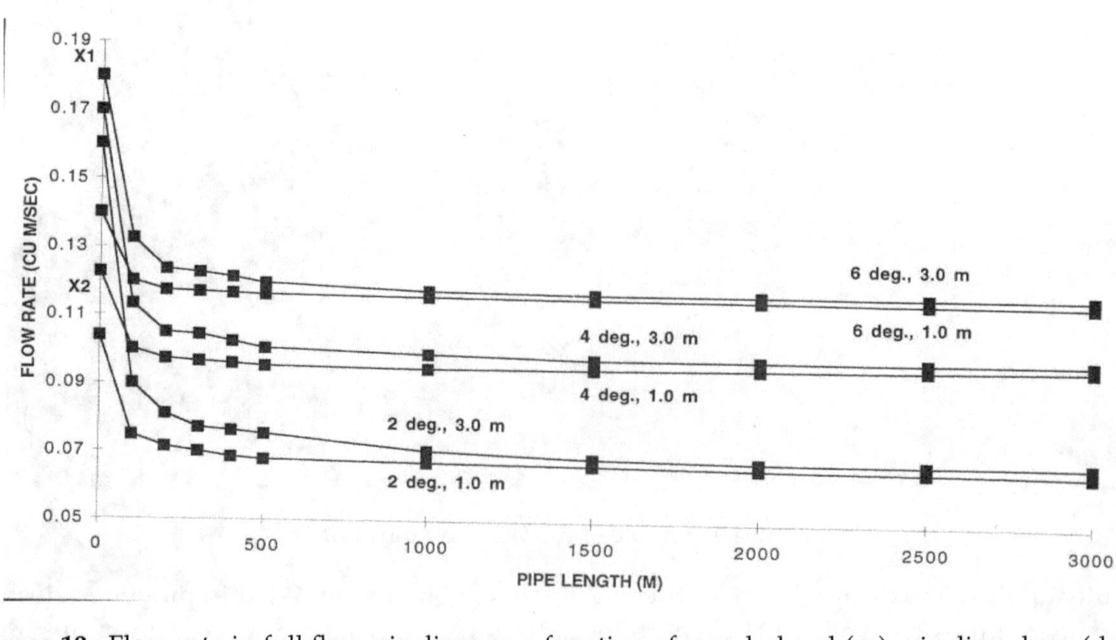

Figure 10. Flow rate in full flow pipelines as a function of supply head (m), pipeline slope (deg.), and pipeline length (M) for a given ε/D value.

The internal pipeline flow rates shown in Figure 10 are based upon a low pipeline wall roughness of ε/D ~0.01; higher ε/D ~0.07 values that account for pipe element connection socket roughness reduce flow rates substantially, leading to Figure 10 curves being shifted downward by a factor of ~3 for typical Re ~10^5 Reynolds number values. On this basis, the observed Siq rough internal wall pipeline at a 2-degree slope permitted a 0.023 m^3/s flow rate and, at a one-degree pipeline slope, permitted a 0.017 m^3/s flow rate corresponding to ~1.5 and ~1.1 m/s mean velocity in the 14-cm internal diameter pipeline with corresponding flow rates of ~80 (0.02 m^3/s) and ~60 m^3/h (~0.02 m^3/s) compared to the Ain Mousa flow rate of 0.01 to 0.04 m^3/s. Thus, even high head values from the Ain Mousa spring to the Siq entrance pipeline had little effect of increasing the flow rate given the long 14-km supply pipeline as Figure 10 implies. The additional ~2-km Outer Siq pipeline lengths to the 14-km Ain Mousa spring to Siq exit length added additional flow resistance due to additional pipeline internal wall roughness resistance; all these elements had to be considered for a substantial fraction of the Ain Mousa spring flow rate to enter the total pipeline length under maximum flow rate conditions without instabilities and leakage problems. For lower flow rate conditions due to Ain Mousa spring seasonal rainfall absorbance conditions, only a trickle flow would go through the Siq pipeline, emphasizing the need for available storage reservoir water supplements to maintain potable water supply for the city's ~20,000 population and caravan trade. Any excess flow beyond the Siq extraction flow rate from the Ain Mousa spring was likely used to recharge upper and lower level reservoir systems for on-demand water usage. Here, normal flow to the Siq may have been temporarily stopped to serve lower reservoir filling—this being done by pipeline blockage—and for upper reservoir filling, pipelines from the Ain Musa spring likely served this purpose. Given the ~10,000 m^3 capacity of the main Zurraba reservoir (Figure 5A) and that of several additional minor reservoirs, the additional Ain Mousa flow rate capacity not used for the total Siq pipeline system served to provide additional reservoir water storage for on-demand and drought remediation use as well as for pipeline maintenance shut-offs where alternate pipeline water supplies were available from reservoirs. In summary, to this point it appears that a sizeable fraction of the Ain Musa spring flow rate could be transmitted through the Siq pipeline with provision for system shut-off to refill lower reservoirs for drought condition water availability, caravan water supply, and pipeline maintenance; however, there is more to this story to demonstrate the engineering that this water system contains.

Along the Siq north-side pipeline, several typical open basin features (Figure 6) are found at intervals along the pipeline. Each basin permits an accumulation of water serving as a head tank for its downstream piping length segment. Figure 10 reveals that hydrostatic head substantially influences flow rate for pipeline lengths less than ~500-m, and more so for shorter pipeline lengths. The head is reset at the origin of each of the piping lengths between basins so that flow through each piping segment is only sensitive to its local head tank head value. This modification led to partial flow (pipeline cross-section partially occupied by water) in Siq pipeline segments reducing overall flow resistance. As pipeline flow extending ~14 km from the Ain Mousa spring to the Siq entry point was mostly subcritical full-flow due to the high pipeline internal wall roughness flow resistance and low slope, the lowered resistance of Siq pipeline segments at ~3-degree hydraulically steep slopes between basins induced partial flow in distal portions of the ~14-km Ain Mousa spring supply pipeline to the Siq pipeline. The slightly lowered flow resistance could help to raise the overall system flow rate due to lowered back pressure and partial flow frictional wall resistance decrease. In simpler terms, lowering the frictional flow resistance in the end regions of a pipeline permitted a higher flow rate to be achieved through the entire pipeline system.

The four Siq north-side open basins (Bellwald, Siq principal excavator, personal communication) additionally served to provide drinking water along the Siq narrow passageway (Figure 2A) and access for cleaning of silt particles entrained in the flow to enhance potable water quality. Atmospheric pressure in the air space above the partial flow water surface in the Siq pipeline segments reduced pipeline joint leakage as would occur under a pressurized, full-flow condition design. To achieve a piping design based on a 14-cm pipe inner diameter at the declination angle of the Siq, ideally partial, critical flow at unit Froude number (Fr = 1) at critical depth (y_c/D~0.8–0.9) would provide the maximum flow rate the piping could convey. As open basins were sequentially lower than their predecessor upstream basins due to the declination slope of the Siq pipeline, each basin water height supplied head for its adjoining downstream piping section. Thus, partial flow existed throughout most of the Siq piping sections where flow entering a ~3-degree slope pipe section from a basin transitioned to a partial-flow, critical depth. This result indicates that the Froude number for the Siq pipeline flow is on the order of unity at close to the permissible ~96 m^3/hr flow rate. As the main purpose of the Siq water supply system was for the maximum flow rate to the urban center, any flow resistance changes that would increase flow rate were part of Nabataean water engineers' design intent. This was apparently achieved to gain a larger portion of the Ain Mousa flow rate to be used to supply the urban center's water needs. The thick sinter build-up in the (now visible) open bottom-most part of Siq piping supports the deposition pattern resulting from partial flow conditions existing in the Siq pipeline segments.

In the area immediately west of the Treasury, the Outer Siq pipeline (Figure 1, LS, B, C, and Figure 8) continued the Siq pipeline and was set at a slightly lower angle than the supply Siq pipeline. Given the Siq pipeline flow rate, a further consideration arises as to the stability of the flow in piping sections within the Siq as well as for water arriving to the Nymphaeum (Figure 1, Appendix A Feature 42) through the further Outer Siq pipeline extension subject to the input Siq flow rate. Any flow resistance from the Outer Siq pipeline extension would have a small effect on the upstream flow rate given the near-critical flow rate sections in the Siq pipeline that limit upstream influence. For water delivery from the Outer Siq portion of the total pipeline, if water delivery was pulsating into open basins within the Nymphaeum, then spillage and sloshing occurs amplifying flow instabilities that would compromise the aesthetic display of fountains and open pools in the Nymphaeum and induce forces within adjoining distribution pipelines promoting joint separation and leakage. For an examination of flow stability within the Outer Siq pipeline extension, CFD models were made for a ~1220-m section of the piping extension with a higher declination slope than that in the Siq pipeline; this length represents the approximate distance from the exit of the Siq pipeline to the Nymphaeum (Feature 42, Figure 1). Previous research [10] has indicated a further south-side Nymphaeum structure across the Wadi Mousa streambed serving the marketplace area with an extension of the south arm of the Siq pipeline; traces of elevated reservoirs providing pressurized water to the south Nymphaeum are described in [25].

FLOW-3D computer solutions (Figure 11) for 0.305(A), 0.610(B), 1.54(C), and 3.05(E) m/s water velocities were made with observed ε/D values typical of the interior surface of piping elements at the higher declination slope of the Outer Siq segment; here water velocity is expected to be higher than in the supply Siq pipeline. Figure 11(D) results are for a smooth interior pipe element wall at 1.54 m/s velocity to demonstrate the effect of wall roughness on flow patterns at the same water velocity as 1.54 m/s (C). For an illustrative example, water at ~96 m^3/hr was accepted into the Outer Siq piping—average pipeline water velocity was then ~1.73 m/s. Given the flow was slightly supercritical at this velocity, the flow approximated critical depth for Cases (B), (C), and (D). For flows close to critical, the normal depth was close to the critical depth [35]. As partial flow existed in the extension section (similar to that in the Siq piping) and had approximately the ~96 m^3/h flow rate at about the steeper declination angle as the Siq piping, near critical flow conditions existed throughout the Outer Siq extension pipeline. Figure 11 shows the transition to lower water depths from the entry boundary condition for all supercritical (B), (C), (E) and (D) cases, primarily due to the steeper declination angle. For the subcritical Case (A), water was at a low level due to low flow velocity. Figure 11, Case (B) at ~1.2 m/s approximated the flow effect at a mean velocity of ~1.75 m/s and indicated that only minor, random height excursions occurred in water flowing within the Outer Siq pipeline—this indicated that the water supply system through the Outer Siq extension to the Nymphaeum was stable, which was advantageous for proper operation of the Nymphaeum water display system. At twice this velocity, Figure 11(C) indicates large height fluctuations in the water surface, indicating unstable flow delivery occurring in the pipeline system. This indicates that the flow rate selected from the Ain Mousa spring by Nabataean engineers was a deliberate design choice to achieve stable flow in the total system. For Case (E), the high flow rate case shows susceptibility to a hydraulic jump creation due to the amplified supercritical flow interaction with pipeline internal roughness. This case serves as limit on the declination angle that could be implemented by Nabataean engineers. Here Cases (A) and (B) likely serve to give the declination angle of the Outer Siq extension pipeline due to the small flow disturbances indicated; Cases (C) to (E) would require a mandatory settling basin design to ensure smooth flow on to city center locations. Such an intermediate settling basin was observed in practice most likely to smooth the passage pipeline flow on to the city center and Nympheaum indicating a safety measure to ensure smooth flow for any of the Cases shown for any of the different declination angles shown that supported higher flow rates.

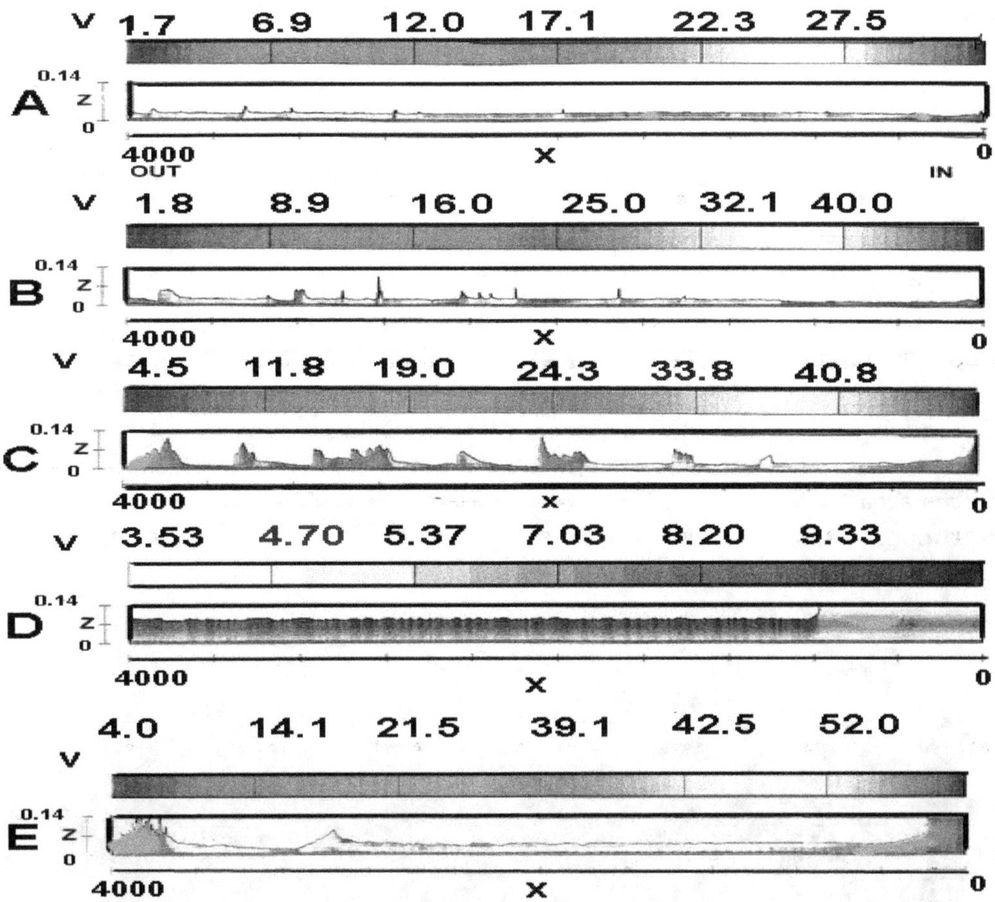

Figure 11. Internal pipeline flow stability results relevant to the Outer Siq pipeline for Cases **A** to **E**; right (IN) to left (OUT) flow direction for all cases.

Comparison of Figure 11, Cases (C) and (D) indicates that internal pipeline wall roughness amplified large-scale turbulent eddies that contributed to flow unsteadiness; this effect was included in the Case (B) result. Tracking the distal end of the Outer Siq extension pipeline revealed it was intercepted by a head reset open basin, indicating a further feature to ensure flow stability for flow continuing onto further pipeline branches and the Nymphaeum.

In summary, the total Siq system was designed to optimally transfer close to the Ain Mousa spring output flow rate of ~96 m^3/hr with partial flow conditions in several pipeline segments. This design largely eliminated large hydrostatic pressure conditions associated with full flow designs that would cause pipeline joint leakage at high hydrostatic pressures. Additionally, open basins along the Siq pipeline permitted removal of sediment debris that would ultimately clog the Siq system and destroy its functioning. In total, the various hydraulic flow regimes designed into the Siq pipeline ($Fr < 1$ subcritical, $Fr > 1$ supercritical, and $Fr = 1$ critical) and its Outer Siq extension necessary to capture a large fraction of the Ain Mousa spring output flow rate indicates an advanced knowledge of hydraulic science.

4. The Wadi Mataha Pipeline System

The water supply systems of Petra employed springs (s, Figure 1) both within (Ain Siyagh) and outside city limits (Ain Dibidbeh, Ain Mousa, Ain Braq, and Ain Ammon) to bring water to city monuments through pipelines. Reservoirs fed by the Ain Mousa (M–Zurraba, Figure 1) as well as other springs and reservoirs (no longer in existence due to modern urban expansion) provided supplemental water on-demand for arriving caravans or other utilitarian purposes. Pipelines composed of short interlocking terracotta pipeline elements, with sealed at joint connections by hydraulic cement, were typical of pipeline construction. Nabataean pipe elements were typically ~0.35-m long with an inner diameter range of ~14 to 23 cm, as observed from sample pipeline sections on display at the Petra on-site museum. As technologies developed over centuries, many different pipeline types developed with improved features to reduce internal pipeline wall frictional resistance. One special design of Nabataean tailored internal wall roughness minimized drag resistance ([25], pp. 268–271) according to modern hydraulic engineering theory and laboratory tests—if intentional from early observations and tests, this hydraulic science development occurred some ~2000 years before its "official" discovery in western hydraulic science.

Later Roman modifications to existing Nabataean water systems in the Great Temple area [20,21] (Figure 12) include small diameter lead piping and Roman standardized pipe diameter designs [22] for Roman extensions of the marketplace and Cardo Street areas, as well as for the Paradeisos Pool areas. The Roman standards for what a city under its control should have (fountains, baths and other water display structures) were an inherent part of Roman reconstructions present in post-106 AD Petra together with standardized piping size elements and lead piping. Excavations in the Cardo area show standardized Roman pipelines grafted onto existing Nabataean pipelines originating from Siq pipeline extensions to this area that expanded the marketplace area with fountains. For the Wadi Mataha pipeline, construction details of the channel on the Jebel al Khubtha mountain face (C-2, Figures 1 and 13, A–C) carry partial pipeline remains permitting analysis of its design and function.

Figure 12. The Great Temple.

The presence of pipeline segments, as opposed to an open channel conveying water (in the Wadi Mataha pipeline for example, Figure 13, A–C), is evident by noting that flow accelerating on a hydraulically steep open channel encountering bends would form surface height excursions and hydraulic jumps, causing channel overflow. For an open channel design, only a trickle flow from the upper reservoir could be contained in the channel without spillage. As water conservation was a prime concern, open channel systems subject to evaporation and spillage were not considered viable compared to pipeline systems. Some outer slab-covered sections of the Wadi Mataha channel are evident from and typical of Nabataean methods to provide a defense against water heating from direct sunlight exposure. Additionally, by surrounding embedded pipelines in channels with stabilizing soil to limit pipeline motion caused by water frictional forces and flow instabilities, leakage could be reduced at pipeline connection joints. Figure 10 governs admissible flow rates for full flow conditions such as would apply to pipeline flows originating from the M–Zurraba reservoir; for partial flow conditions as indicated in Figure 11, theoretical means are available for flow rate estimates [35,36]. Note also that open channels carrying water from distant springs were not a design option due to evaporation loss of scarce water supplies.

The pipeline along the eastern face of Jebel al Khubtha originated from a pipeline branch originating from the M–Zurraba reservoir complex. Typical full flow maximum flow rate estimates from Figure 10 are on the order of ~0.05 m^3/s, which would satisfy on-demand water supply requests; lesser flow rates would be possible by using a flow control valve originating from the reservoir. Piping from the high-level reservoir source proceeds to a lower reservoir elevated above the Sextus Florentinus Tomb area(Appendix A Feature 22, Figure 14), then proceeds southward toward the royal compound area [19] (Figure 15, Figure 1, 2-C), then on to monumental royal tomb architecture (Figure 15) near the Palace Tomb and a nearby fountain [10], then further on to the Nymphaeum for further pipeline water distribution to other parts of the city together with multiple basin/reservoir areas created by multiple dams along the Wadi Mataha streambed (-d, Figure 1).

(A)

Figure 13. *Cont.*

(B)

(C)

Figure 13. (**A**) The Wadi Mataha pipeline channel on the Jebel al Khubtha mountainside. (**B**) Continuation of the Wadi Mataha pipeline above the Sextus Florentinus Tomb (Figure 14). (**C**) Continuation of the Wadi Mataha channel is above workshop areas.

Figure 14. The Sextus Florentinus Tomb below the Wadi Mataha pipeline channel.

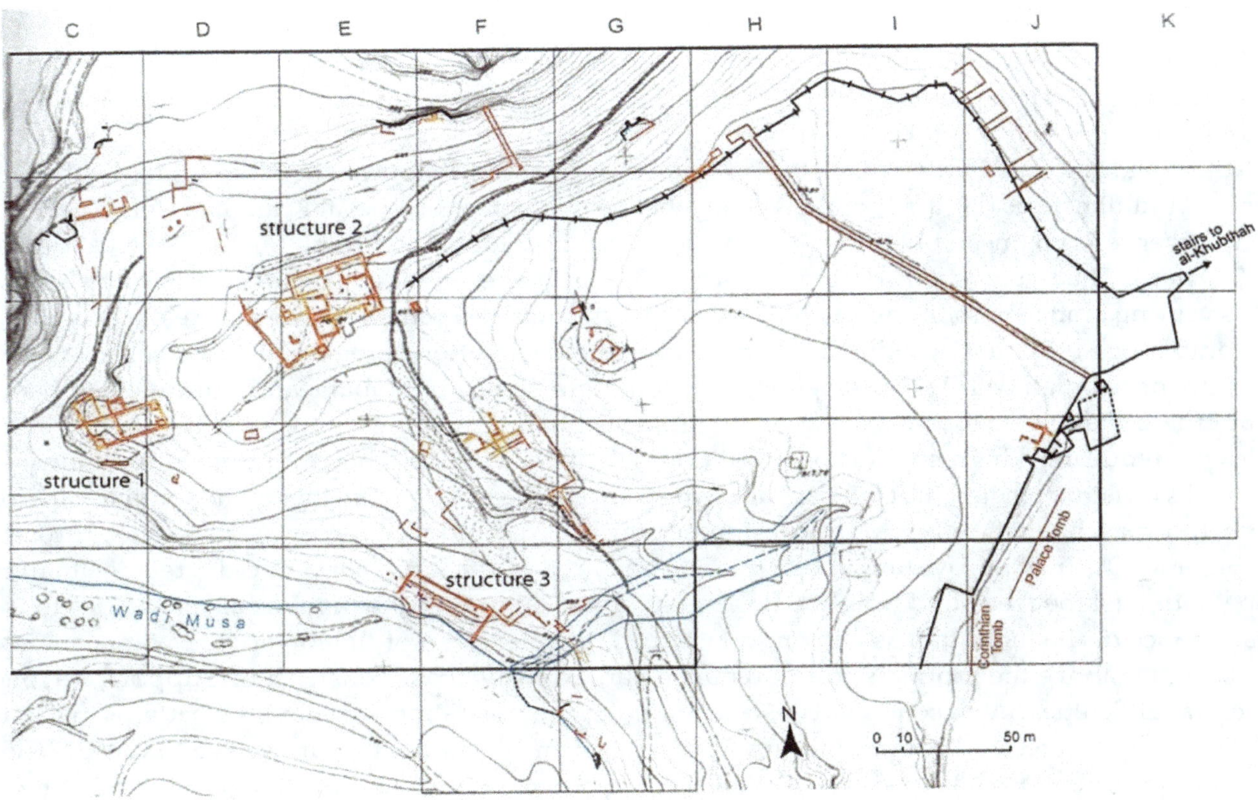

Figure 15. Royal compound structures 1, 2, and 3 in the Wadi Mataha area [18].

This flow network provided water to cisterns associated with ceramic and metal workshop areas and for celebratory triclinium functions associated with monumental royal tomb structures (Figure 1, Appendix A 62, 63, 64, and 65; **B** in region 2-B) and additionally made available upper reservoir water as a drought remediation measure. Recent research [18,19] indicates that several royal compounds exist (Figure 15) in the terminal pipeline area supplied by the Wadi Mataha pipeline (Figure 1, **B**, region 2-C). In conjunction with water from the Outer Siq pipeline extension directed into this area, ample water supply to royal compounds, even in drought periods, was assured as royal compound water needs for both consumption and aesthetic water displays dictated. Pipeline connections to the multiple royal compounds await excavation, but given nearby multiple water sources from the Wadi Mataha and Ain Mousa springs through the Outer Siq pipeline extension, as well as the upper reservoir water supply, abundant water supplies from multiple sources would be available to serve multiple palace structures. Water continued by pipeline past the Nymphaeum southward to city center areas and ultimately was transferred by a pipeline across a bridge to the commercial marketplace and fountain area district to supplement Siq pipeline water to the marketplace area. Final water discharge from all pipeline sources was ultimately led to the Wadi Siyagh natural drainage path. The present-day bridge in the Q'asar al Bint area contains visible ancient pipeline elements remaining after modern modifications to the earlier Nabataean bridge; this pipeline permitted water to cross over the bridge to sites on the south-side marketplace, Great Temple (Figure 12) and al Bint (Figure 2A) areas. Multiple pipeline water supplies to the same site areas would permit pipeline cleaning outages given redundant water supplies from other parts of the city, as well as a pedestrian transit path across the Wadi Mataha streambed.

Given the slope (~2.8-degrees) and construction detail of the entire Wadi Mataha pipeline, the question arises as to the engineering behind the selection of a pipeline slope that would yield the maximum flow rate given a variety of possible design options. Given the necessity to construct an elevated basin carrying the pipeline on the near vertical face of the Jebel al Khubtha mountainside (Figure 13, A–C) and the surveying required for the pipeline slope over a considerable length, a preliminary engineering analysis was necessary preceding construction. Analysis of alternate pipeline slope choices that have the potential to maximize the flow rate, while minimizing leakage from the many connecting joints between piping elements, was vital to a successful design. Details of flow patterns within the pipeline connecting the upper to the lower reservoir were analyzed by computer simulation to provide insight into the decision-making rationale underlying the Nabataean water engineers' choice of the most efficient pipeline slope.

Starting from an elevation of ~910 masl (meters above sea level) near the Mughur al Mataha and Jebel al Mudhlim areas (Figure 1, point **R**), the pipeline on the face of Jebel al Khubtha extends ~1.5 km to the lower reservoir near, but offset and above, the Sextus Florentinus Tomb (Figure 14) area located at ~770 masl. The upper water source is from the M–Zurraba reservoir connection supplied by a Wadi Mousa spring branch possibly augmented by a local spring. A ~15 m lower reservoir elevation above the tomb base at 770 masl yields an observed pipeline declination slope of ~2.8 degrees measured from the horizontal. While this slope was the ultimate choice of Nabataean hydraulic engineers, an alternate shallower slope choice of one-degree declination would require a yet higher elevation reservoir, requiring a high support base built onto the near-vertical Jebel el Khubtha mountainside. While this construction had the benefit of allowing the pipeline to extend further to the south to provide water to urban habitation areas, this design option was redundant given water supplies through the Siq and Outer Siq providing water to the same area of the city. For a one-degree declination slope option (slope measured from the horizontal origin point **R** of the pipeline as it switches to the western face of the mountain as shown in Figure 1), the elevation of the lower reservoir would be much higher above the ground surface, requiring an extensive base construction support. Beyond structural difficulties involved in this construction, computer modeling of water flow patterns in a long pipeline at one-degree declination slope revealed a problem. To expose this problem, a FLOW-3D CFD model (Figure 16, Case A) consisted of a left-most upper reservoir connected to a lower right-most reservoir by a pipeline at a one-degree declination slope; CFD calculations qualitatively duplicated hydraulic phenomena present in the pipeline flow from the upper reservoir **UR** to the lower reservoir **LR**. Figure 16, Case A results demonstrated the pipeline cross-section was fully occupied by water (full flow) given the submerged pipeline exit condition into the lower reservoir. A design with the pipeline exit above the lower reservoir water level would still support mostly full-flow conditions over most of its length, but be subject to ingested air slugs moving counter to the flow direction ingested at the pipeline exit, causing unstable flow conditions [34]. A volumetric flow rate of ~0.05 m^3/s entered the pipeline from the upper reservoir; this was the permissible full-flow rate for ε/D ~0.1 pipeline internal roughness over long pipeline distances (Figure 10) at a one-degree declination slope. Increasing the upper reservoir head did not result in a proportionate increase in volumetric flow rate for long pipelines over 400–500 m length as indicated by Figure 10. CFD results for Figure 16, Case A indicated full-flow for a submerged outlet pipeline under high hydrostatic pressure near the end reaches of the pipeline—this increased the likelihood of pipeline joint leakage due to full flow conditions under high hydrostatic pressure. Based on the combination of leakage problems and the difficulty of building a high-elevation reservoir with a steep connecting pipeline to the Nymphaeum, hydraulic engineers rejected the one-degree declination slope pipeline design as impractical.

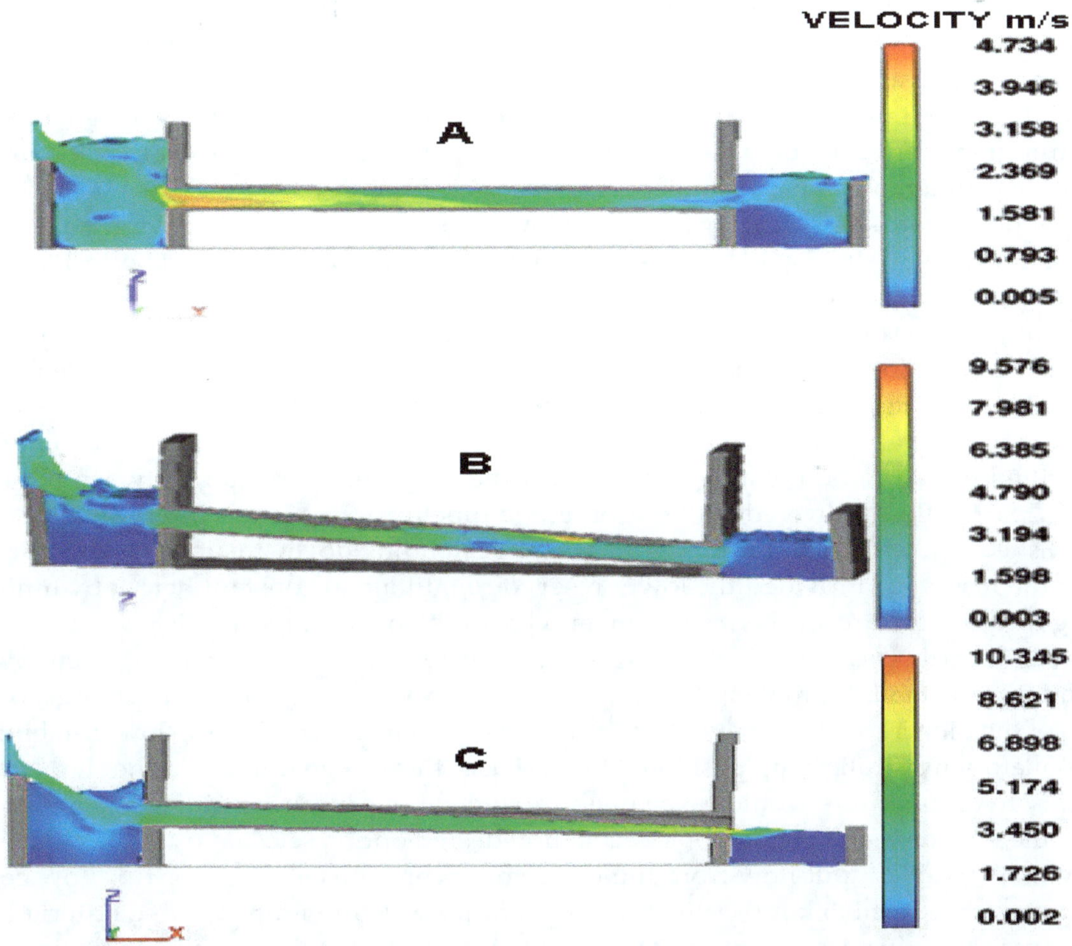

Figure 16. FLOW-3D computational fluid dynamics (CFD) results for Cases **A**–**C**. The hydraulic jump shown(directly under **B**) in Case **B** causes pressure oscillations, inducing separation forces at pipeline connection joints amplifying the potential for joint leakage and does not support the highest flow rate possible despite its steeper slope.

A further design option involved a steeper pipeline slope choice of four degrees with a volumetric flow rate adjusted to 0.08 m^3/s. The intuitive (but incorrect) observation that the steeper the pipeline slope, the faster the water velocity will be serving the on-demand rapid water delivery maximum flow rate requirement best, is an observation that may have guided the steeper slope choice. The steeper slope design option would result in a lower reservoir height from ground level but would require an additional length of piping to carry water on to the tomb structures (Figure 1, Appendix A site numbers 62, 63, 64, and 65) as water would be delivered to ground level cisterns distant from these structures. Further piping to the (North) Nymphaeum and the nearby sites including the Temple of the Winged Lions (Figure 9) required pipeline construction which, at low slope, may have further impeded the flow on to further water distribution centers across the lower bridge to south-side sites. Based upon the easier design option construction of a low-level reservoir on the el Khubtha mountainside, this option appears to have had positive benefits for workshop areas north of the monumental tomb structures; however, hydraulic analysis of this design option reveals a problem. For a steep pipeline slope transitioning from a mild slope, the high velocity supercritical internal flow developed a hydraulic jump with an upstream partial vacuum above the partial flow water surface (Figure 16, Case B). The flow height approached normal depth [35] for supercritical flow in a hydraulically steep pipeline. This result, approximated as flow resistance, was amplified by the high partial flow velocity that converted flow from supercritical ($Fr > 1$) to subcritical ($Fr < 1$) through a hydraulic jump within the pipeline as Figure 16, Case B calculation shows. This effect was largely caused by pipeline wall roughness and the high-water velocity used for Case B. The hydraulic jump is indicated below **B** in Case B. Froude number (Fr) was defined as $Fr = V/(g\ D)^{1/2}$, where V is the average flow velocity, g the gravitational constant ($9.82\ m/s^2$), and D is the hydraulic depth [36]. While partial-flow in the upper supercritical portion of the pipeline helped eliminate pressurized full flow conditions that promoted leakage, the full-flow, post-hydraulic jump region within the downstream reaches of the pipeline promoted air entry into the pipeline exit to counter the partial vacuum region that promoted unstable flow pressure variations that caused pipeline joint leakage. Since water in the lower reservoir may have undergone sloshing and transient height changes during this process, the hydraulic jump position could be oscillatory due to sporadic air ingestion slugs countering the partial vacuum region from the pipeline exit downstream of the hydraulic jump. This effect contributed to sloshing in the lower reservoir and pipeline forces that compromised connection pipeline joint stability. While a design option existed to have the pipeline outlet above the lower reservoir water level under free fall conditions, this only amplified air ingestion to counter the partial vacuum region and flow instability. The computer solution shown in Figure 16, Case B, verified the formation of a hydraulic jump showing the interaction between the incoming supercritical partial flow delivered by the submerged pipeline to the lower reservoir. The hydraulic jump established a pressurized, full-flow water region extending downstream from its location in the pipeline to the reservoir. Provided the lower reservoir maintained constant height by limiting flow into piping directed toward the Nymphaeum and basin regions, the hydraulic jump position could be stabilized; nevertheless, the full-flow, pressurized water region constituted a potential pipeline leakage and flow instability problem. Given problems associated with the steep-piping slope choice (regardless of the pipeline exit submerged or with a free-fall delivery condition), flow instabilities arose from an oscillatory hydraulic jump position. On this basis, this design option was not implemented as an oscillatory hydraulic jump would cause pressure fluctuations inducing pipeline joint leakage.

Given the ~2.8-degree pipeline slope declination design option selected by Nabataean engineers and given an allowable input flow from the upper reservoir of $0.062\ m^3/s$ for full-flow conditions into a long stretch of the 3.5 km pipeline upstream of point **R** in Figure 1, three criteria had to be met for acceptance of this design option:(1) An easy-to-build and maintain low reservoir that would provide observation of pipeline exit flows to check for flow problems; (2) a stable, high volume flow rate with an atmospheric pressure airspace above the (partial flow) water surface to eliminate hydrostatically pressurized connection joint leakage; and (3) easy access to the tomb structures for ritual events. Regarding these conditions, Figure 16, Case C provides insight into design criteria that satisfies conditions (1) to (3).

Theoretically, a flow approximating critical flow at $Fr \approx 1$ would yield the highest flow rate possible with flow occupying a fraction of the cross-section of the pipe with an air space above the water surface. Calculations for a 0.062 m^3/s full-flow volumetric flow rate based on a ~2.8-degree slope and upper reservoir hydrostatic head values from one to three meters indicated that average velocity in the pipeline was on the order of V ~1.5 m/s with $(g\ D)^{1/2} \approx 1.5$ m/s, leading to $Fr \approx 1.0$, indicating that a near critical flow condition was approximated at the ~2.8-degree declination slope. Given the near-critical flow observed at a~2.8-degree declination slope (corresponding to the observed slope) and observing computer results (Figure 16, Case C) that indicate that this condition would produce an air space over the water free surface approximating $y_c/D \approx 0.8$, an approximate critical flow condition existed at the ~2.8-degree declination slope. The significance of this is that the 0.062 m^3/s volumetric flow rate was the maximum value that this long pipeline design could sustain due to pipeline internal wall frictional effects and that the critical flow water height in the pipeline, although close to full-flow conditions, left an atmospheric pressure airspace over the water surface over a long length of the pipeline to limit pressurized leakage. Additionally, the presence of a critical flow eliminated the upstream influence from downstream disturbances thus promoting a stable flow within the critically sloped pipeline. Therefore, among the choices of the pipeline slope available to Nabataean engineers, the ~2.8-degree declination slope was selected. This slope option produced the maximum transport volumetric flow rate possible, which is what an on-demand water system is set to accomplish.

5. The Ain Braq and Theater Area Pipeline Continuance System

The upper reaches of Ain Braq water system (Figure 1) consisted of a channel cut into bedrock and surface soils that earlier supported a pipeline; its continuance to the plateau north of Wadi Farasa (Figure 1) by a supply line continues to the B-3 area shown in Figure 1 and supplied water to sites 12 to 18 (Appendix). A branch of this system may have extended to the eastern reaches of the housing district south of the Great Temple, as further excavations may indicate. A pipeline originating from the theater frontage area (Figure 17), given the land contours shown in ([37], p.136), would support a canal path to this area and sites along its path. The plateau shown in the Weis reference (Abb.5, p.136) [37] indicates many water structures consistent with this water supply system.

Figure 17. Theater with a seating capacity of 5000 to 6000 persons. Frontal pipelines extend to elevated basins above the south bank marketplace area to support fountains with (a probable) pipeline extension to areas above the Great Temple.

A survey of the area from the Ain Braq, Ain Ammon springs (Figure 1) revealed channels cut into mountain sides that supported a pipeline water supply from the two main springs to sites shown in Figure 1. Some traces of an attempt to bridge a large valley by a siphon are indicated in Figure 18. This design choice originated from a spring originating above the Wadi Mousa town that was likely abandoned early due to the necessity of constructing a large siphon across a deep valley to reach city precincts. The main Ain Braq and Ain Ammon springs supported a pipeline that sent water to the Lion Fountain (Figure 19, Appendix A Feature 14) and sites 12 to 18.

Figure 18. Remains of an abandoned early pipeline channel that would employ a large siphon to cross a valley originating from the spring source located north of the modern town.

Figure 19. The Lion Fountain.

Figures 1 and 20 show surface water distribution features in the Wadi Farasa area (B-1, Figure 1), together with the subterranean pipeline basin pit **A** connecting to the distribution junction **E**. Further exploration of possible contour pipeline paths threaded through mountainous areas to lower sites is necessary to establish the water supply source to **D**, the Lion Fountain (Figure 19), the Soldier's Tomb area, and associated water basin structures sources by a branch of the Ain Braq pipeline. The contour map of the hill area south of the Great Temple (Figure 12) area indicated the possibility of a water supply from a Southern Siq channel passing by the theater frontage (Figure 16), but erosion changes in the landscape over millennia and multiple earthquake landscape disturbances have made the water supply connection to **D** difficult to establish.

Figure 20. Details of pipeline components located above the Great Temple area. Pipeline (**A**) is immediately adjacent and connected to (**E**) and empties into a pit basin.

Of interest is the water supply to the Great Temple area. Starting from the hilltop junction (Figure 1) above the Great Temple, a pipeline component assemblage exists (Figure 20). While likely sourced by a Siq extension pipeline from the theater frontage area, there may have been a branch pipeline originating from the main Ain Braq, Ain Ammon pipeline as indicated in the B-3 area of Figure 1. Independent of confirmation of this water source to the hill area above the Great Temple, the assemblage of water transfer components (Figure 20) indicated water transfer to downhill housing areas with a probable pipeline water transfer to the Great Temple area. Combining excavated details of the Figure 20 components, a FLOW-3D computer model is surmised in Figure 21. One of the pipelines **B** from **E** led water to the lower Zhanthur housing district ([37], p. 136), while another pipeline branch

C also served the same purpose; a possible branch of the **C** pipeline may have led to the upper reaches of the Great Temple and lower Paradeisos area [8,20,21] as further excavations may verify. On site conversations with Brown University personnel during excavation of the Great Temple area indicated that a pipeline element was found heading downhill to the Great Temple, but was now covered over from excavation soil from the Great Temple renovation activity. Independent of verification of this pipeline water supply to the Great Temple's upper levels, much can be learned from existing Figures 20 and 21 components that give insight into the hydraulic engineering knowledge base of Nabataean water engineers.

The FLOW-3D computer model (Figure 21) composed of Figure 20 components utilized an input water velocity consistent with Figure 10 for full flow conditions into pressurized pipeline **D**. The water supply input into **D** was from the theater district (or from an Ain Braq branch pipeline) with a level portion of the pipeline existing before pipeline branches **E–C** and **E–B** continued water flow on to steeper terrain to supply downhill housing destinations and possibly the Great Temple upper level areas. A basin pit **A** shows a connection pipeline adjacent to and connected into the **E** pipeline.

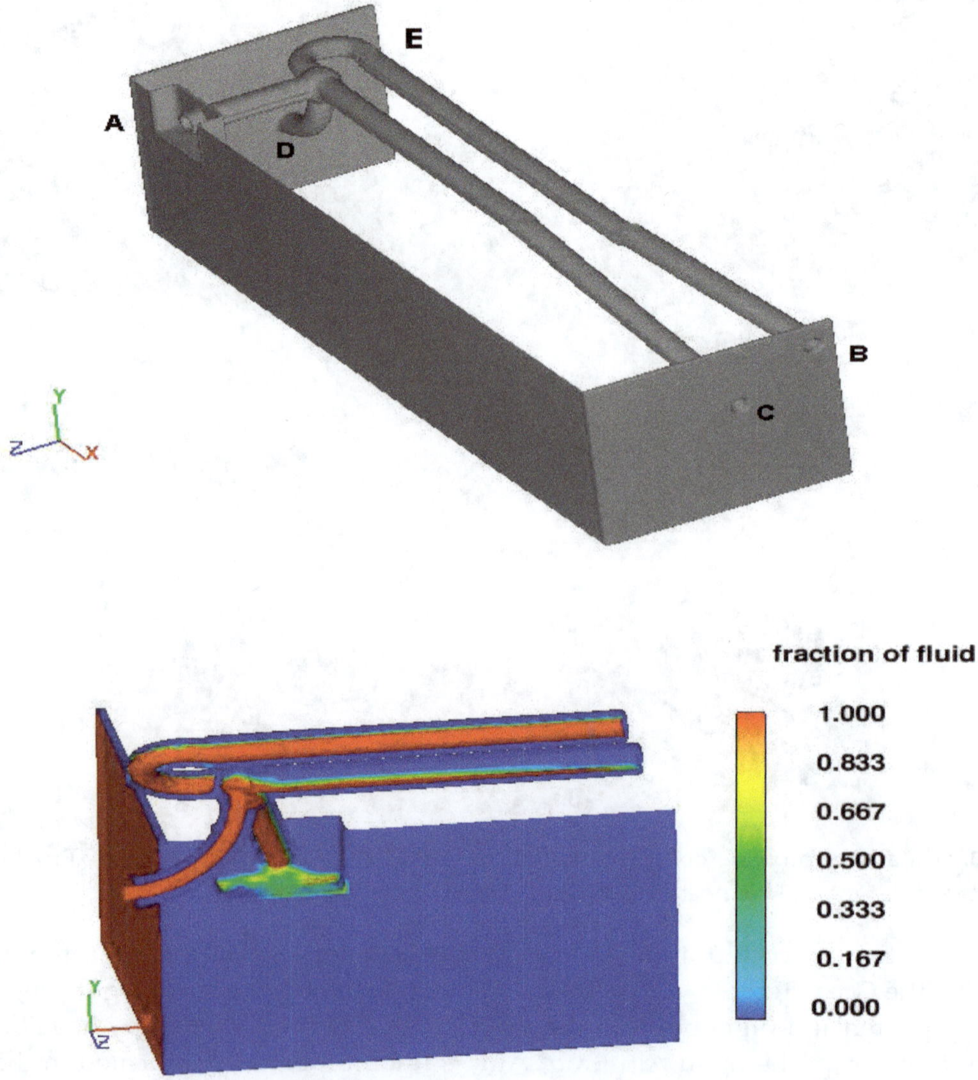

(A)

Figure 21. *Cont.*

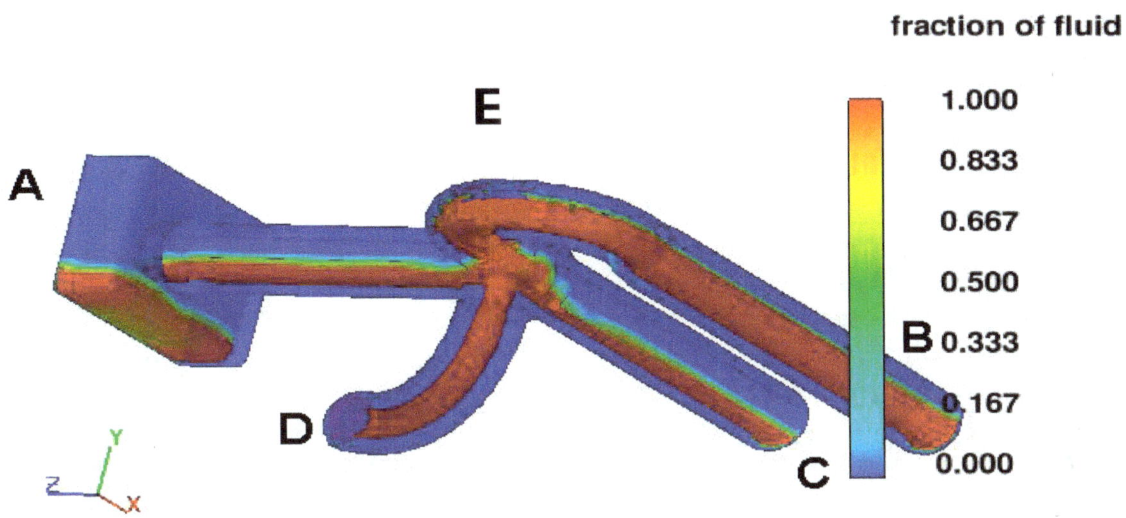

(B)

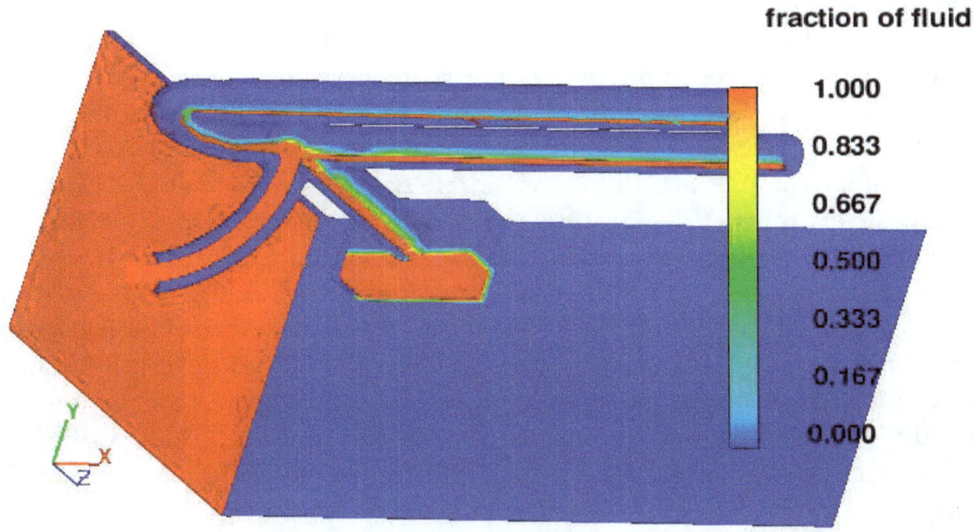

(C)

Figure 21. FLOW-3D CFD model of the pipeline connections constructed from Figure 20 components and downhill exposed pipeline elements. (**A**) Incipient filling of pit basin **A** due to hydraulic jump instability (or an **E–B, E—C** blockage) causing transient back flow conditions that are damped by water flow into the **A** pit basin. (**B**) Partial filling of pit basin **A** after pipeline instabilities are damped; note partial flow in high slope pipelines **E–B** and **E–C**. (**C**) Alternate view of Figure 21**B** conditions.

For one hydraulic function, the basin pit pipeline **A** (Figure 21) served to divert flow away from the dual downhill pipeline **E–B** and **E–C** branches if these were blocked during repair or if the input flow at **D** had flow surges past what the remaining downstream pipelines could accommodate. The blockage then diverted water to basin pit **A** in Figure 21. The flow into the basin pit continued until a pressure balance of water accumulated in basin pit **A** balances with the pressure in water supply pipeline **D** promoting an automatic system shut-off if the lower exit ports **B** and **C** (Figure 21) were blocked. The more sophisticated hydraulic engineering use of the Figure 21 configuration originated from the observation that subcritical full flow conditions existed in the lengthy, low slope supply pipeline **D**. Due to the hydraulically steep declination angle in dual pipelines **E–B** and **E–C**, input full flow past the junction area was converted to supercritical partial flow in both steep pipeline branches. This flow transition induced a local partial vacuum region above the partial flow water surface. As the branch pipeline exits were submerged into housing area reservoirs at their destination locations, then air was drawn into pipeline exit openings toward the upstream partial vacuum regions by flow patterns described in [34]. The intermittent air bubble stream proceeded upstream into the dual steep declination pipelines' branches to counter the partial vacuum region leading to flow intermittency and oscillatory hydraulic jumps in both the dual pipeline branches. This hydraulic phenomenon affected the flow stability of the entire system, leading to pulsating discharges into the downstream reservoirs as well as affecting the stability of input flow into **D**. The pulsating flows induced internal pressure changes that could affect pipeline joint connections causing leakage. If pipeline exit flow into housing area reservoirs was free-fall, then the flow instability condition was even more severe as there was easier access of air going upstream to partial vacuum regions. A further cause for pipeline flow instability was the creation of a hydraulic jump within the pipelines due to internal pipeline wall roughness when subject to upstream supercritical flow. Many situations arose then that caused flow instabilities, leading to pulsating flow delivery tohousing area reservoirs and pressure variations that induced pipeline joint leakage.

A cure to unstable flow conditions was anticipated by Nabataean water engineers when air was supplied through basin pit pipeline **A** as shown in Figure 21. Air led into the partial vacuum region where it occurred canceled the partial vacuum region and air bubble transfer from pipeline exits and subsequent flow instabilities in the dual subterranean pipeline branches. The hydraulic jumps still occurred in **E–B** and **E–C** pipeline branches, but with the partial vacuum regions eliminated, the hydraulic jumps and flow within pipelines remained stable. In Figure 21 **A–C**, the fluid fraction ff = 1 denotes water, ff = 0 denotes air, and intermediate ff values denote evaporative moisture levels. These figures indicate the transfer of air into the upstream reaches of **E** pipelines through the basin pit pipeline **A**. In summary, the basin pit pipeline connection served to minimize transient water surges originating from air pulses entering partial vacuum regions by providing an air passageway to **E** and elimination of transient water instabilities in the downstream high declination slope dual pipelines. This was done by transferring backup water caused by transient hydraulic jump motion into the pit basin, thus preventing its effect on flow input stability. Figure 21**A** shows incipient water transfer into the pit basin **A** due to pipeline blockage. Figure 21**C** shows continued filling of the pit basin until a pressure balance was achieved, shutting down further input water flow from **D**. Figure 21B shows that with air input from the connecting pipeline in **A** to the main **E–B** and **E–C** pipelines, the partial vacuum regions were eliminated, producing stable partial flow in these pipeline branches.

Excavation of the Great Temple area over many years by the Brown University team under the direction of Martha Joukowski [20,21] revealed that rainfall seepage from a hillside wall south of the temple was collected into a drainage channel [25] that led water to an underground reservoir on the east part of the site. A further subterranean channel starting from the surface entrance to the reservoir led water to the front platform part of the Great Temple site and likely served as an overflow drain channel activated when the reservoir was full. These additional water transfer features indicate that water supply to the Great Temple (and Paradeisos area) may have had an additional pipeline source from another branch of the Figure 21 assemblage, which is yet to be discovered. Given Great Temple water supplies from the extension of the theater district piping at lower levels, possible higher level water supplies originated from the collection trench above the Great Temple north wall ([25], pp. 265–267), as well as a possible connection from the Figure 21**C** channel as previously noted. Previously, this trench was assigned a drainage collection channel role from rainfall seepage to protect the foundation stability of the Great Temple, but other uses were possibilities. Given the elaborate upper level community assemblage structure that likely represented [20,21] a site leadership meeting place, access to water at higher Great Temple levels appeared a necessity. Given that discussions with Brown University personnel involved in the excavation indicated that a pipeline existed coming down from the hill above the temple, now covered with excavation soil, gives credence to the water supply from a branch canal from pipeline **C** shown in Figure 21 or an extension pipeline from the theater system (Figure 1). Future excavations are necessary to establish these connections to show the source of enigmatic structures that may have association with water transfer at upper levels of the Great Temple.

The Monastery Temple (Figure 22) at high elevation above the urban city center had no apparent pipeline water supply system due to the absence of high-level springs, but rather a series of nearby water collection basins that collected rainfall runoff.

Figure 22. The Monastery Temple.

6. Conclusions

Three pipeline water supply systems supplying urban Petra required different hydraulic engineering approaches to overcome problems to optimize the water supply to Petra's city center.

In each case, problems were overcome by engineering designs requiring empirical knowledge of fluid mechanics principles known in a prescientific format predating western science's "official" discovery of similar principles some ~2000 years later. Given the constraints from limited excavation data and given that the Nabataean hydraulics knowledge base is not known from surviving literature, its breadth may be gauged by use of CFD recreations of solutions that inform of thought processes and technical means available to Nabataean hydraulic engineers. As many of their solutions exhibit modern approaches, one can now attribute a role in the history of hydraulic science to Nabataean engineers.

Appendix A (Figure 1 Site Captions)

1	Zurraba–M reservoirs
2	Petra rest house
3	Park entrance
4	Early B-C-D Siq Covered Floor Channel
5	Dijn monument
6	Obelisk tomb and Triclinium
7	Siq entrance elevated arch remnants
8	Flood bypass tunnel and dam
9	Eagle Monument
10	Siq passageway
11	Treasury (El Kazneh)
12	High Place sacrifice center
13	Dual obelisks
14	Lion fountain monument
15	Garden Tomb
16	Roman Soldier Tomb
17	Renaissance Tomb
18	Broken pediment Tomb
19	Roman theater
20	Uneishu Tomb
21	Royal Tombs (62, 63, 64,65)
22	Sextus Florentinus Tomb
23	Carmine façade
24	House of Dorotheus
25	Colonnade Street
26	Winged LionsTemple
27	Pharaoh's column
28	Great Temple
29	Q'asar al Bint
30	Museum
31	Quarry
32	Lion Triclinium
33	El Dier (Monastery)
34	468 Monument
35	North city wall
36	Turkamaniya Tomb
37	Armor Tomb
38	Wadi drainage
39	Aqueduct
40	Al Wu'aira crusader castle
41	Byzantine tower

42	North Nymphaeum
43	Paradeisos, Temenos Gate area
44	Wadi Mataha major dam
45	Bridge abutment
46	Wadi Thughra tombs
47	Royal tombs
48	Jebel el Khubtha high place
49	El Hubtar necropolis
50	Block Tombs
51	Royal Tombs
52	Obelisk Tomb
53	Columbarium
54	Conway Tower
55	Tomb complex
56	Convent tomb
57	Additional Tomb complex
58	Pilgrim's spring
59	Jebel Ma'aiserat high place
60	Snake monument
61	Zhanthur mansion
62	Palace Tomb
63	Corinthian Tomb
64	Silk Tomb
65	Urn Tomb
66	Rectangular platform/temple
67	South Nymphaeum
—	Pipeline
- - -	Buried channel
B	Dual pipeline supplied basin area
d	Catchment dam
c	Cistern
s	Spring
D	Multi-level dam
A	Kubtha aqueduct
T	Water distribution tank
-d	Major dam

References

1. Browning, I. *Petra*; Chatto and Windus Publishers, Ltd.: London, UK, 1982.
2. Glueck, N. *Deities and Dolphins*; Strauss and Cutahy Publishers: New York, NY, USA, 1965.
3. Guzzo, M.G.; Schneider, E. *Petra*; University of Chicago Press: Chicago, IL, USA, 2002.
4. Hammond, P. *The Nabataeans: Their History, Culture and Archaeology*; Studies in Mediterranean Archaeology 37; Astrom Publishers: Gothenburg, Sweden, 1973.
5. McKenzie, J. *The Archaeology of Petra*; Oxford Press: London, UK, 1990.
6. Ossorio, F.A.; Porter, B.A. *Petra-Splendors of the Nabataean Civilization: Highlights*; White Star Publishers: Vercelli, Italy, 2009.
7. Akasheh, T.S. Ancient and Modern Water Management at Petra. *Near East. Archaeol.* **2002**, *65*, 220–224. [CrossRef]
8. Bedal, L.-A. *The Petra Pool Complex: A Hellenistic Paradeisos in the Nabataean Capital*; Gorgias Dissertations, Near Eastern Studies 4; Gorgias Press: Piscataway, NJ, USA, 2002.
9. Bedal, L.-A.; Gleason, K.L.; Schryver, J.G. The Petra Garden and Pool Complex, 2003–2005. *Annu. Dep. Antiq. Jordan* **2007**, *51*, 151–176.

10. Bellwald, U. The Hydraulic Infrastructure of Petra-A Model for Water Strategies in Arid Lands. *Historia* **2007**, *19*, 94–100.
11. Bellwald, U.; Al-Huneidi, M.; Salihi, A.; Naser, R. *The Petra Siq: Nabataean Hydrology Uncovered*; Petra National Trust Publication: Amman, Jordan, 2002.
12. Oleson, J.P. The Origins and Design of Nabataean Water-Supply Systems. In *Studies in the History and Archaeology of Jordan*; Department of Antiquities: Amman, Jordan, 1995; Volume 5.
13. Oleson, J.P. Nabataean Water Supply: Irrigation and Agriculture. In *Proceedings of the International Conference on the World of Herod and the Nabataeans*; Politis, K., Ed.; Franz Steiner Verlag: Stuttgart, Germany, 2002; Volume 2.
14. Ortloff, C.R. The Water Supply and Distribution System of the Nabataean City of Petra (Jordan), 300 BC–AD 300. *Camb. Archaeol. J.* **2005**, *15*, 93–109. [CrossRef]
15. Ortloff, C.R. Water Engineering at Petra (Jordan): Recreating the Decision Process underlying Hydraulic Engineering at the Wadi Mataha Pipeline. *J. Archaeol. Sci.* **2014**, *44*, 91–97. [CrossRef]
16. Ortloff, C.R. Three Hydraulic Engineering Masterpieces at 100 BC–AD 300 Nabataean Petra. In *Proceedings of the De Aquaductuatque Aqua Urbium Lyciae Phamphyliae Pisidiae: The Legacy of Julius SextusFrontinus*; Wiplinger, G., Ed.; Peeters Publishing: Leuven, Belgium, 2014; ISBN 978-90-429-3361-3.
17. Ortloff, C.R.; Crouch, D. The Urban Water Supply and Distribution System of the Ionian City of Ephesus in the Roman Imperial Period. *J.Archaeol. Sci.* **2002**, *28*, 843–860. [CrossRef]
18. Schmid, S.G. Die Wasserversorgung des Wadi Farasa Ost in Petra. In *Cura Aquarum in Jordanien, Proceedings of the 13th International Conference on the History of Water Management and Hydraulic Engineering in the Mediterranean Region*; DWhG Publications: Berlin, Germany, 2007; pp. 95–117.
19. Schmid, S.G.; Bienkowski, Z.P.; Ziema, T.; Kolb, B. The Palaces of the Nabataean Kings at Petra. In *The Nabataeans in Focus: Current Archaeological Research at Petra*; Nehme, L., Wadeson, L., Eds.; Supplement to Proceedings of the Seminar for Arabian Studies 42; Archaeopress: Oxford, UK, 2012; pp. 73–98.
20. Joukowsky, M. *Petra Great Temple, 1993–1997*; Brown University Press: Providence, RI, USA, 1998; Volume 1, pp. 265–270.
21. Joukowsky, M. The Petra Great Temple Water Supply. In *Proceedings of the 9th International Conference on the History and Archaeology of Jordan- Cultural Interaction throughout the Ages; Abstracts 25*; Department of Antiquities, Al-Hussein Bin Talal University Publication: Str Ma'an, Jordan, 2004.
22. Hodge, A.T. *Roman Aqueducts & Water Supply*; Bristol Classical Press: London, UK, 1992.
23. Laureano, P. *The Water Atlas, Traditional Knowledge to Combat Desertification*; Laia Libros: Barcelona, Spain, 2002.
24. Mays, L. Ancient Water Technologies. Springer Publishers: Dordrecht, The Netherlands, 2007; pp. 21–22.
25. Ortloff, C.R. *Water Engineering in the Ancient World: Archaeological and Climate Perspectives on Ancient Societies of South America, the Middle East and South East Asia*; Oxford University Press: Oxford, UK, 2010.
26. Ortloff, C.R. *The Hydraulic State: Science and Society in the Ancient World*; Routledge Press: London, UK, 2020.
27. Parr, P. La Date du Barrage du Sîq á Petra. *Rev. Biblique* **1967**, *74*, 45–49.
28. Parcak, S.; Tuttle, C.A. Hiding in Plain Sight: The Discovery of a New Monumental Structure at Petra, Jordan, Usung WorldView-1 Satellite Imagery. *Bull. Am. Sch. Orient. Res.* **2016**, *375*, 35–51. [CrossRef]
29. Markoe, G. *Petra Rediscovered: Lost City of the Natataeans*; H. Abrams Publishers, Inc.: New York, NY, USA, 2003.
30. Flow Science. *FLOW-3D/V.10.1 Users' Manual*; Flow Science, Inc.: Santa Fe, NM, USA, 2019.
31. Heinrichs, K.; Azzam, R. Investigation of Salt Weathering on Stone Monuments by use of a Modern Wire Sensor Network exemplified for Rock-Cut Monuments at Petra. *Int. J. Herit. Digit. Era* **2012**, *1*, 191–216. [CrossRef]
32. Strabo. *Geography, Books 15–16*; Harvard University Press: Cambridge, MA, USA, 2000; Volume 164, pp. 21–22.
33. Urban, T.M.; Alcock, S.E.; Tuttle, C.A. Virtual Discoveries at a Wonder of the World: Geophysical Investigations and Ancient Plumbing at Petra, Jordan. *Antiquity* **2012**, *86*, 331.
34. Lugt, J. *Vortex Flow in Nature and Technology*; John Wiley & Sons Publishers: Hoboken, NJ, USA, 1983.
35. Morris, H.; Wiggert, M. *Applied Hydraulics in Engineering*; The Ronald Press: New York, NY, USA, 1972.
36. Chow, V.T. *Open-Channel Hydraulics*; McGraw-Hill Book Company: New York, NY, USA, 1959.

37. Weis, L. Das Wasser der Nabataer: Zwishen Lebensotwendigkeit und Lexus: The Northwestern Project. In *Proceedings of the De Aquaductuatque Aqua Urbium Lyciae Phamphyliae Pisidiae: The Legacy of Julius Sextus Frontinus*; Wiplinger, G., Ed.; Peeters Publishing: Leuven, Belgium, 2014.

2

Investigation of Free Surface Turbulence Damping in RANS Simulations for Complex Free Surface Flows

Arun Kamath[1,*], Gábor Fleit [2] and Hans Bihs [1]

1 Department of Civil and Environmental Engineering, Norwegian University of Science of Technology, NTNU, 7491 Trondheim, Norway; hans.bihs@ntnu.no
2 Department of Hydraulic and Water Resources Engineering, Budapest University of Technology and Economics, Moegyetem rkp. 3, 1111 Budapest, Hungary; fleitg@gmail.com
* Correspondence: arun.kamath@ntnu.no

Abstract: The modelling of complex free surface flows over weirs and in the vicinity of bridge piers is presented in a numerical model emulating open channel flow based on the Reynolds Averaged Navier-Stokes (RANS) equations. The importance of handling the turbulence at the free surface in the case of different flow regimes using an immiscible two-phase RANS Computational Fluid Dynamics (CFD) model is demonstrated. The free surface restricts the length scales of turbulence and this is generally not accounted for in standard two-equation turbulence modelling approaches. With the two-phase flow approach, large-velocity gradients across the free surface due to the large difference in the density of the fluids can lead to over-production of turbulence. In this paper, turbulence at the free surface is restricted with an additional boundary condition for the turbulent dissipation. The resulting difference in the free surface features and the consequences for the solution of the flow problem is discussed for different flow conditions. The numerical results for the free surface and stream-wise velocity gradients are compared to experimental data to show that turbulence damping at the free surface provides a better representation of the flow features in all the flow regimes and especially in cases with rapidly varying flow conditions.

Keywords: free surface flow; embankment weir; bridge piers; hydraulic jump; turbulence; CFD

---

1. Introduction

The interface between air and water or the free water surface is one of the most significant features of open channel flow. The motion of the free surface and the flow properties such as velocities and the pressure in the flow are strongly related. A change in the geometry of the channel, such as in the case of flow around hydraulic structures, steep pressure gradients occur due to flow constriction. According to Bernoulli's equation, under ideal conditions the total energy in the flow is conserved through transfer of energy between potential and kinetic energy. In addition, energy is lost to fluid-fluid and fluid-solid interface friction. This results in a change in the free surface location and the flow features, and governs the design of hydraulic structures.

The free surface features resulting from different flow conditions such as hydraulic jumps, shock waves, and standing waves can have negative consequences on the hydraulic structure and should be considered during design. Several researchers have proposed different empirical formulae to estimate different features such as the height of a shock front or deflection angles [1–4]. Physical model tests investigating different flow conditions and the resulting flow features can provide more details regarding the flow phenomena [5–8].

Numerical modelling of the open channel flow phenomena can yield further detail into the flow physics. Due to the strong relationship between the flow hydrodynamics and the free surface, the correct representation of the free surface along with the boundary conditions are essential to accurately represent the flow in a numerical model. Over recent decades many different methods for the prediction of the free water surface elevation have been developed. They reach from simple empirical estimations with 1D numerical tools, e.g., Brunner [9] to more advanced 3D models, e.g., Olsen [10].

The application of Computational Fluid Dynamics (CFD) methods to study free surface flows is an interesting avenue where different flow features can be calculated in detail with the solution of the Navier-Stokes equations, where the Reynolds Averaged Navier-Stokes (RANS) framework is employed [11,12]. Further several recent studies in literature present eddy-resolving frameworks such as Kirkil et al. [13], Kara et al. [14], Kang and Sotiropoulos [15] and McSherry et al. [16]. Eddy-resolving simulations provide a higher level of detail regarding the coherent structures in a turbulent flow and a better resolution of the flow compared to the RANS models [17]. On the other hand, the potential of the RANS models to evaluate complex free surface flows without resolving details regarding the turbulent eddies is not yet fully used. Unsteady RANS simulations offer a balance between computational efficiency and resolution of the flow, which can be used for many engineering applications. A challenge with two-phase RANS simulations is that the effect of the free surface is not accounted for in the standard RANS turbulence models. The turbulence production is determined by the strain rate in fluids and at the interface, the abrupt large difference in the fluid density leads to a high turbulence production. In reality, the presence of the free surface results in the elongation of the turbulent eddies parallel to the flow and dissipation of the turbulent energy in the direction normal to the free surface. This turbulence dissipation at the free surface must be included using additional boundary conditions in RANS turbulence models.

Several authors have presented different approaches to overcome this limitation in the RANS equations for two-phase flow. Jacobsen et al. [18] proposed an alternative method to calculate the turbulence production in the two-equation turbulence models for two-phase CFD models using the vorticity in the flow instead of the strain rate to avoid the over-production of turbulence around the free surface. Devolder et al. [19] demonstrated excessive damping of the free surface due to the over-production of turbulence around the free surface with the $k-\omega$ SST model in OpenFOAM and proposed a buoyancy modification term to account for this problem. Furthermore, Larsen and Fuhrman [20] showed that the overestimation of turbulence around the free surface is inherent in the classical RANS turbulence modelling approach and proposed a new closure term to avoid the exponential growth of the eddy viscosity.

In the current study, the relationship shown by Naot and Rodi [21], following the concept of wall function dissipation adjusted to free surface conditions based on experimental observations is implemented to address turbulence over-production around the free surface. This provides a boundary condition for the turbulence dissipation around the free surface based on the flow physics. The boundary condition is implemented in the open-source CFD model, REEF3D [22] and used to study complex flow features downstream of different hydraulic structures for different flow conditions. The model has been previously applied to various problems in which the representation of the complex deformation of the free surface is essential such as wave trapping between cylinders [23], breaking wave forces [24,25], breaking wave kinematics [26] and floating bodies on waves [27].

The influence of the free surface turbulence boundary condition is investigated for various flow cases in this paper to demonstrate the applicability of the method in a universal manner to complex free surface flows. Numerical simulations are carried out for controlled and free stream outflow conditions to obtain different flow conditions such as sub-critical flow throughout the domain, change from sub-critical to super-critical flow regime, transitional flow, and change from sub-critical to super-critical and back to sub-critical. These changing flow conditions are investigated for free surface flow over hydraulic structures such an embankment weir, a broad-crested weir and in the vicinity of bridge piers with focus on the boundary condition for the turbulence dissipation at the free surface. The numerical results for the free surface and the stream-wise velocity profiles over the water depth are compared to experimental data. The change in the distribution of the eddy viscosity around the free surface and the stream-wise velocity profiles in the presence and absence of the free surface turbulence boundary condition is also analyzed to investigate the influence and importance of turbulence damping at the free surface in the different cases.

2. Numerical Model

REEF3D uses the incompressible RANS equations together with the continuity equation to solve the immiscible two-phase fluid flow problem:

$$\frac{\partial u_i}{\partial x_i} = 0 \tag{1}$$

$$\frac{\partial u_i}{\partial t} + u_j \frac{\partial u_i}{\partial x_j} = -\frac{1}{\rho}\frac{\partial p}{\partial x_i} + \frac{\partial}{\partial x_j}\left[(\nu + \nu_t)\left(\frac{\partial u_i}{\partial x_j} + \frac{\partial u_j}{\partial x_i}\right)\right] + g_i \tag{2}$$

where u is the time averaged velocity, ρ is the density of the fluid, p is the pressure, ν is the kinematic viscosity, ν_t is the eddy viscosity and g the acceleration due to gravity. The equations are solved simultaneously for the two phases and the location of the free surface determines the material properties of the fluid cell such as density and viscosity.

The projection method [28] is used for pressure treatment and the resulting Poisson pressure equation is solved with a geometric multigrid preconditioned BiCGStab solver [29] provided by the high-performance solver library, HYPRE [30]. Turbulence modelling is carried out using the two-equation $k - \omega$ model [31]:

$$\frac{\partial k}{\partial t} + u_j \frac{\partial k}{\partial x_j} = \frac{\partial}{\partial x_j}\left[\left(\nu + \frac{\nu_t}{\sigma_k}\right)\frac{\partial k}{\partial x_j}\right] + P_k - \beta_k k \omega \tag{3}$$

$$\frac{\partial \omega}{\partial t} + u_j \frac{\partial \omega}{\partial x_j} = \frac{\partial}{\partial x_j}\left[\left(\nu + \frac{\nu_t}{\sigma_\omega}\right)\frac{\partial \omega}{\partial x_j}\right] + \frac{\omega}{k}\alpha P_k - \beta \omega^2 \tag{4}$$

where k is the turbulent kinetic energy, ω is the specific turbulent dissipation rate, P_k is the turbulent production rate, the coefficients have the values $\alpha = \frac{5}{9}$, $\beta_k = \frac{9}{100}$, $\beta = \frac{3}{40}$, $\sigma_k = 2$ and $\sigma_\omega = 2$.

This turbulence model is chosen as it is the most suitable for two-phase flow calculations with unsteady flow situations as in the cases presented in this study. This is because $k - \omega$ model tends to be numerically more stable due to the linear relationship between k and ω, $\nu_t = k/\omega$. In comparison, the $k - \epsilon$ model defines this relationship with a quadratic term as in $\nu_t = \rho\, C\, k^2/\epsilon$, which can lead to issues with numerical stability for areas of low levels of turbulence intensity.

To avoid the over-production of eddy viscosity outside the boundary layer in simulations of highly strained flow, eddy-viscosity limiters proposed by [32] are used:

$$\nu_t = \min\left(\frac{k}{\omega}, \sqrt{\frac{2}{3}}\frac{k}{|\mathbf{S}|}\right) \tag{5}$$

where $|\mathbf{S}|$ is the magnitude of the strain tensor.

The free surface influences the turbulence in such a way that the components normal to the surface are damped whereas the components parallel to the surface are enhanced. This is included in the turbulence model through an additional boundary condition at the free surface as shown by [21]. Here, the specific turbulence dissipation term at the free surface ω_s is defined as:

$$\omega_s = \frac{c_\mu^{-\frac{1}{4}}}{\kappa} k^{\frac{1}{2}} \cdot \left(\frac{1}{y'} + \frac{1}{y^*}\right) \tag{6}$$

where $c_\mu = 0.07$ and $\kappa = 0.4$. The variable y' is the virtual origin of the turbulent length scale, and was empirically found to be 0.07 times the mean water depth [33]. Including the distance y^* from the nearest wall gives a smooth transition from the free surface value to the wall boundary value of ω. The damping is carried out only around the interface using the Dirac delta function $\delta(\phi)$:

$$\delta(\phi) = \begin{cases} \frac{1}{2\epsilon}\left(1 + \cos\left(\frac{\pi\phi}{\epsilon}\right)\right) & if\ |\phi| < \epsilon \\ 0 & else \end{cases} \tag{7}$$

where $\epsilon = 1.6dx$ is the width over which the delta function is applied and dx is the grid size. The value is chosen to ensure that at least one grid cell is included around the interface in each direction and the boundary condition is applied just around the interface.

2.1. Free Surface

The free surface is determined with the level set method. The zero-level set of a signed distance function $\phi(\vec{x},t)$ is used to represent the interface between air and water [34]. Moving away from the interface, the level set function gives the shortest distance from the interface. The sign of the function distinguishes between the two fluids across the interface as shown in Equation (8):

$$\phi(\vec{x},t) \begin{cases} > 0 & \text{if } \vec{x} \text{ is in the air phase} \\ = 0 & \text{if } \vec{x} \text{ is at the interface} \\ < 0 & \text{if } \vec{x} \text{ is in the water phase} \end{cases} \tag{8}$$

The level set function is moved under the influence of the velocity field u_j obtained from the solution of the RANS equations to represent the change in the free surface as the simulation progresses, with the convection equation in Equation (9):

$$\frac{\partial \phi}{\partial t} + u_j \frac{\partial \phi}{\partial x_j} = 0 \tag{9}$$

The kinematic and dynamic boundary conditions at the free surface are implicitly enforced by the level set function. The discontinuity in the material properties such as the density and viscosity are handled using an interpolation with the Heaviside function around the interface over a distance of $2.1dx$ [22]. The level set function loses its signed distance property on convection and is reinitialized after every iteration using a partial differential equation-based reinitialization procedure [35] to regain its signed distance property.

2.2. Discretization Schemes

The fifth-order conservative finite difference Weighted Essentially Non-Oscillatory (WENO) scheme [36] is applied to discretize the convective terms of the RANS equations. The level set function, turbulent kinetic energy and the specific turbulent dissipation rate are discretized using the Hamilton-Jacobi formulation of the WENO scheme [37]. The WENO scheme is a minimum third-order accurate and numerically stable even in the presence of large gradients. Time advancement for the momentum and level set equations is carried out using a Total Variation Diminishing (TVD) third-order Runge-Kutta explicit time scheme [38]. Adaptive time stepping is employed to satisfy the CFL criterion based on the maximum velocity in the domain. This ensures numerical stability throughout the simulation with an optimal value of time step size. An implicit scheme is used for the time advancement of the turbulent kinetic energy and the specific turbulent dissipation, as these variables are mostly source term driven with a low influence of the convective terms. Diffusion terms of the velocities are also subjected to implicit treatment to remove them from the CFL criterion. The numerical model has been validated in detail by Bihs et al. [22] and the convergence and order of the model presented. The model is found to be monotonically convergent with the convergence ratio $R \in [0, 1]$. The combination of the higher-order schemes results in a combined order of about 2.7 for the model.

The numerical model uses a Cartesian grid for the spatial discretization together with the Immersed Boundary Method (IBM) to represent the irregular boundaries in the domain. Berthelsen and Faltinsen [39] developed the local directional ghost cell IBM to extend the solution smoothly in the same direction as the discretization, which is adapted to three dimensions in the current model. REEF3D is fully parallelized using the domain decomposition strategy and Message Passing Interface (MPI).

3. Numerical Setup

The free surface flow features in open channel flow vary with the flow conditions. Depending on the boundary conditions, the flow around hydraulic structures can stay sub-critical throughout, change from a sub-critical to a super-critical flow or change from sub-critical to super-critical and then back to a sub-critical flow. An overview of the flow conditions and cases investigated is provided in Table 1. The details regarding numerical setup and boundary conditions are provided in this section.

Table 1. Overview of the flow conditions and cases investigated in the study.

No.	Flow Condition	Case	Re	Fr_{max}
1.	sub-critical flow	embankment weir	4.5×10^4	0.33
		bridge piers	2.0×10^5	0.40
2.	sub-critical to super-critical flow	broad-crested weir	4.6×10^4	3.98
3.	transitional flow	embankment weir	6.4×10^4	2.63
4.	sub-critical to super-critical to sub-critical flow	embankment weir	7.8×10^4	4.16
		bridge piers	1.95×10^5	2.09

3.1. General Boundary Conditions

In all the cases presented in this study, a logarithmic velocity profile is prescribed at the inflow for the water phase. The air velocities are not prescribed and are a part of the solution. The outflow boundary condition for the simulations with an embankment weir is a zero-gradient or free stream outflow condition, where no resistance is offered to the flow. The required tailwater level in the different cases is maintained using a rectangular obstacle placed towards the end of the channel to represent the outflow gate in the experiments. The free stream outflow boundary condition is also applied in the simulation of the rectangular weir where the flow transitions from sub-critical to super-critical. In the case of the bridge piers, the tailwater level is prescribed at the outflow. The side walls in the two-dimensional numerical channel are treated with a zero-gradient Neumann condition, whereas a no-slip wall boundary condition is used in the three-dimensional channel. A no-slip wall boundary condition is enforced at the bottom of the channel and a zero-gradient Neumann condition is applied to the top of the numerical channel in all the simulations. An initial water depth is prescribed to initialize the free surface in the channel at $t = 0$ and a discharge given in the experiments is also prescribed. A Cartesian grid is employed in all the simulations with $dx = dy = dz$ for all the cases.

3.2. Sub-Critical Flow

The first case for this flow condition is the experiment carried out for submerged overflow on an embankment weir by Fritz and Hager [5]. The experiments are carried out in a channel 7 m long, 0.499 m wide and 0.700 m high. The embankment weir is 0.3 m high and 1.5 m long, with a slope of 1V: 2H on both the upstream and downstream faces. The two-dimensional numerical channel setup is similar to the experimental setup and is illustrated in Figure 1 showing the water and air phases within the outline of the channel. The channel has a length of 12 m and a height of 0.7 m, and the water depth is initialized as 0.454 m and a discharge of 0.038 $m^3/s/m$ is specified. The tailwater level in the channel is maintained at $d_t = 0.450$ m using a rectangular obstacle of height 0.355 m and length 0.5 m placed at 11.0 m in the channel. Stream-wise velocity profiles are calculated at four locations in the channel, upstream of the weir at 1.94 m, on the upstream face at 2.54 m, over the downstream toe at 3.54 m and downstream of the weir at 3.94 m.

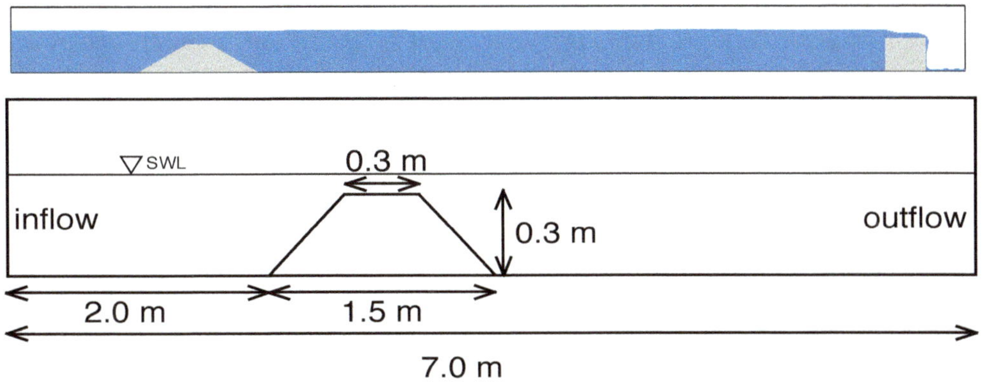

Figure 1. Numerical channel setup for the cases with an embankment weir following Fritz and Hager [5].

The second case for this flow condition is the flow around a pair of bridge piers presented by Szydłowski [40]. The channel in the experiments is 15 m long, 0.62 m wide and a pair of bridge piers of diameter 0.11 m are placed 2 m downstream of the inflow section equidistant from each other and the channel walls. The numerical setup is 4.0 m long, 0.62 m wide and 1.0 m high. The water depth is initialized to 0.48 m at $t = 0$ s and the tailwater level is prescribed to be 0.48 m along with a discharge of 0.114 m^3/s in the channel. The piers are placed 1.30 m from the inflow region, equidistant from the channel walls and from each other with a distance of 0.134 m. Stream-wise velocity profiles are calculated at four locations, upstream of the piers at 0.8 m, in between the piers at 1.30 m, downstream of the piers at 1.63 m and 3.0 m. The numerical setup is illustrated in Figure 2 showing the water and air phases within the outline of the numerical channel and the piers.

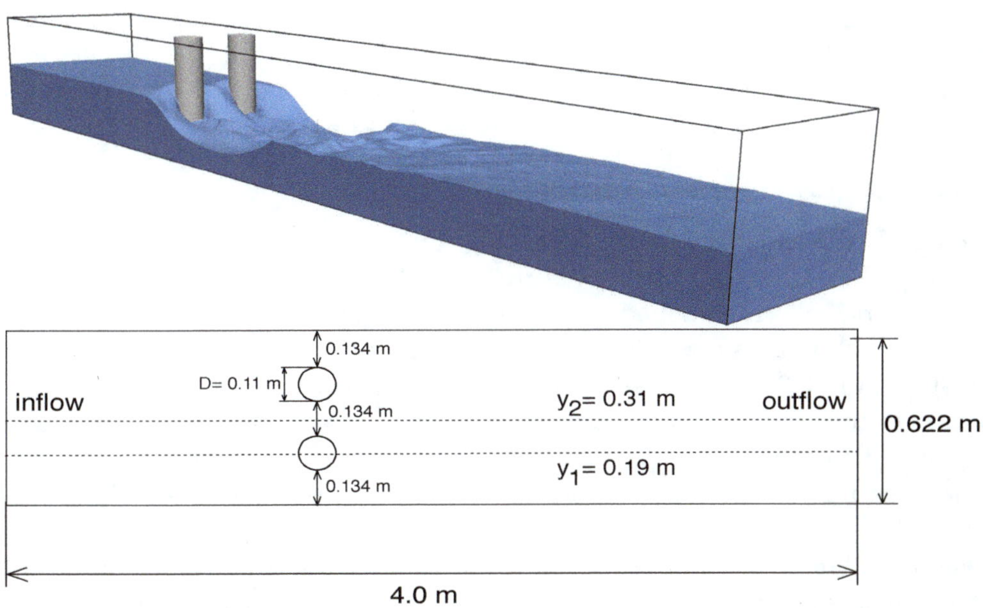

Figure 2. Numerical channel setup with bridge piers following Szydłowski [40].

3.3. *Sub-Critical to Super-Critical Flow*

The experiment presented by Sarker and Rhodes [41] for flow over a rectangular broad-crested weir is simulated for this flow condition. The experimental channel is 3.162 m long and 0.105 m wide. A rectangular weir with sharp edges 0.4 m long and 0.1 m high is placed in the channel with a discharge of 0.004684 m^3/s. The numerical setup is illustrated in Figure 3 showing the outline of the numerical channel, the water, and air phases and the weir. The numerical domain is 3.16 m long, 0.105 m wide and 0.4 m high. The broad-crested weir is 0.4 m long and 0.1 m high as in the experiments and placed at 1.20 m from the inlet. The free surface elevation at the inflow and the discharge are prescribed to be 0.194 m and 0.004684 m^3/s respectively, using the values from the experimental data.

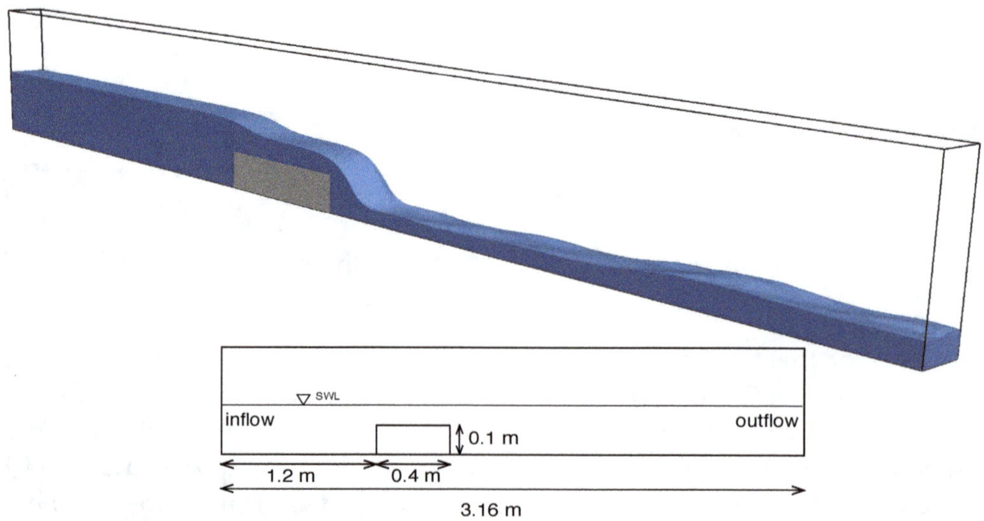

Figure 3. Numerical channel setup for broad-crested weir following Sarker and Rhodes [41].

3.4. *Transitional Flow*

For this flow condition, the plunging jet flow resulting from a 1% submergence of the embankment weir presented by Fritz and Hager [5] is presented. The numerical setup is similar to that illustrated in Figure 1, with the height of the rectangular obstacle set to 0.19 m to maintain a tailwater level of $d_t = 0.301$ m. A discharge of 0.054 $m^3/s/m$ is specified in the two-dimensional channel. The water depth is initialized at 0.40 m at $t = 0$ s in the simulation. Stream-wise velocity profiles are calculated at four locations in the channel, upstream of the weir at 1.94 m, on the upstream face at 2.54 m, over the downstream toe at 3.54 m and downstream of the weir at 3.94 m.

3.5. *Sub-Critical to Super-Critical to Sub-Critical Flow*

The first case for this flow condition is the free overflow on an embankment weir presented by Fritz and Hager [5]. The two-dimensional numerical channel is the same as that illustrated in Figure 1. The tailwater level is maintained at $d_t = 0.159$ m using a rectangular obstacle of height 0.06 m. A discharge of 0.054 $m^3/s/m$ is prescribed in the two-dimensional channel and the water depth is initialized at 0.40 m at $t = 0$ s. Stream-wise velocity profiles are calculated at four locations in the channel, upstream of the weir at 1.95 m, on the upstream face at 2.54 m, over the downstream toe at 3.54 m and downstream of the weir at 3.94 m.

The second case for this flow condition is the flow around a pair of bridge piers presented by Szydłowski [40], using the numerical setup illustrated in Figure 2, but with a height of 0.5 m. A tailwater level of 0.19 m, a discharge of 0.114 m^3/s is prescribed and the water depth initialized at 0.19 m at $t = 0$ s. Stream-wise velocity profiles are calculated at four locations, upstream of the piers at 0.8 m, in between the piers at 1.30 m, downstream of the piers at 1.63 m and 3.0 m.

4. Results and Discussion

The different flow conditions analyzed in this study have special free surface features associated with them. The change of the flow conditions from a super-critical condition to a sub-critical condition involves major energy losses, formation of special free surface features such as a hydraulic jump and can be especially challenging to represent correctly in numerical simulations. In the following sections, the numerical results for different scenarios with different flow conditions around an embankment weir, a broad-crested weir, and a pair of bridge piers are presented. The numerical results for the free surface elevation in all the cases are compared to experimental data. The velocity profiles at various locations in the numerical channel and the difference in the eddy viscosity in the numerical channel with and without the application of the free surface turbulence damping (FSTD) treatment is presented. In addition, the extent of the recirculation zones from Fritz and Hager [5] is presented along with the numerical results for the embankment weirs.

4.1. Sub-Critical Flow

4.1.1. Flow over an Embankment Weir

In the first case, submerged overflow on an embankment weir with a surface jet flow and a nearly horizontal free surface is studied. A grid convergence study is carried out for this case with FSTD using $dx = 0.02$ m, 0.01 m and 0.005 m. The grid convergence for the free surface elevation is presented in Figure 4 and for the stream-wise velocity profiles in Figure 5. The numerical results for the free surface and the velocity profiles are seen to be similar for $dx = 0.01$ m and $dx = 0.005$ m and the simulation with a grid size of $dx = 0.01$ m is used for further analysis of this flow condition.

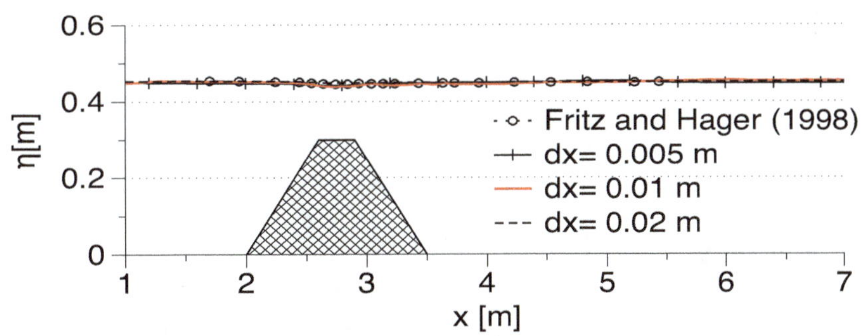

Figure 4. Grid convergence study with FSTD for the free surface elevation in the case of submerged flow with a surface jet over an embankment weir.

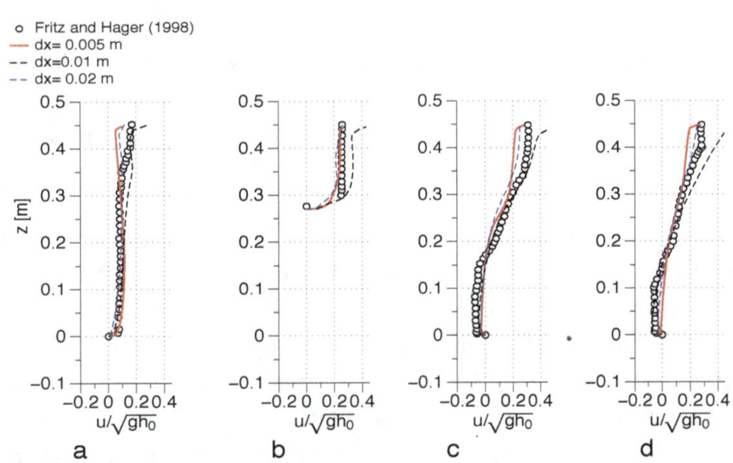

Figure 5. Grid convergence study with FSTD for the horizontal velocity profiles over the water depth at various locations for submerged flow over an embankment weir: (**a**) $x = 1.95$ m, (**b**) $x = 2.54$ m, (**c**) $x = 3.54$ m, (**d**) $x = 4.04$ m.

The numerical results for the free surface elevation with and without FSTD are compared to the experimental data in Figure 6. A good agreement is seen between the numerical result with FSTD and the experimental data. The numerical free surface elevation without FSTD in Figure 6 is seen to be similar but slightly lower just downstream of the weir between 2.6 m and 4.5 m. Due to the sub-critical flow condition throughout the channel, the free surface has a relatively simple, almost horizontal form and as a result the two simulations show similar results for the free surface elevation. The stream-wise velocity profiles in the simulations with and without FSTD are presented in Figure 7. A large deviation is seen in the velocities closer to the free surface in Figure 7b–d, indicating that the recirculation zones in the flow are calculated differently in the two simulations. The velocity profiles with FSTD show a reasonable agreement with the experimental data.

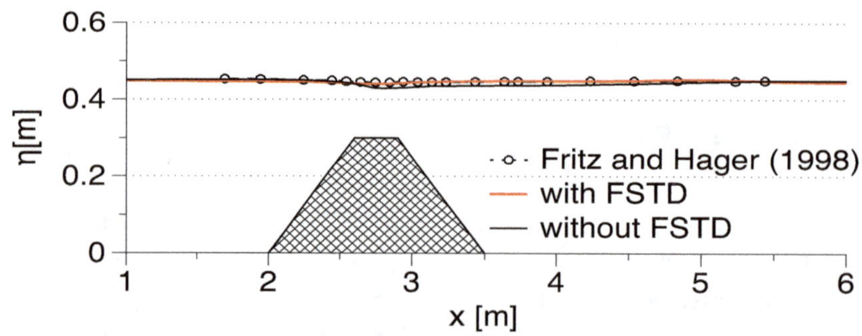

Figure 6. Comparison of the experimental data [5] and the numerical results with and without FSTD for the free surface elevation in the case of submerged flow with a surface jet over an embankment weir.

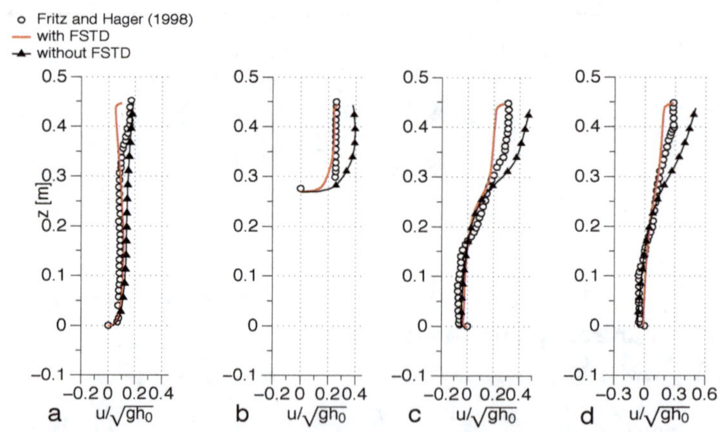

Figure 7. Horizontal velocity (u) profiles over the water depth at various locations along the channel with and without FSTD for submerged flow with a surface jet over an embankment weir: (**a**) $x = 1.95$ m, (**b**) $x = 2.54$ m, (**c**) $x = 3.54$ m, (**d**) $x = 4.04$ m.

The flow field in the numerical model for this scenario with and without FSTD is presented in Figure 8. The lower limit of the recirculation zone presented by Fritz and Hager [5] is overlaid as white squares on the streamlines to provide a comparison of the recirculation zone calculated in the simulations and the experiments. The streamlines in the water phase in Figure 8a show that while the surface jet is represented, the recirculation zone is over-estimated in the simulation without FSTD. In the simulation with FSTD, the surface jet and the extent of the recirculation zone below the free surface downstream of the embankment weir is correctly represented, as seen from Figure 8b. Thus, despite the numerical free surface elevations in the two simulations being similar, the flow features agree better with the experimental observation in the simulation with FSTD.

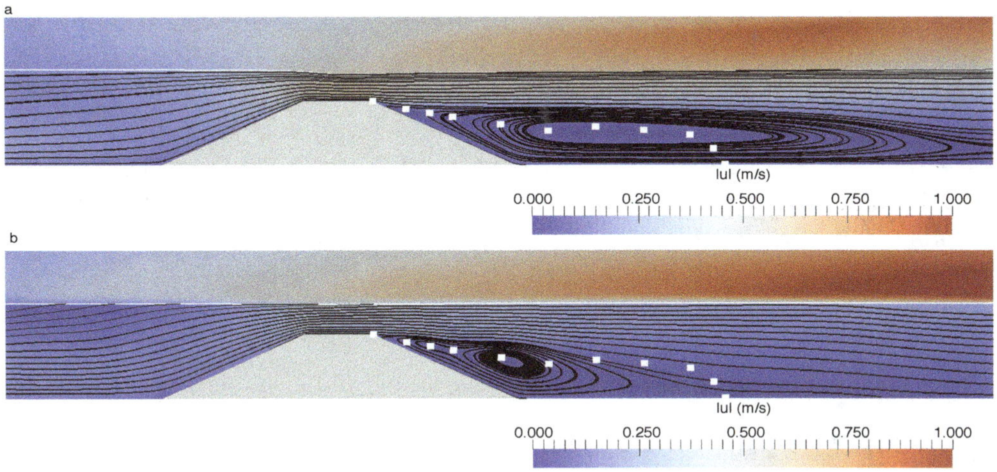

Figure 8. Free surface features for submerged flow over an embankment weir without and with FSTD, the white dots denote the lower limit of the mixing layer presented in Fritz and Hager [5]: (**a**) without FSTD, (**b**) with FSTD.

Furthermore, the eddy viscosity in the numerical channel is presented in Figure 9. In the simulation without FSTD, Figure 9a shows that the eddy viscosity has large values around the free surface downstream of the weir, both in the air and water phases. Whereas, in Figure 9b with FSTD, the eddy-viscosity values are much lower around the free surface and higher around the recirculation zone under the free surface and further downstream. Thus, it is seen that even in the simplest case of sub-critical flow throughout the channel, the damping of the turbulence at the free surface significantly aids in the better representation of the flow, even though no significant changes are seen in the calculation of the free surface elevation.

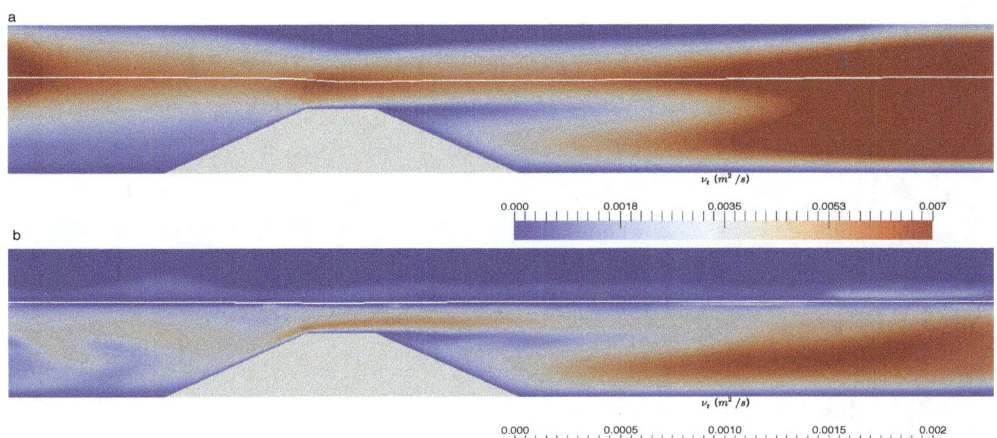

Figure 9. Eddy-viscosity contours for submerged flow over an embankment weir: (**a**) without FSTD, high values of eddy viscosity around the free surface both in air and water phases, (**b**) with FSTD, the reduction in eddy viscosity around the free surface is clearly seen.

4.1.2. Flow around a Pair of Bridge Piers

In the second case with sub-critical flow throughout the domain, the flow around a pair of bridge piers is presented. The grid convergence study for the free surface elevation along a line through a pier and along the center of the channel is shown in Figure 10a,b and is seen that the numerical results are similar for all the grid sizes. The results from the simulation with $dx = 0.01$ m are used for further analysis. The free surface calculated in simulations with and without FSTD treatment in Figure 11 show that the results are similar. There is a small deflection in the free surface downstream of the piers along the center of the channel which is slightly over-estimated in the simulation without

FSTD treatment. The velocity profiles upstream of the piers in Figure 12a at 0.8 m and Figure 12b at 1.30 are identical, indicating that the FSTD treatment does not influence the flow upstream of the cylinder. The velocity profile in the vicinity of the deflection at 1.63 m in Figure 12c shows a difference in the flow field corresponding to the small difference in the location of the free surface at this location. Further downstream from the piers, the velocity profiles at 3.0 m in Figure 12d, are identical again for the simulations with and without FSTD treatment. The results show that the FSTD treatment has only minor implications for the flow in this case as well.

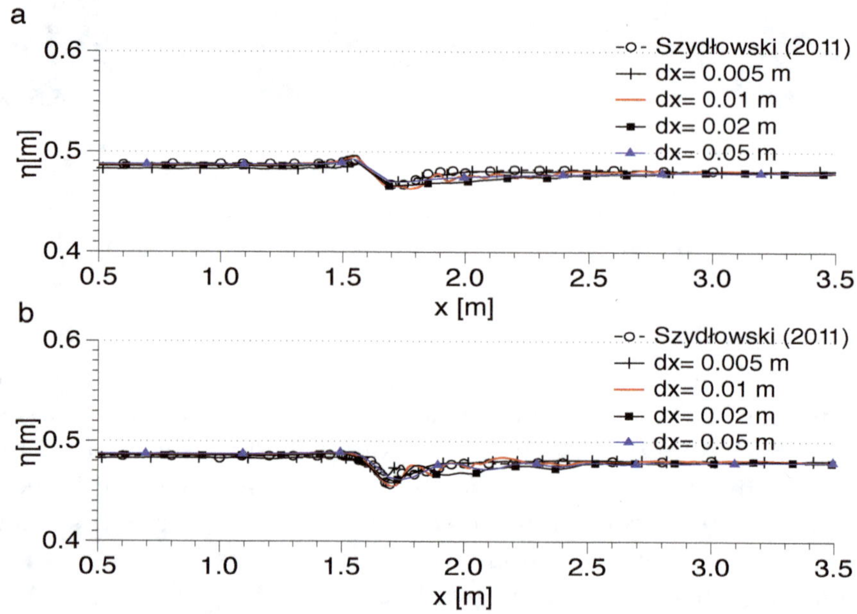

Figure 10. Grid convergence study with FSTD for the numerical free surface elevation for flow around a pair of bridge piers: (**a**) along the center line through a pier $y_1 = 0.19$ m, (**b**) along the center of the channel $y_2 = 0.31$ m.

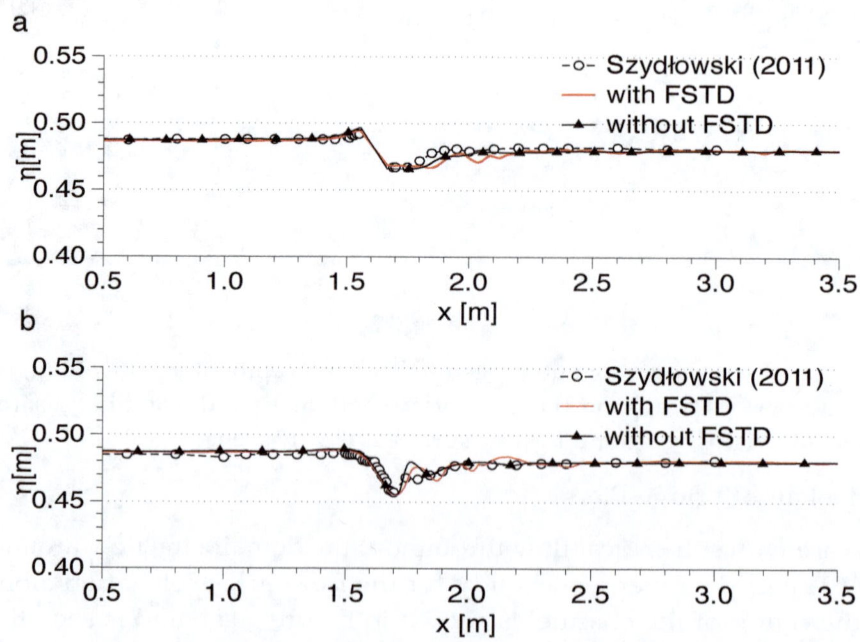

Figure 11. Comparison of the experimental data [40] and the numerical results for the free surface elevation with and without FSTD: (**a**) along the center line through a pier $y_1 = 0.19$ m, (**b**) along the center of the channel $y_2 = 0.31$ m.

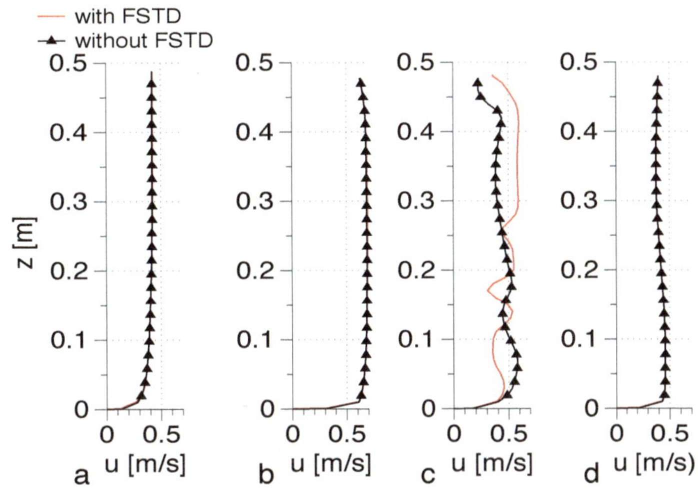

Figure 12. Horizontal velocity (u) profiles over the water depth at various locations along the channel with and without FSTD: (**a**) $x = 0.80$ m, (**b**) $x = 1.30$ m, (**c**) $x = 1.63$ m, (**d**) $x = 3.00$ m.

From the results presented for sub-critical flow throughout the channel with an embankment weir and a pair of bridge piers, it is seen that the representation of the free surface is similar in simulations with and without FSTD treatment. The flow pattern in the vicinity of the structures, on the other hand, can be influenced to some extent. The velocity profiles in the case of the embankment weir with a shallower water depth show a significant difference and consequently the recirculation pattern is not represented correctly without FSTD treatment. In the case of the bridge piers with a higher water depth, the velocity profiles show only a minor difference in the vicinity of the piers.

4.2. Sub-Critical to Super-Critical

In this scenario, the flow condition changes from sub-critical to super-critical after passing the hydraulic structure and stays super-critical until the outflow region. A grid convergence study is carried out for this case with FSTD using $dx = 0.025$ m, 0.01 m, 0.005 m and 0.0025 m. Figure 13 shows that the most coarse grid with $dx = 0.025$ m results in a spurious wave-like flow, which disappears on refining the grid to $dx = 0.01$ m. The results for the free surface elevation presented in Figure 13 show that the solution converges at grid size $dx = 0.005$ m and this is selected for further analysis of this flow condition.

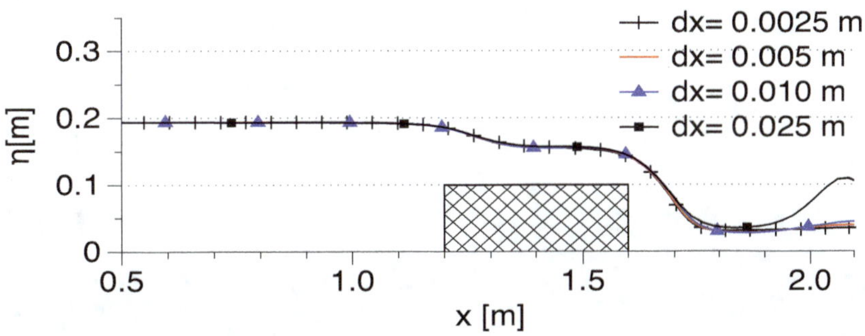

Figure 13. Grid convergence study with FSTD for the free surface elevation in the case of flow over a broad-crested weir with a free stream outflow condition.

The numerical results with and without FSTD for the free surface elevation are compared to the experimental data from Sarker and Rhodes [41] in Figure 14. The numerical result with FSTD agrees well with the experimental data. Also, the difference in the numerical results for the free surface elevation with and without FSTD is not very significant.

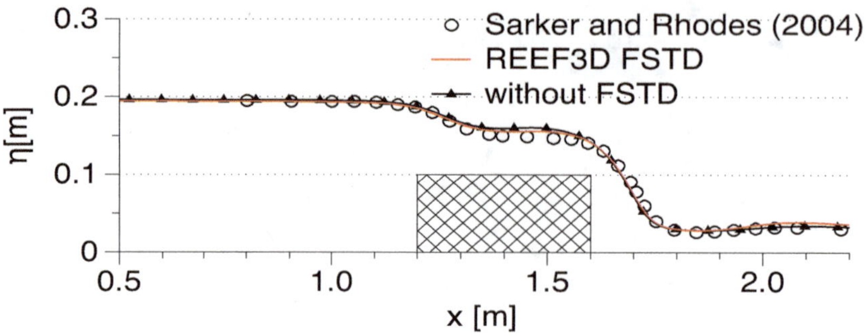

Figure 14. Comparison of the experimental data [41] and the numerical results with and without FSTD for the free surface elevation over a broad-crested weir with free stream outflow condition.

The streamlines and the velocity magnitude contours in the two simulations are shown in Figure 15. It is seen that the extent of the recirculation zone is similar in the two simulations whereas the maximum velocities in the numerical channel are higher in the simulation without FSTD. The distribution of the eddy viscosity along the center of the numerical channel is presented in Figure 16. The higher values of eddy viscosity are distributed around the free surface, in the region with super-critical flow downstream of the weir in Figure 16a for the simulation without FSTD. Whereas in the simulation with FSTD shown in Figure 16b, the higher values of eddy viscosity are found below the free surface and in the recirculation zone downstream of the weir and a region of reduced eddy viscosity around the free surface is clearly seen.

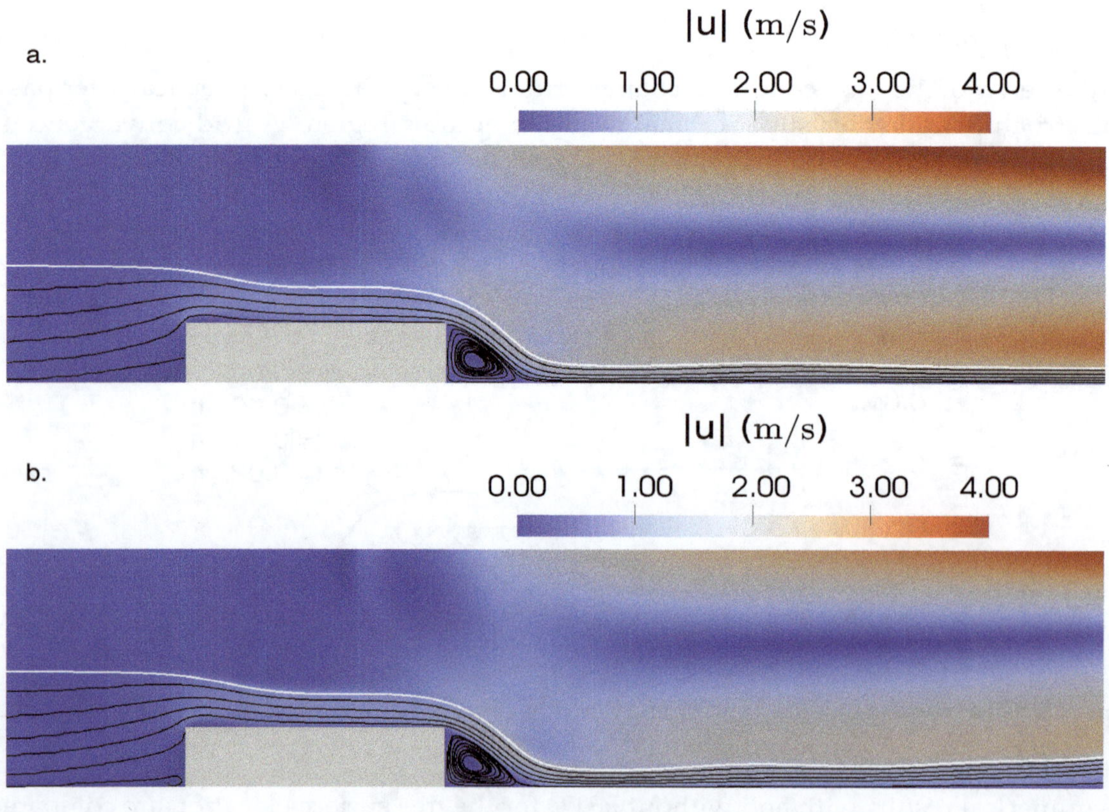

Figure 15. Free surface features around the broad-crested weir with and without FSTD showing streamlines and velocity magnitude contours: (**a**) without FSTD, (**b**) with FSTD.

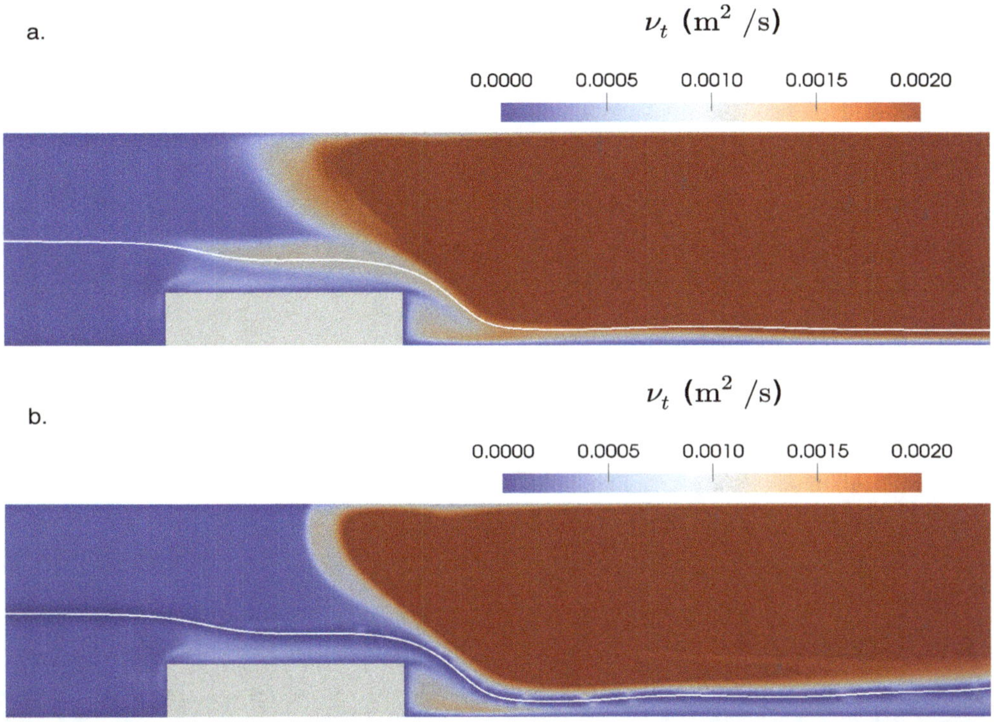

Figure 16. Eddy viscosity around the broad-crested weir: (**a**) without FSTD, high values of eddy viscosity around the free surface in both air and water phases, (**b**) with FSTD, the reduction in the eddy viscosity around the free surface is clearly seen.

Thus, in the scenario where the flow condition changes from sub-critical to super-critical and stays super-critical until the outlet, the free surface elevation is not changed significantly by FSTD. Some changes are seen in the distribution of the eddy viscosity and the maximum velocity in the channel.

4.3. Transitional Flow

In this section, the flow condition with a plunging jet flow on an embankment weir is presented. The grid convergence study for the free surface elevation in Figure 17 shows that for a coarse grid with $dx = 0.02$ m, the plunging jet flow is not represented. With finer grids of $dx = 0.01$ m and $dx = 0.005$ m the numerical free surface elevation agrees with the experimental data. This is also reflected in the grid convergence study for the stream-wise velocity profiles presented in Figure 18, where the results for $dx = 0.02$ m does not agree with the experimental observation downstream of the weir. The velocity profiles for $dx = 0.01$ m and $dx = 0.005$ m are similar and the results from $dx = 0.01$ m are used for further analysis of the flow situation.

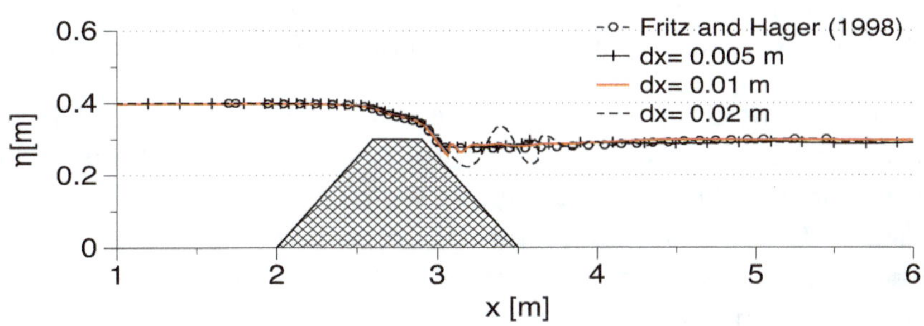

Figure 17. Grid convergence study with FSTD for the free surface elevation in the case of plunging jet flow on an embankment weir.

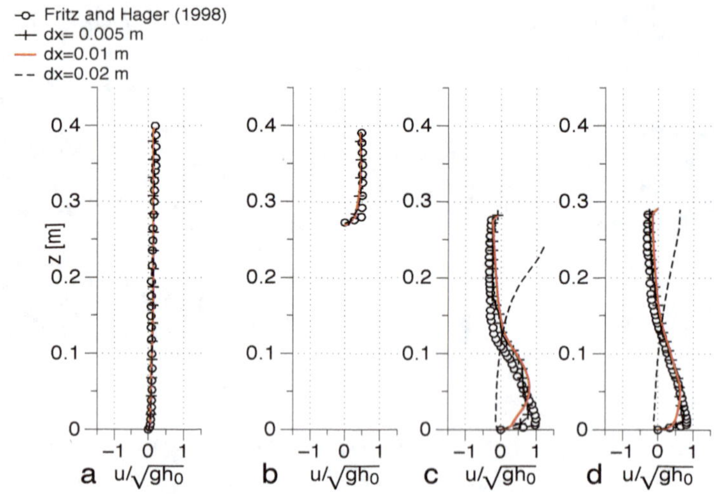

Figure 18. Grid convergence study with FSTD for the horizontal velocity profiles over the water depth at various locations for plunging jet flow on an embankment weir: (**a**) $x = 1.94$ m, (**b**) $x = 2.54$ m, (**c**) $x = 3.54$ m, (**d**) $x = 3.94$ m.

The free surface elevations in the simulations with and without FSTD treatment are presented in Figure 19. In the simulation without FSTD, the deflection in the free surface at the location of the plunging jet is over-estimated, whereas in the simulation with FSTD, the free surface agrees with the experimental data. In addition, the free surface further downstream of the weir is underestimated in the simulation without FSTD treatment. The stream-wise velocity profiles in Figure 20a,b show that the calculated flow field is identical upstream of the weir at $x = 1.94$ m and on the upstream face of the weir at $x = 2.54$ m is identical and in agreement with the experimental data. For the velocity profiles on the downstream face of the weir at $x = 3.54$ m and further downstream at $x = 3.94$ m in Figure 20c,d respectively, the numerical results with FSTD agree with the experimental data, whereas without FSTD, the velocities close to the free surface are over-estimated.

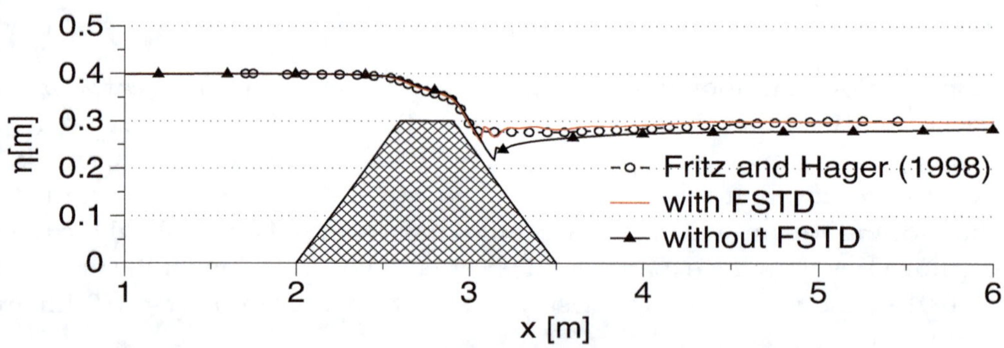

Figure 19. Comparison of the experimental data [5] and the numerical results with and without FSTD for the free surface elevation in the case of plunging jet flow on an embankment weir.

The flow in the numerical channel with and without FSTD treatment is investigated using streamlines and the extent of the recirculation zone is compared with the observations in the experiments by Fritz and Hager [5] in Figure 21. While the plunging jet flow is seen in both simulations from the velocity magnitude contours, the recirculation pattern is different in the two simulations. In the simulation without FSTD treatment in Figure 21a, the extent of the recirculation zone is smaller than in the experiments but is correctly represented in the simulation with FSTD as seen in Figure 21b. The eddy viscosity in the simulation without FSTD treatment is seen to have very high values around the free surface in Figure 22a. The FSTD treatment results in lower turbulent production around the interface as seen in Figure 22b.

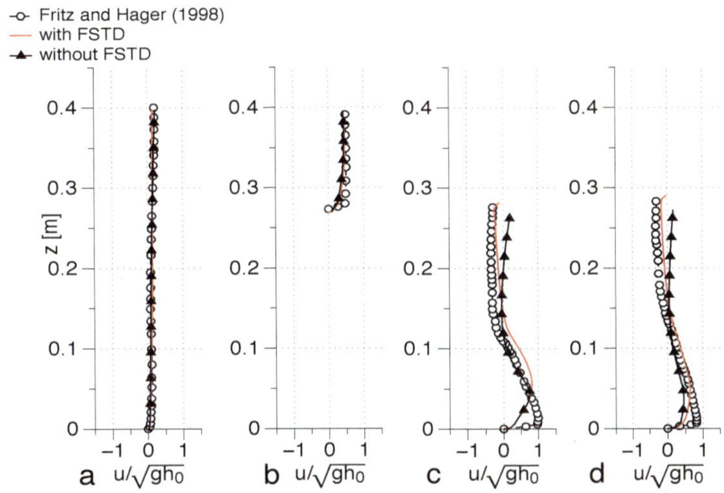

Figure 20. Horizontal velocity (u) profiles over the water depth for plunging jet flow at various locations along the channel with and without FSTD (FSTD): (**a**) x = 1.94 m, (**b**) x = 2.54 m, (**c**) x = 3.54 m, (**d**) x = 3.94 m.

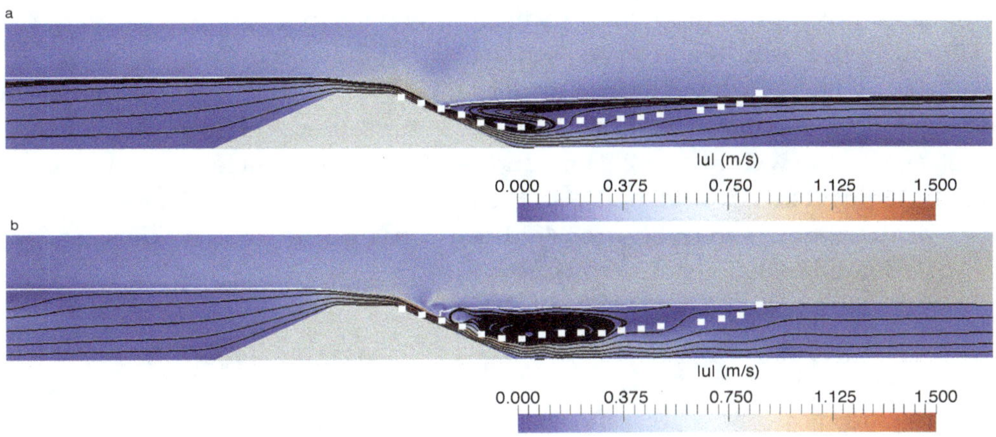

Figure 21. Free surface features for plunging jet flow on an embankment weir without and with FSTD, the white dots denote the lower limit of the mixing layer presented in Fritz and Hager [5]: (**a**) without FSTD, (**b**) with FSTD.

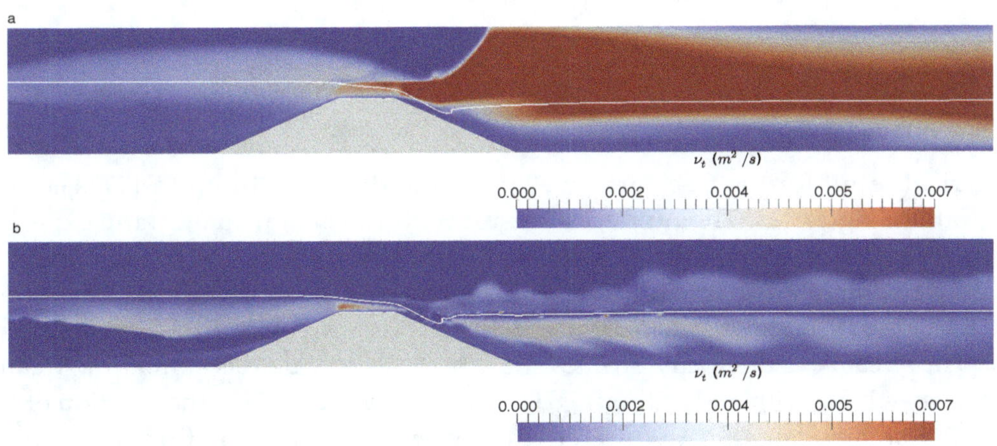

Figure 22. Eddy-viscosity contours for plunging jet flow on an embankment weir: (**a**) without FSTD, high values of eddy viscosity are noticed around the interface in both air and water phases, (**b**) with FSTD, the reduction in the eddy viscosity around the free surface is clearly seen.

From the results for the transitional flow regime with a plunging jet flow on an embankment weir, it is seen that the difference in the calculation of the free surface and the recirculation pattern is significantly influenced by the over-production of turbulence around the free surface in the absence of the FSTD algorithm. This contrasts with the flow scenarios with sub-critical flow throughout the channel and sub-critical to super-critical flow transition presented in the previous sections, where the difference in the numerical results are generally smaller with and without the application of FSTD.

4.4. Sub-Critical to Super-Critical to Sub-Critical

This is the most complex flow scenario simulated in this study where the flow conditions change from sub-critical to super-critical and back to sub-critical in the vicinity of the hydraulic structure. Two cases are presented for this scenario, the first case is the free overflow over an embankment weir presented by Fritz and Hager [5] and the second case is the flow around a pair of bridge piers presented by Szydłowski [40].

4.4.1. Flow over an Embankment Weir

A grid convergence study is carried out with FSTD using $dx = 0.03$ m, 0.02 m, 0.01 m and 0.005 m. The results for the free surface elevation shown in Figure 23. For the coarse grid with $dx = 0.03$ m, the hydraulic jump is not represented, and additional waves are seen downstream of the weir for $dx = 0.02$ m. The numerical results improve with grid refinement and the free surface elevation with $dx = 0.01$ m and $dx = 0.005$ m are seen to be similar and agree with the experimental data. The grid convergence study for the stream-wise velocity profile is presented in Figure 24. The velocity profiles are identical upstream of the weir at $x = 2.94$ m, and on the upstream face of the weir at $x = 2.54$ m for all the grid sizes in Figure 24a,b, respectively. At the location of the hydraulic jump at $x = 3.54$ m and further downstream at $x = 3.94$ m in Figure 24c,d respectively, the velocity profiles for $dx = 0.01$ m and $dx = 0.005$ m are similar. The results from $dx = 0.01$ m are used for further analysis of the case.

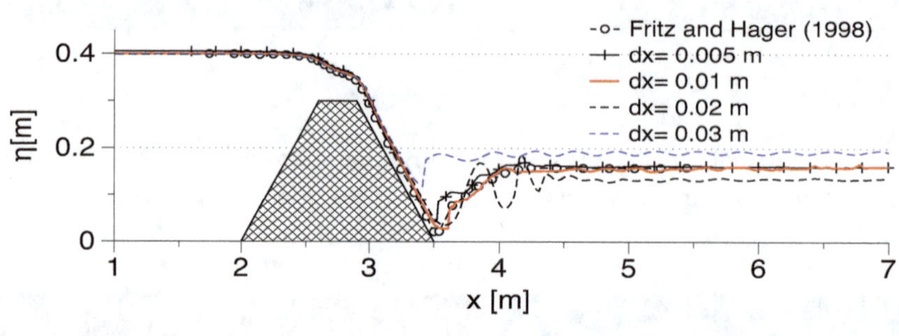

Figure 23. Grid convergence study with FSTD for the free surface elevation in the case of free overflow on an embankment weir.

The numerical result for the free surface elevation with and without FSTD is compared to the experimental data in Figure 25. The numerical result with FSTD agrees with the experimental data, including the representation of the hydraulic jump. The difference between the calculated free surface elevations in the simulations with and without FSTD is clearly seen. The free surface elevation returns to the tailwater level sharply downstream of the weir and has a constant free surface in the simulation without FSTD. The stream-wise velocity profiles show that the flow field is similar upstream of the weir in Figure 26a,b respectively with and without FSTD. The velocity profile at the location of the hydraulic jump in Figure 26c shows a significant difference between the two simulations corresponding to the difference in the free surface location. The velocity profile in the simulation with FSTD follows the observations from the experiments, though a slight overestimation of the free surface location by about 0.01 m and a corresponding deviation from the experimental observation is seen in Figure 26c. Further downstream of the weir at $x = 3.94$ m in Figure 26d, the over-estimated horizontal free surface in the simulation without FSTD corresponds to a different velocity profile, while the result with FSTD treatment is closer to the experimental observation.

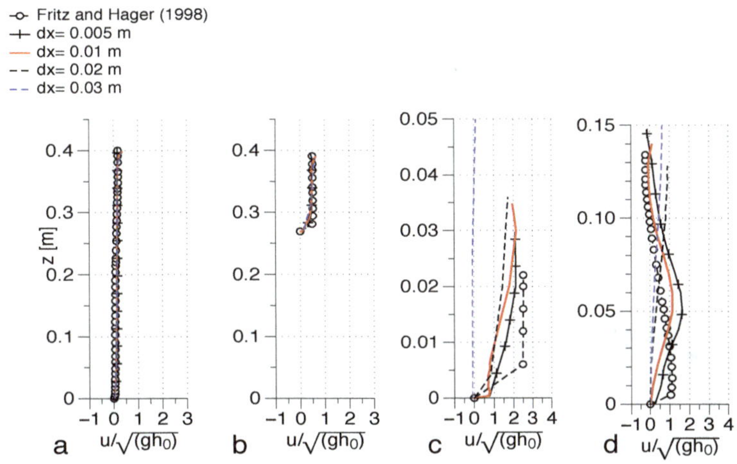

Figure 24. Grid convergence study with FSTD for the horizontal velocity profiles over the water depth at various locations for free overflow on an embankment weir: (**a**) $x = 1.94$ m, (**b**) $x = 2.54$ m, (**c**) $x = 3.54$ m, (**d**) $x = 3.94$ m.

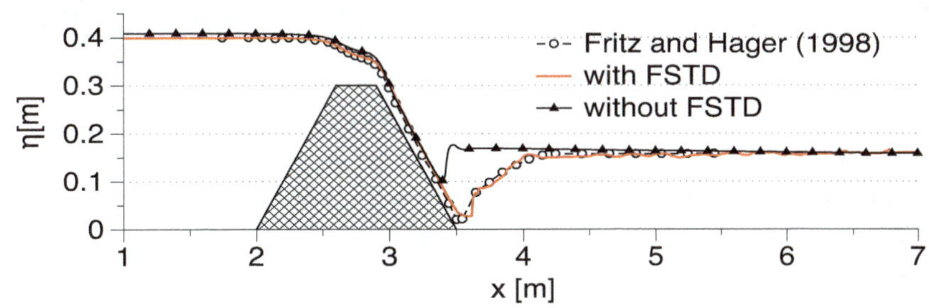

Figure 25. Comparison of the experimental data [5] and the numerical results with and without FSTD for the free surface elevation in the case of free overflow on an embankment weir.

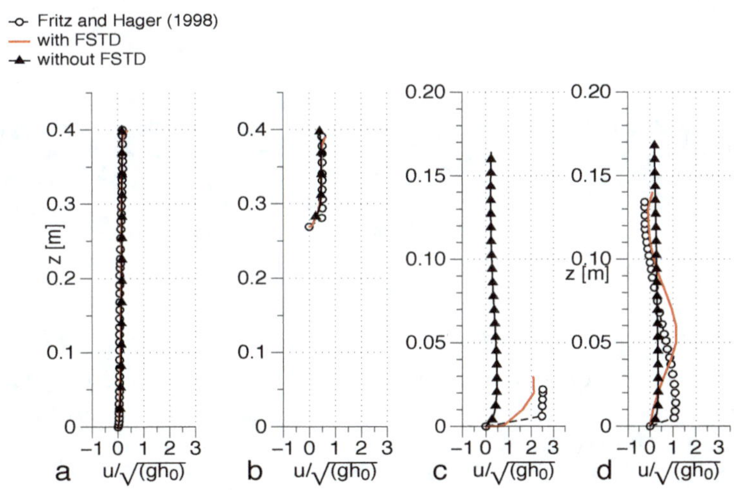

Figure 26. Horizontal velocity (u) profiles over the water depth for free overflow on an embankment weir at various locations along the channel with and without FSTD: (**a**) $x = 1.94$ m, (**b**) $x = 2.54$ m, (**c**) $x = 3.54$ m, (**d**) $x = 3.94$ m.

The lower limit of the recirculation zone presented by Fritz and Hager [5] is overlaid as white squares on the streamlines to provide a comparison of the recirculation zone calculated in the simulations in Figure 27. The streamlines and the velocity magnitude contours in Figure 27a show that the recirculation zone and the submerged flow are not calculated in the simulation without FSTD. The simulation with FSTD in Figure 27b shows that the calculated recirculation region agrees well with the lower limit of the recirculation zone in the experiments. The distribution of the eddy viscosity in the channel for the two simulations is presented in Figure 28. In the simulation without FSTD, the higher values of eddy viscosity are located around the free surface, in the region of the hydraulic jump. This has a damping effect on the flow in the region and the hydraulic jump is not represented correctly as seen in Figure 28a. In the simulation with FSTD in Figure 28b, the values of eddy viscosity are the lowest just around the free surface.

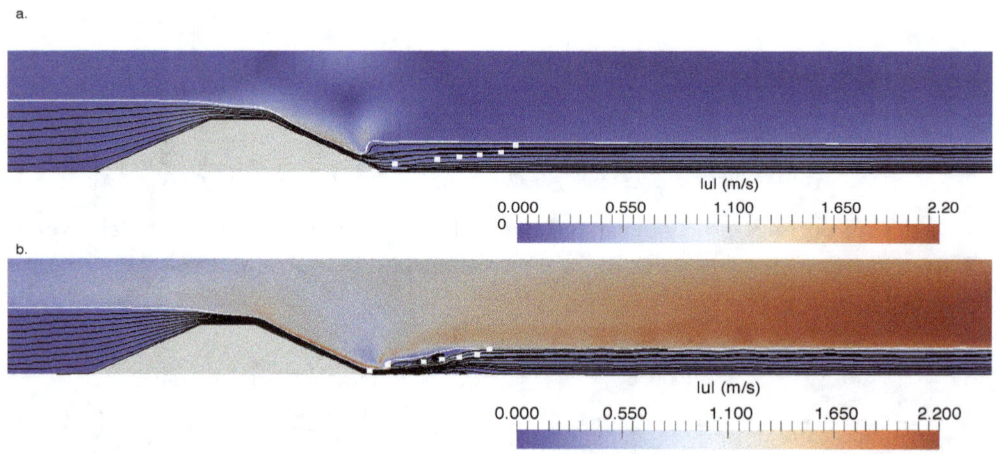

Figure 27. Free surface features for free overflow on an embankment weir without and with FSTD, the white dots denote the lower limit of the mixing layer presented in Fritz and Hager [5]: (**a**) without FSTD, (**b**) with FSTD.

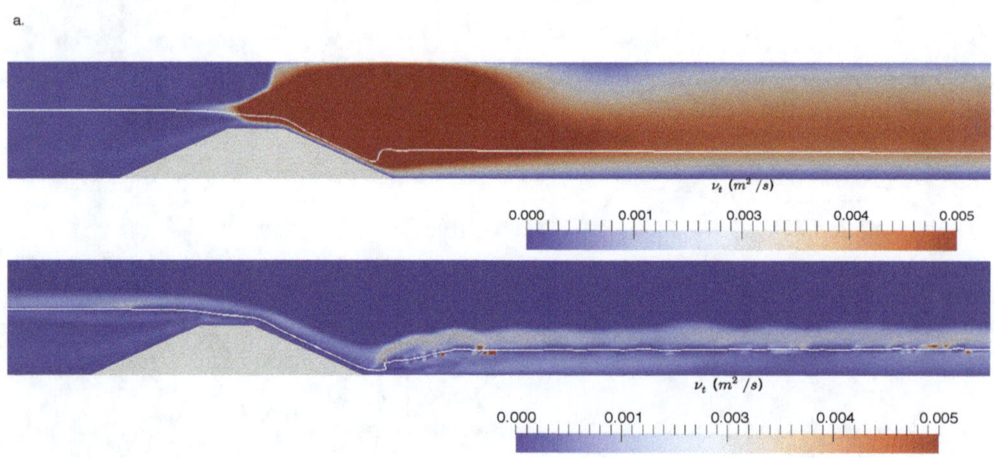

Figure 28. Eddy-viscosity contours for free overflow on an embankment weir: (**a**) without FSTD, high values of eddy viscosity are noticed around the interface in both air and water phases, (**b**) with FSTD, the reduction in the eddy viscosity around the free surface is clearly seen.

Thus, it is clearly seen from this case that the FSTD treatment has a significant effect on the calculation of the free surface elevation and the flow features involved in a scenario where the flow conditions undergo rapid changes from sub-critical to super-critical and back to sub-critical. The representation of the free surface elevation and the recirculation zone is more realistic in the simulations with FSTD as seen from the comparisons with the experimental results.

4.4.2. Flow around a Pair of Bridge Piers

A grid convergence study is carried out in this case using grid sizes $dx = 0.00625$ m, 0.01 m, 0.025 m, 0.05 m and 0.10 m using FSTD in the simulations. The results for the free surface along the center of the channel ($y_2 = 0.31$m) and along the center line through a pier ($y_1 = 0.19$ m) are presented in Figure 29. The free surface elevations agree with the experimental data for $dx = 0.01$ m and $dx = 0.00625$ m with the hydraulic jump correctly represented in the numerical result. The grid convergence study for the stream-wise velocity profiles in Figure 30 shows that the numerical results improve with grid refinement and similar values are obtained with $dx = 0.01$ m and $dx = 0.00625$ m. The results obtained with the grid size of $dx = 0.01$ m are used in the analysis of this flow condition.

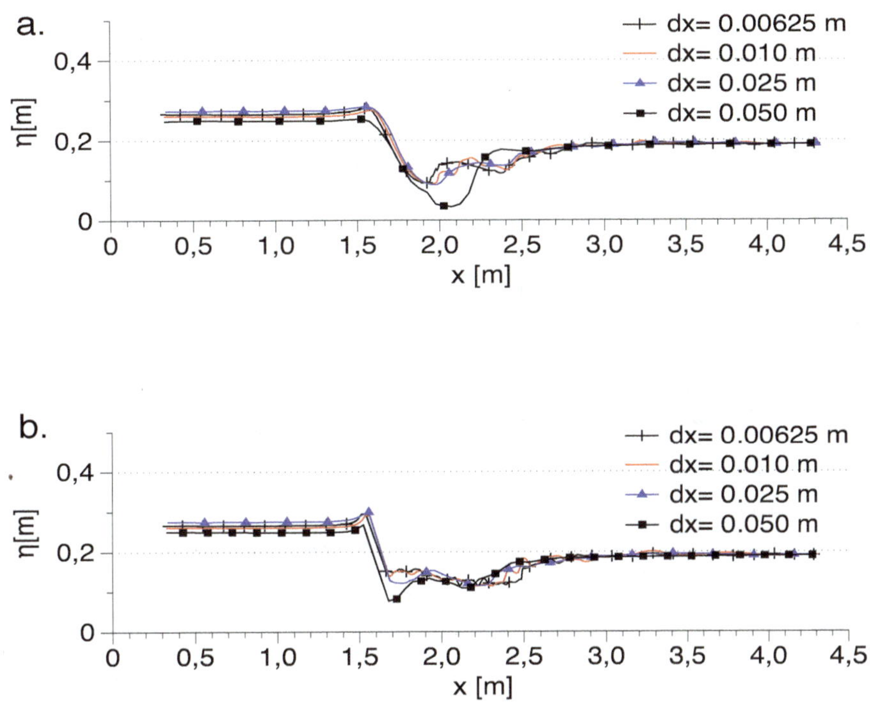

Figure 29. Grid convergence study with FSTD for the numerical free surface elevation for flow around a pair of bridge piers: (**a**) along the center line through a pier $y_1 = 0.19$ m, (**b**) along the center of the channel $y_2 = 0.31$ m.

The obstruction due to the bridge piers leads to super-critical flow downstream of the piers and the flow regime changes back to sub-critical near the outflow. The comparison of the numerical results with experimental data in this case with and without FSTD along the center line through a pier ($y_1 = 0.19$ m) is shown in Figure 31a and along the center of the channel ($y_2 = 0.31$ m) is presented in Figure 31b. These results show the importance of FSTD in complex flow situations. The free surface downstream of the piers is elevated and the free surface in the super-critical region is not represented correctly both along the center of the channel and along the center line through one of the piers without FSTD. The free surface downstream of the piers slopes downwards towards the outlet to meet the controlled outflow boundary condition. This together with the elevated free surface upstream of the piers indicates an unphysical energy loss in the numerical channel due to the over-production of eddy viscosity around the free surface. On the other hand, in the numerical results with FSTD follow the experimental data all along the channel including the hydraulic jump and the slope of the downstream free surface does not have a distinct slope towards the outlet.

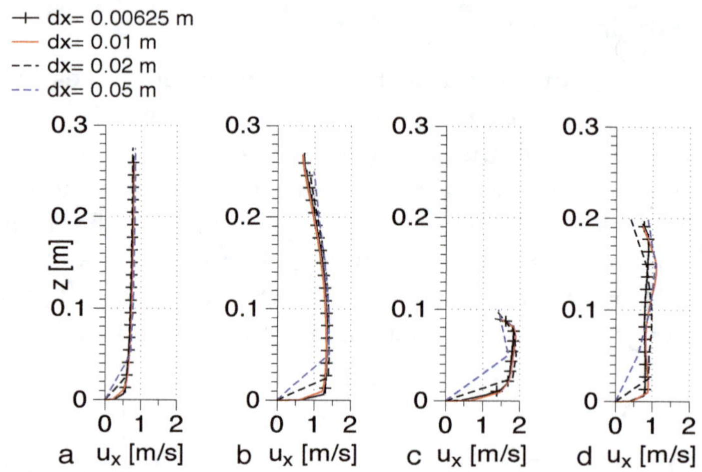

Figure 30. Grid convergence study with FSTD for the horizontal velocity profiles over the water depth at various locations in the numerical channel: (**a**) $x = 0.8$ m, (**b**) $x = 1.30$ m, (**c**) $x = 1.63$ m, (**d**) $x = 3.0$ m.

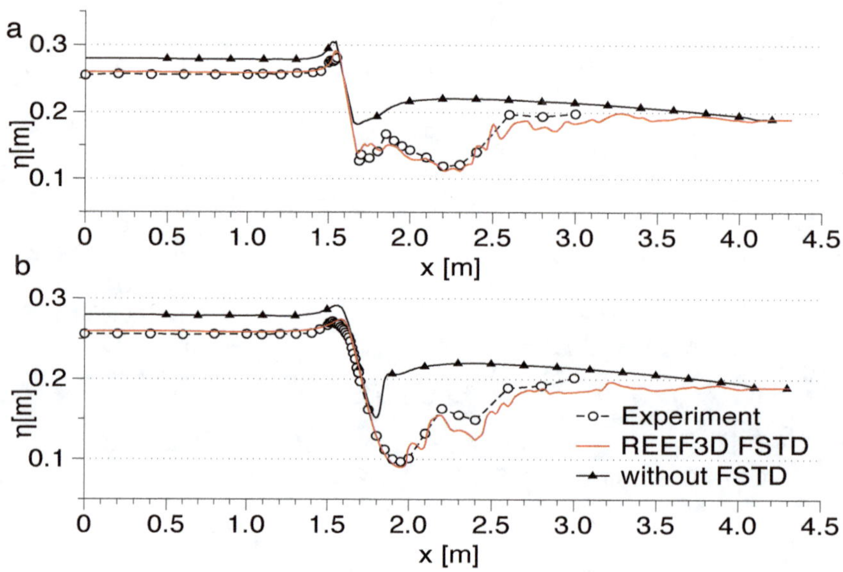

Figure 31. Comparison of the experimental data [40] and the numerical results for the free surface elevation with and without FSTD: (**a**) along the center line through a pier $y_1 = 0.19$ m, (**b**) along the center of the channel $y_2 = 0.31$ m.

The stream-wise velocity profiles with and without FSTD at various locations in the channel show that the flow field is similar upstream of the piers at $x = 0.8$ m and between the piers at $x = 1.30$ m in Figure 32a,b, respectively. The major difference is seen in Figure 32c at the trough of the hydraulic jump at $x = 1.63$ m. The application of FSTD results in a more physical distribution of the horizontal velocity along the water depth and the maximum horizontal velocity is seen just under the free surface. In the absence of FSTD, the free surface is higher, and the maximum horizontal velocity occurs much below the free surface. Further downstream at $x = 3.0$ m in Figure 32, the velocity profiles differ only in the vicinity of the free surface, corresponding to the small difference in the free surface elevation.

The difference between the flow regimes in the two simulations is further demonstrated with the eddy viscosity in the channel. Figure 33 shows the eddy-viscosity distribution around the free surface without and with FSTD. Without FSTD, high values are seen for the eddy viscosity around the free surface especially in the region of the hydraulic jump behind the piers in Figure 33a. The strain in the flow at this location is the highest in the channel due to the occurrence of the hydraulic jump. The highly strained flow results in an unphysical high value of eddy viscosity in the region, resulting in a damped out free surface. On the other hand, with FSTD, the eddy viscosity around the interface is reduced in Figure 33b and the free surface resembles the experimental observations. The flow in the channel and the free surface features in the three-dimensional numerical channel without FSTD is shown in Figure 34a shows the abrupt jump in the free surface downstream of the piers and the horizontal free surface further downstream. In the simulation with FSTD in Figure 34b, the region downstream of the piers shows a rapid decrease in free surface elevation with a trans-critical flow, the formation of a hydraulic jump and the flow returns to a sub-critical flow further downstream.

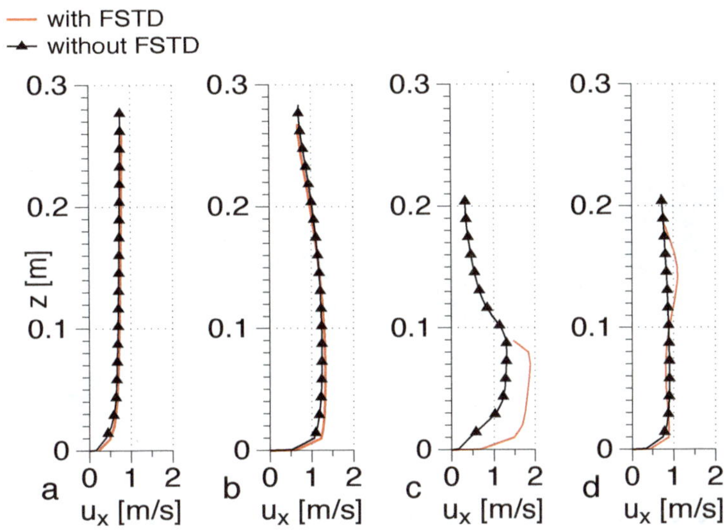

Figure 32. Horizontal velocity (u) profiles over the water depth at various locations along the center of the channel in simulations with and without FSTD: (**a**) $x = 0.80$ m, (**b**) $x = 1.30$ m, (**c**) $x = 1.63$ m, (**d**) $x = 3.00$ m.

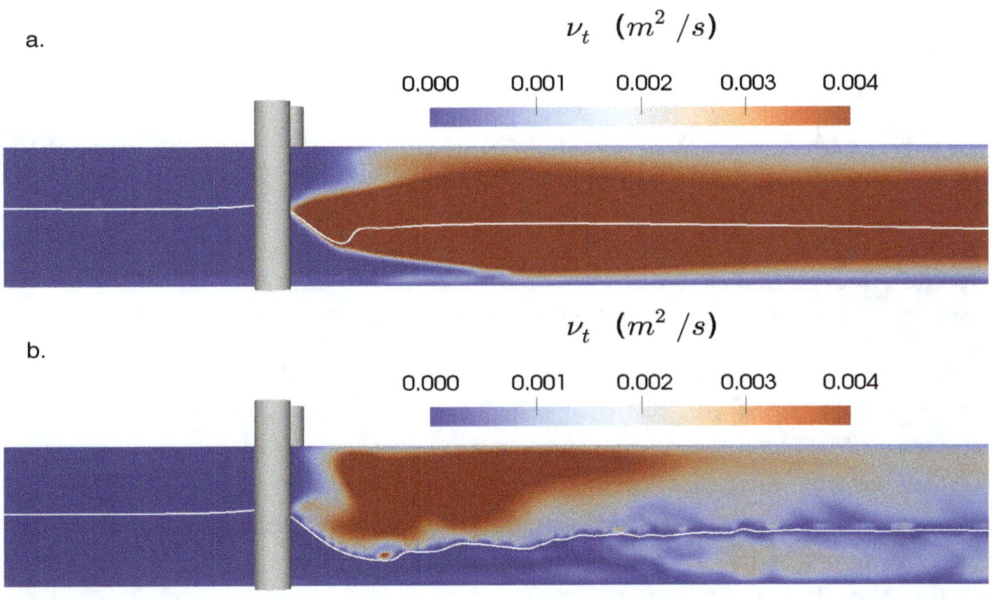

Figure 33. Eddy viscosity along the center of the channel with bridge piers: (**a**) without FSTD, high values of eddy viscosity are seen around the free surface in both air and water phases, (**b**) with FSTD, reduction of the eddy viscosity around the interface is clearly seen.

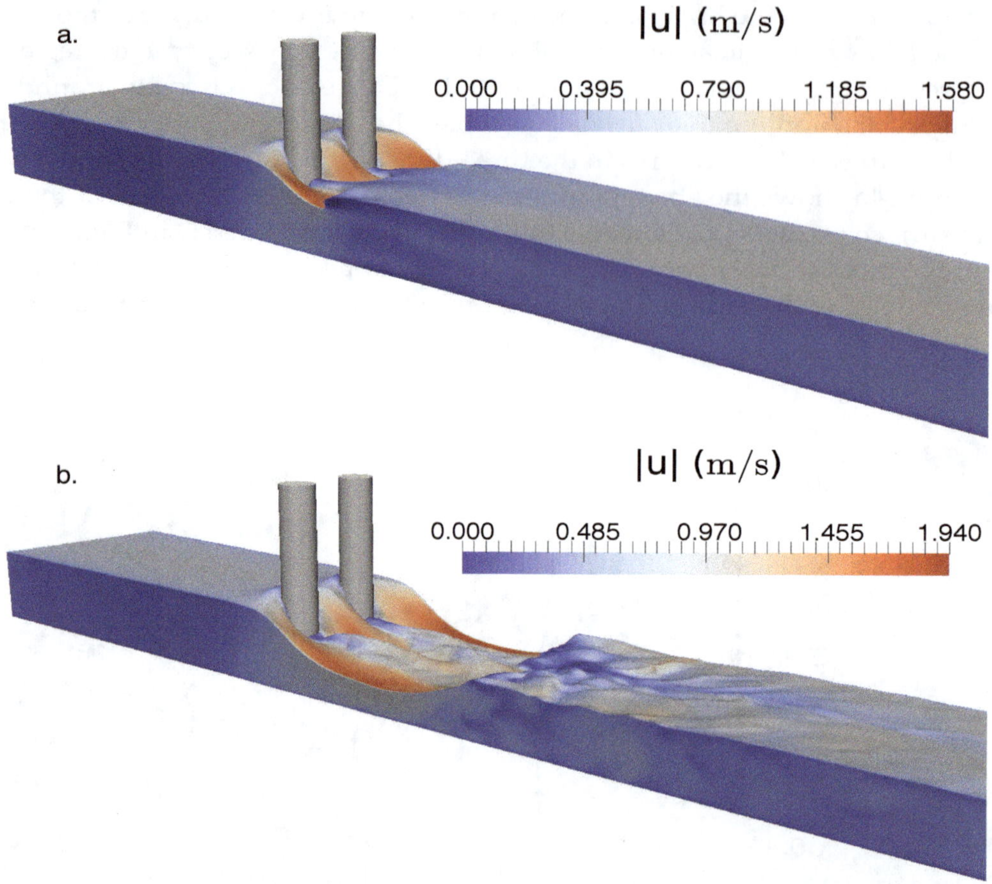

Figure 34. Free surface features in the simulations with and without FSTD along with the velocity magnitude contours throughout the channel: (**a**) without FSTD, (**b**) with FSTD.

The results presented above for a rapidly varying flow situation from sub-critical to super-critical and back to sub-critical with the free surface elevations, velocity profiles, the recirculation zone and the distribution of the eddy viscosity show that the application of a FSTD algorithm is essential to obtain realistic and physical results in such cases using immiscible two-phase RANS equations.

5. Conclusions

The open-source CFD model REEF3D is used to simulate free surface flows around different hydraulic structures with different flow conditions using immiscible two-phase RANS equations. The distribution of the eddy viscosity, the velocity magnitude, and the recirculation zones in the numerical channel are analyzed for the different cases and compared to experimental data and correlated with the effect of FSTD treatment. From the investigation of the different flow scenarios presented in this study, the following conclusions can be drawn:

- When the flow conditions remain sub-critical throughout the domain, the calculated free surface elevations are similar with and without FSTD but the submerged recirculation zone is better represented when FSTD is used.
- When the flow conditions change from sub-critical to super-critical, the calculated free surface elevation and the flow features such as the recirculation zone are similar with and without FSTD.
- In the case of transitional flow, the differences in the calculation of the free surface and the recirculation pattern become more apparent.
- When the flow conditions change from sub-critical to super-critical and back to sub-critical, the calculated free surface elevation and the flow features such as the hydraulic jump and the recirculation zone are represented correctly only in the simulation with FSTD.

- In general, simulations with FSTD provide a more physical representation of the recirculation zone. The calculation of free surface elevation is significantly affected in the complex case of flow conditions changing from sub-critical to super-critical to sub-critical.

The results demonstrate the importance of FSTD treatment in immiscible two-phase RANS simulations of flow around hydraulic structures. The implementation also addresses an important issue of over-production of turbulence at the free surface in two-phase RANS modelling during complex free surface flow using an empirical approach that is based on the flow physics. The proposed boundary condition is shown to be numerically stable in all the flow situations studied and has varying degree of influence on the results, depending in the flow situation. It is also demonstrated that two-phase RANS modelling that does not resolve the detailed eddies in turbulent flow can be used to obtain reasonably accurate representation of complex flow conditions. In the case of flows with hydraulic jumps, the internal structures of the flow and the air entrainment are not presented, but the bulk quantities that influence engineering design of hydraulic structures such as the free surface and the velocity profile over depth are well represented. This provides an alternative tool to analyze engineering design of hydraulic structures at a computational expense that is much lower than that required by eddy-resolving models.

Author Contributions: Lead Author A.K. was responsible for the design of the simulations, compilation of the results and drafting the manuscript. G.F. carried out initial testing and validation. Concept development, coding and academic supervision by H.B.

Acknowledgments: The authors thank Prof Michał Szydłowski for providing the experimental data for the free surface elevations for the flow scenario with the bridge piers. GF was supported through the ÚNKP-18-3-I New National Excellence Program of the Ministry of Human Capacities. This research was supported in part with computational resources at the Norwegian University of Science and Technology (NTNU) provided by NOTUR (No. NN2620K), http://www.notur.no.

References

1. Rouse, H.; Bhoota, B.; Hsu, E.Y. Design of Channel Expansion. *Proc. Am. Soc. Civ. Eng. (ASCE)* **1951**, *75*, 1369–1385.
2. Ippen, A.T.; Dawson, J.H. Design of channel contractions. *Proc. Am. Soc. Civ. Eng. (ASCE)* **1951**, *75*, 1348–1368.
3. Hager, W.H. Supercritical flow in channel junctions. *J. Hydraul. Eng.* **1989**, *115*, 595–616. [CrossRef]
4. Mazumder, S.K.; Hager, W.H. Supercritical expansion flow in Rouse modified and reversed transitions. *J. Hydraul. Eng.* **1993**, *119*, 201–219. [CrossRef]
5. Fritz, H.M.; Hager, W.H. Hydraulics of embankment weirs. *J. Hydraul. Eng.* **1998**, *124*, 963–971. [CrossRef]
6. Tullis, B.; Neilson, J. Performance of submerged ogee-crest weir head-discharge relationships. *J. Hydraul. Eng.* **2008**, *134*, 486–491. [CrossRef]
7. Murzyn, F.; Chanson, H. Free-surface fluctuations in hydraulic jumps: Experimental observations. *Exp. Therm. Fluid Sci.* **2009**, *33*, 1055–1064. [CrossRef]
8. Chachereau, Y.; Chanson, H. Free-surface fluctuations and turbulence in hydraulic jumps. *Exp. Therm. Fluid Sci.* **2011**, *35*, 896–909. [CrossRef]
9. Brunner, G.W. HEC-RAS (River Analysis System). In Proceedings of the North American Water and Environment Congress & Destructive Water, Anaheim, CA, USA, 22–28 June 1996; pp. 3782–3787.
10. Olsen, N. *A Three Dimensional Numerical Model for Simulation of Sediment Movements in Water Intakes with Multiblock Option: Version 1.1 and 2.0 User's Manual*; Department of Hydraulic and Environmental Engineering, The Norwegian University of Science and Technology: Trondheim, Norway, 2003; p. 138.
11. Haun, S.; Olsen, N.R.B.; Feurich, R. Numerical modeling of flow over trapezoidal broad-crested weir. *Eng. Appl. Comput. Fluid Mech.* **2011**, *5*, 397–405. [CrossRef]
12. Fleit, G.; Baranya, S.; Bihs, H. CFD Modeling of Varied Flow Conditions Over an Ogee-Weir. *Period. Polytech. Civ. Eng.* **2017**, 1–7. [CrossRef]

13. Kirkil, G.; Constantinescu, S.G.; Ettema, R. Coherent structures in the flow field around a circular cylinder with scour hole. *J. Hydraul. Eng.* **2008**, *134*, 572–587. [CrossRef]
14. Kara, S.; Kara, M.C.; Stoesser, T.; Sturm, T.W. Free-surface versus rigid-lid LES computations for bridge-abutment flow. *J. Hydraul. Eng.* **2015**, *141*, 04015019. [CrossRef]
15. Kang, S.; Sotiropoulos, F. Large-eddy simulation of three-dimensional turbulent free surface flow past a complex stream restoration structure. *J. Hydraul. Eng.* **2015**, *141*, 04015022. [CrossRef]
16. McSherry, R.J.; Chua, K.V.; Stoesser, T. Large eddy simulation of free-surface flows. *J. Hydrodyn. Ser. B* **2017**, *29*, 1–12. [CrossRef]
17. Viti, N.; Valero, D.; Gualtieri, C. Numerical Simulation of Hydraulic Jumps. Part 2: Recent Results and Future Outlook. *Water* **2019**, *11*, 28. [CrossRef]
18. Jacobsen, N.G.; Fuhrman, D.R.; Fredsøe, J. A wave generation toolbox for the open-source CFD library: OpenFOAM. *Int. J. Numer. Methods Fluids* **2012**, *70*, 1073–1088. [CrossRef]
19. Devolder, B.; Rauwoens, P.; Troch, P. Application of a buoyancy-modified $k-\omega$ SST turbulence model to simulate wave run-up around a monopile subjected to regular waves using OpenFOAM®. *Coast. Eng.* **2017**, *125*, 81–94. [CrossRef]
20. Larsen, B.E.; Fuhrman, D.R. On the over-production of turbulence beneath surface waves in Reynolds-averaged Navier-Stokes models. *J. Fluid Mech.* **2018**, *853*, 419–460. [CrossRef]
21. Naot, D.; Rodi, W. Calculation of secondary currents in channel flow. *J. Hydraul. Div. ASCE* **1982**, *108*, 948–968.
22. Bihs, H.; Kamath, A.; Alagan Chella, M.; Aggarwal, A.; Arntsen, Ø.A. A new level set numerical wave tank with improved density interpolation for complex wave hydrodynamics. *Comput. Fluids* **2016**, *140*, 191–208. [CrossRef]
23. Kamath, A.; Bihs, H.; Alagan Chella, M.; Arntsen, Ø.A. Upstream-Cylinder and Downstream-Cylinder Influence on the Hydrodynamics of a Four-Cylinder Group. *J. Waterw. Port Coast. Ocean Eng.* **2016**, *142*, 04016002. [CrossRef]
24. Kamath, A.; Alagan Chella, M.; Bihs, H.; Arntsen, Ø.A. Breaking wave interaction with a vertical cylinder and the effect of breaker location. *Ocean Eng.* **2016**, *128*, 105–115. [CrossRef]
25. Bihs, H.; Kamath, A.; Alagan Chella, M.; Arntsen, Ø.A. Breaking-Wave Interaction with Tandem Cylinders under Different Impact Scenarios. *J. Waterw. Port Coast. Ocean Eng.* **2016**, *142*, 04016005. [CrossRef]
26. Alagan Chella, M.; Bihs, H.; Myrhaug, D.; Muskulus, M. Breaking solitary waves and breaking wave forces on a vertically mounted slender cylinder over an impermeable sloping seabed. *J. Ocean Eng. Mar. Energy* **2017**, *3*, 1–19. [CrossRef]
27. Bihs, H.; Kamath, A. A combined level set/ghost cell immersed boundary representation for floating body simulations. *Int. J. Numer. Methods Fluids* **2017**, *83*, 905–916. [CrossRef]
28. Chorin, A. Numerical solution of the Navier-Stokes equations. *Math. Comput.* **1968**, *22*, 745–762. [CrossRef]
29. Ashby, S.F.; Falgout, R.D. A Parallel Mulitgrid Preconditioned Conjugate Gradient Algorithm for Groundwater Flow Simulations. *Nucl. Sci. Eng.* **1996**, *124*, 145–159. [CrossRef]
30. Center for Applied Scientific Computing. *HYPRE High Performance Preconditioners—User's Manual*; Lawrence Livermore National Laboratory: Livermore, CA, USA 2006.
31. Wilcox, D.C. *Turbulence Modeling for CFD*; DCW Industries Inc.: La Canada, CA, USA, 1994.
32. Durbin, P.A. Limiters and wall treatments in applied turbulence modeling. *Fluid Dyn. Res.* **2009**, *41*, 1–18. [CrossRef]
33. Hossain, M.S.; Rodi, W. Mathematical modeling of vertical mixing in stratified channel flow. In Proceedings of the 2nd Symposium on Stratified Flows, Trondheim, Norway, 24–27 June 1980.
34. Osher, S.; Sethian, J.A. Fronts propagating with curvature-dependent speed: Algorithms based on Hamilton-Jacobi formulations. *J. Comput. Phys.* **1988**, *79*, 12–49. [CrossRef]
35. Peng, D.; Merriman, B.; Osher, S.; Zhao, H.; Kang, M. A PDE-based fast local level set method. *J. Comput. Phys.* **1999**, *155*, 410–438. [CrossRef]
36. Jiang, G.S.; Shu, C.W. Efficient implementation of weighted ENO schemes. *J. Comput. Phys.* **1996**, *126*, 202–228. [CrossRef]
37. Jiang, G.S.; Peng, D. Weighted ENO schemes for Hamilton-Jacobi equations. *SIAM J. Sci. Comput.* **2000**, *21*, 2126–2143. [CrossRef]

38. Shu, C.W.; Osher, S. Efficient implementation of essentially non-oscillatory shock capturing schemes. *J. Comput. Phys.* **1988**, *77*, 439–471. [CrossRef]
39. Berthelsen, P.A.; Faltinsen, O.M. A local directional ghost cell approach for incompressible viscous flow problems with irregular boundaries. *J. Comput. Phys.* **2008**, *227*, 4354–4397. [CrossRef]
40. Szydłowski, M. Numerical Simulation of Open Channel Flow between Bridge Piers. *TASK Q.* **2011**, *15*, 271–282.
41. Sarker, M.A.; Rhodes, D.G. Calculation of free-surface profile over a rectangular broad-crested weir. *Flow Meas. Instrum.* **2004**, *15*, 215–219. [CrossRef]

3

Three-Dimensional Aerators: Characteristics of the Air Bubbles

Shuai Li [1,2], Jianmin Zhang [3,*], Xiaoqing Chen [1,2] and Jiangang Chen [1,2]

1. Key Laboratory of Mountain Hazards and Earth Surface Processes, Chinese Academy of Sciences, Chengdu 610041, China; lishuai@imde.ac.cn (S.L.); xqchen@imde.ac.cn (X.C.); chenjg@imde.ac.cn (J.C.)
2. Institute of Mountain Hazards and Environment, Chinese Academy of Sciences and Ministry of Water Conservancy, Chengdu 610041, China
3. State Key Laboratory of Hydraulics and Mountain River Engineering, Sichuan University, Chengdu 610065, China

* Correspondence: zhangjianmin@scu.edu.cn.

Abstract: Three-dimensional aerators are often used in hydraulic structures to prevent cavitation damage via enhanced air entrainment. However, the mechanisms of aeration and bubble dispersion along the developing shear flow region on such aerators remain unclear. A double-tip conductivity probe is employed in present experimental study to investigate the air concentration, bubble count rate, and bubble size downstream of a three-dimensional aerator involving various approach-flow features and geometric parameters. The results show that the cross-sectional distribution of the air bubble frequency is in accordance with the Gaussian distribution, and the relationship between the air concentration and bubble frequency obeys a quasi-parabolic law. The air bubble frequency reaches an apex at an air concentration (C) of approximately 50% and decreases to zero as $C = 0\%$ and $C = 100\%$. The relative location of the air-bubble frequency apex is 0.210, 0.326 and 0.283 times the thickness of the layers at the upper, lower and side nappes, respectively. The air bubble chord length decreases gradually from the air water interface to the core area. The air concentration increases exponentially with the bubble chord length. The air bubble frequency distributions can be fit well using a "modified" gamma distribution function.

Keywords: three-dimensional aerator; air concentration; air bubble frequency; air bubble chord length

---

1. Introduction

Cavitation erosion caused by high velocity flows is a common phenomenon in spillways or chutes with high-head dams. A commonly adopted counter-measure to reduce or prevent cavitation damage is aeration [1–6]. Three-dimensional aerators are used to enhance air entertainment into water and to prevent cavitation erosion on spillways. Full interfacial aeration is commonly observed at the free surface of the upper, lower and side nappe along the free jet (Figure 1). The effect of air entrainment is based on the jet disintegration process due to turbulence and secondary interactions with the surrounding atmosphere.

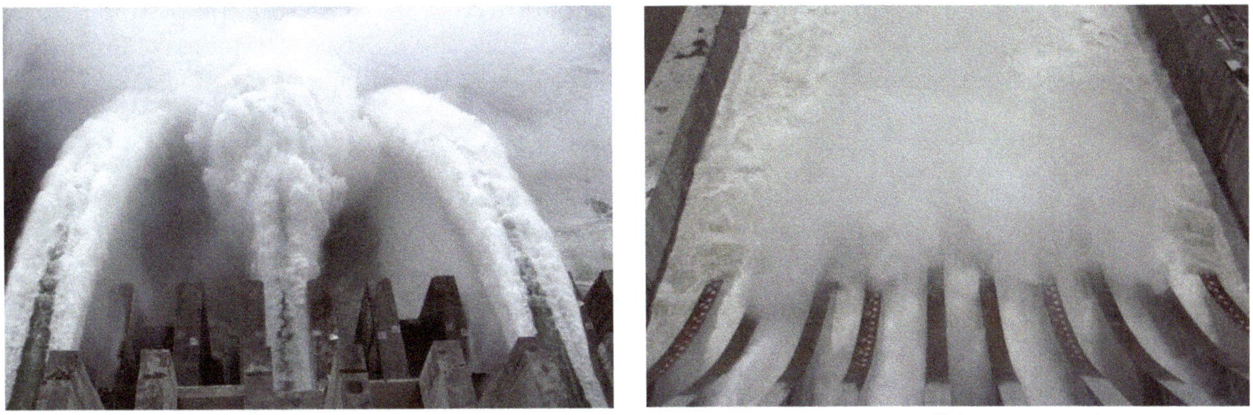

(a) (b)

Figure 1. Air water flow in three-dimensional aerators: (**a**) Three Gorges Project; (**b**) Xiang Jia-Ba Project.

Air entrainment at the interfacial area is a fundamental parameter in numerical models for simulating two-phase flows. Likewise, it is also an essential consideration when designing optimum aerators. During the last few decades, numerous studies have been conducted to investigate the air concentration and free-surface aeration on an aerator. A series of experiments were conducted to study the free-surface aeration and diffusion characteristics of the bottom aeration devices [7–11]. Results indicate that the quantities and geometric scales of air bubbles are important parameters in characterizing the air concentration in the interfacial area. Chanson [12] described the flow structure of the developing aerated region in a flat chute and found that the maximum mean air content in different cross-sections was 12%. Toombes [13] showed that the maximum bubble count rates typically correspond to air concentration between 40% and 60%. The distribution of velocity indices at the different nappes, which stays at the downstream of an expanding chute aerator, were found to be different [14] and a solution was also found to improve both the bottom and lateral cavities' length [15]. The longitudinal position of the upper trajectory apex was found to be further downstream than that of the lower trajectory apex [16]. Kramer and Hager [17] found that both the entrainment rate and dimensionless bubble count rate increased with jet length. The mean air concentration is affected by aerators, but the air concentration at the bottom is limited [18]. Furthermore, Toombes and Chanson [19] studied flows past a backward-facing step, where the void fraction distribution, bubble count rate, local air distribution and water chord size were indicated. The results showed that the relationship between bubble count rate and void fraction can be defined by a quasi-parabolic function, while probability density functions of local chord size exhibited a quasi-log-normal shape. Some researchers [20–24] also established analytical equations to predict the air concentration distributions at the aerators' downstream. Numerical models such as the lattice Boltzmann method were found to successfully simulate two-phase flows and free-surface flows [25–27]. These models may be used to complement physical models and experimental measurements after careful validation and verification of model parameters such as the void fraction distribution, velocity, and air-bubble chord sizes. Chanson [28] reviewed the recent progress in turbulent free-surface flows and the mechanics of aerated flows and highlighted that physical modelling tests are efficient methods to validate phenomenological, theoretical and numerical models.

Although numerous studies regarding air–water flows have been conducted over the years, only a few researchers, however, have systematically investigated the air bubble count rate and bubble size of high velocity air–water flows downstream of the pressure outlet joined by a three-dimensional aerator (sudden vertical drop and lateral enlargement) in the tunnel. There is also a lack of knowledge regarding the influence of the bubble count rate and bubble size on the air concentration in two-phase flows, which is necessary for improving turbulence models. Thus, further insight is needed regarding the details of air entrainment in high-speed flows and the related core area features. In this study, a series of experimental investigations were systematically conducted to further the knowledge of air water flows among the upper, lower and side nappes downstream of a three-dimensional aerator (Figure 2). The experimental data and equations obtained from this study provide novel information and a quantitative description of the aeration flow structure at the high-speed interfacial area.

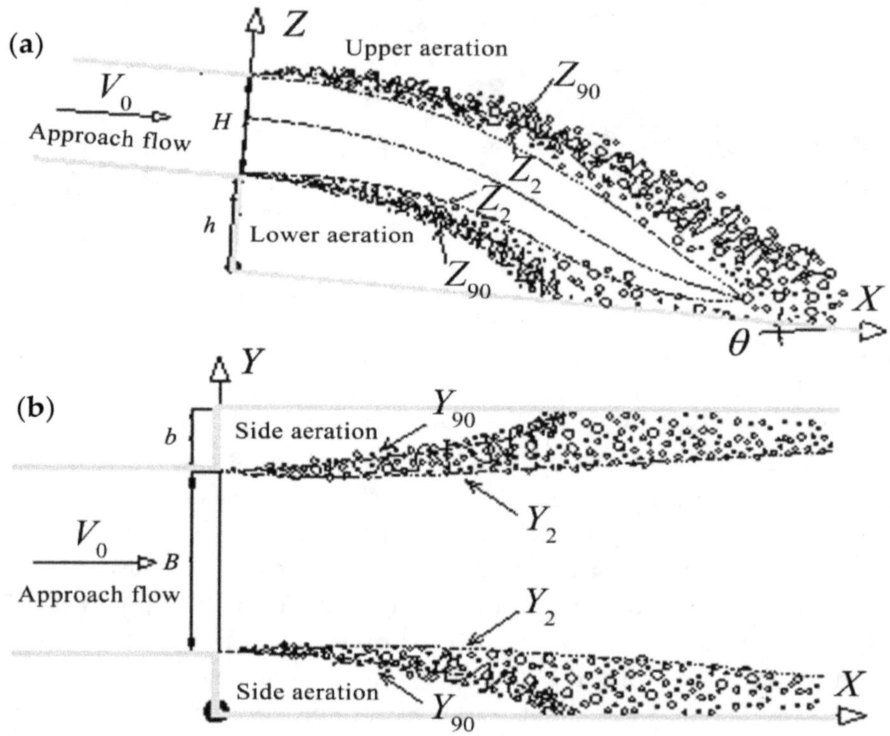

Figure 2. Schematic with the relevant parameters: (**a**) side view and (**b**) plan view.

2. Experimental Setup

Experiments were performed at the State Key Laboratory of Hydraulic and Mountain River Engineering, Sichuan University, China. The experimental setup (Figure 3) includes an upstream reservoir, a pressure section, a sudden vertical drop and lateral enlargement aerator, a free flow section, a tail-water section, and an underground reservoir. Water is discharged (can be up to 0.5 m^3/s) by a continuous and stable water supply system (2.5 m wide, 3.5 m long, and 6.3 m high). The components of the pressure inlet were assembled with a smooth steel plate. The pressure section is a rectangular pipe (0.25 m wide, 0.15 high and 2.0 m long) with a variable inclination angle varied with the downstream chute. The sudden fall-expansion aerator (height $h \times$ width b) was installed at the end of the pressure section. The free-flow section was fabricated from polymethyl methacrylate (PMMA) with a roughness height of 0.008 mm. The upper, lower, and side nappes were exposed to atmospheric pressure. The constant water levels corresponding to a particular head were maintained in the upstream reservoir.

Experiments were carried out for mean velocities (V_0) of 6 m/s, 7 m/s, 8 m/s, and 9 m/s; chute slopes of 0%, 10%, and 25%; and aerator sizes (h, b) of 0.025m, 0.045m, and 0.065 m. The test cases are summarized in Table 1 ($0.9 < q_w < 1.35$, $0.89 < R_e < 1.34$). It should be noted that some combinations (series 7–10, and 13 in Table 1) pertain to cases involving three-dimensional (vertical drop and lateral enlargement) aerators. The other cases involve only either vertical drop aerators or lateral enlargement aerators. Upstream flows were supercritical ($4.95 < Fr < 7.42$) for all of the investigated flow conditions. Downstream of the aerator, a free jet, with the use of a ventilated air cavity below and/or beside it, was formed.

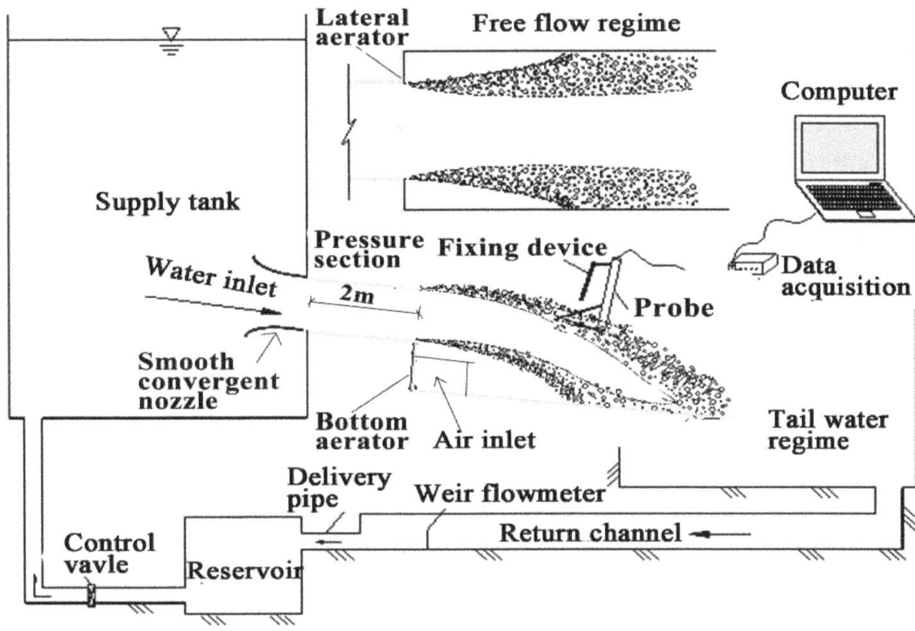

Figure 3. Sketch of the experimental model and test equipment.

Table 1. Summary of the operating conditions for the experiments.

Series	(h, b)/m	θ
1	(0.025, 0.000)	10%
2	(0.045, 0.000)	10%
3	(0.065, 0.000)	10%
4	(0.000, 0.025)	10%
5	(0.000, 0.045)	10%
6	(0.000, 0.065)	10%
7	(0.025, 0.025)	10%
8	(0.045, 0.045)	10%
9	(0.065, 0.065)	10%
10	(0.045, 0.045)	0%
11	(0.000, 0.045)	0%
12	(0.045, 0.000)	0%
13	(0.045, 0.045)	25%
14	(0.000, 0.045)	25%
15	(0.045, 0.000)	25%

The air concentration, bubble count rate, and chord length were recorded using a CQY—Z8a measurement instrument [14,29,30] with a double-tip conductivity probe (Figure 4). Different voltage indices between air and water were measured by the platinum tip. The probe consisted of two identical tips, including an external stainless steel electrode with 0.7-mm-diameter and an internal concentric platinum electrode with 0.1-mm-diameter. The tips are aligned in the flow direction and the distance between the two tips is 12.89 mm. Both tips were connected to an electronics device with a response time less than 10 μs. The vertical translation of the probe was dominated by a fixed device with an accuracy of 1 mm. The probe measurements were taken at regular intervals of 5 mm along the vertical direction from the chute bottom to the free surface (or side nappe surface) and 100 mm along the flow direction. At each location, signals were recorded at a scan rate (f) of 100 kHz per channel for a scan period (T) of 10 s. The scan period was based on the findings of Toombes [13] which indicated that a scanning period of 10 s was long enough to provide a reasonable representation of the flow characteristics while maintaining realistic time and data storage constraints.

Figure 4 presents a typical aeration structure detected by a probe at a fixed location along the aeration flow. Assuming each segment is either air or water, the air concentration can be presented as the probability of any discrete element being air. The air concentration is computed as the encountering air's probability at the leading tip of the probe.

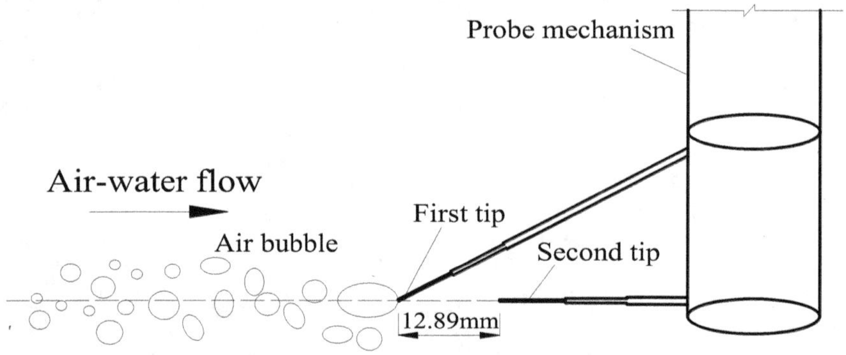

Figure 4. Sketch of the double-tip conductive probe.

The number of data (N) can be obtained by multiplying the frequency (f) of the measurement instrument with the sampling time T, i.e., $N = f \times T$. Those data were classified according to two categories, air and water signals. The air concentration (C) was computed as the encountering air's probability at the leading tip of the probe, which can be expressed as [26]

$$C = \frac{\sum_{i=1}^{N} R_i}{N} \quad (1)$$

where R_i represents the ratio of bubbles measured all throughout the whole measurement time. When the probe tip is completely immersed in air bubbles $R_i = 1$, otherwise $R_i = 0$.

The length of bubble chord is defined as the straight distance between the two intersections of the interface [12]. Note that an air bubble is defined as a volume of air (i.e., air entity), which can be detected by the leading tip of the probe between two continuous air–water interface events. The bubble chord length is calculated by

$$d_i = \frac{v \times n_i}{f} \quad (2)$$

where n_i is the number of detected bubbles recorded; and v is the bubble velocity equal to the local mean aeration velocity (i.e., no slip between the air and water phases).

3. Results

The air bubble frequency and chord length are two important parameters used to characterize the air concentration of aerated flows. We first describe air bubble frequency among the upper, lower and side air–water mixed layers in the experimental flows with various conditions. In this section, the relationships between air concentration with air bubble frequency, and with the relative location at which the maximum air bubble frequency are also discussed. Next, we identify the effects of initial flow velocity on air bubble chord length. In addition, the relationships between the air concentration and air bubble chord length, and between air bubble count rate and air bubble size are also evaluated.

3.1. Air Bubble Frequency

The air bubble frequency characterizes the flow fragmentation, which is proportional to the specific area of the interface between two phases. The air bubble frequency, or air bubble count rate, is a function of bubbles' shape and size, surface tension, and shear forces of the fluid. Predicting how air concentration affects the bubble frequency is a complex problem [19]. A simplified analogy would be to consider on aeration flows past a fixed probe, as a series of discrete one-dimensional air and water elements (Figure 4). Here, the air bubble frequency (Fa) is defined as the number of air-structures (Na) per unit time (t) detected by the leading tip of the probe sensor (i.e., $Fa = Na/t$).

Typical air bubble frequency distributions at each cross-section along the upper, lower, and side nappes are presented in Figure 5. The air bubble frequency distributions exhibit unitary self-similarity along the thickness of the mixed layer and the horizontal distance. The trend of the cross-sectional distribution tends to initially increase and then decrease from the aeration interface to the inside of the water following a Gaussian distribution that is defined by the following functions:

$$\frac{F_a}{(F_a)_{\max}} = \begin{cases} \frac{1}{\sqrt{2\pi}\sigma_0} e^{-\frac{\left(\frac{Z-Z_2}{Z_{90}-Z_2}-\mu_0\right)^2}{2\sigma^2}} & \text{the upper nappe} \\ \frac{1}{\sqrt{2\pi}\sigma_0} e^{-\frac{\left(\frac{Z-Z_{90}}{Z_2-Z_{90}}-\mu_0\right)^2}{2\sigma^2}} & \text{the lower nappe} \\ \frac{1}{\sqrt{2\pi}\sigma_0} e^{-\frac{\left(\frac{Y-Y_{90}}{Y_2-Y_{90}}-\mu_0\right)^2}{2\sigma^2}} & \text{the side nappe} \end{cases} \quad (3)$$

where, $F_a/(F_a)_{\max}$ is the dimensionless air bubble frequency and $(F_a)_{\max}$ is the maximum bubble frequency in the cross-section; The characteristic value of μ_0 reflects the depth corresponding to the air bubble frequency; σ_0 represents the degree of deviation between the relative depth and its mean value.

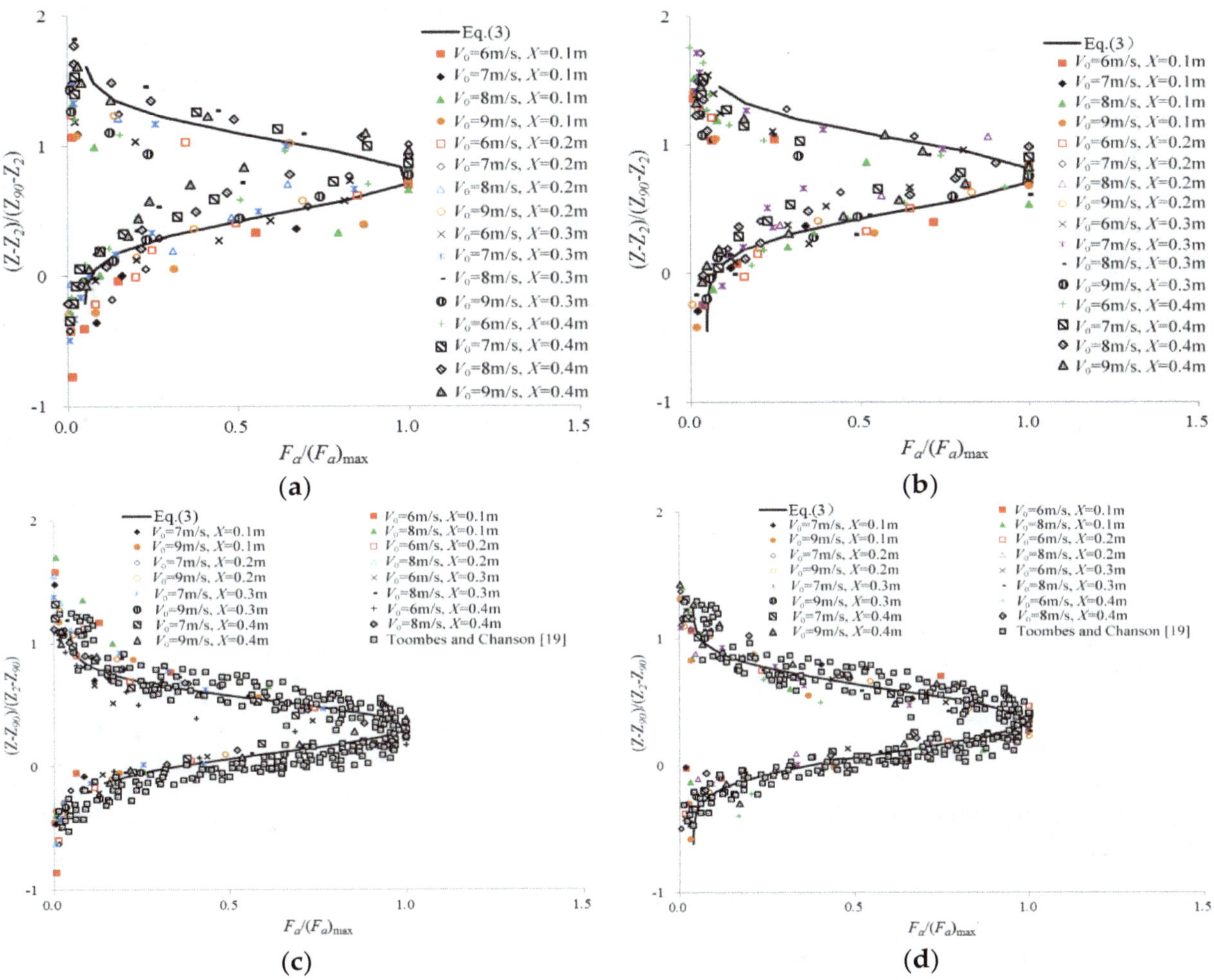

Figure 5. *Cont.*

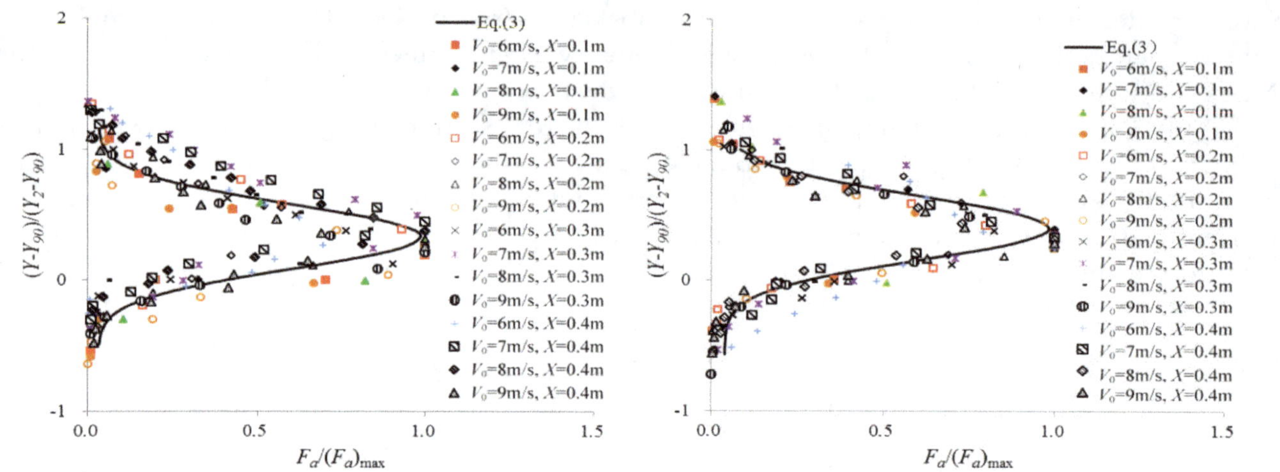

(e) (f)

Figure 5. Air bubble frequency distributions compared with the theoretical distribution obtained from Equation (3). Upper nappe (**a**) (h, b) = (4.5 cm, 4.5 cm), θ = 25%. (**b**) (h, b) = (6.5 cm, 6.5 cm), θ = 10%; lower nappe (**c**) (h, b) = (4.5 cm, 4.5 cm), θ = 25%. (**d**) (h, b) = (6.5 cm, 6.5 cm), θ = 10%; side nappe (**e**) (h, b) = (4.5 cm, 4.5 cm), θ = 25%. (**f**) (h, b) = (6.5 cm, 6.5 cm), θ = 10%.

For all experimental results, the air bubble frequency distributions correspond well with the data reported by Toombes and Chanson [19], as illustrated in Figure 5. Figure 6 presents a plot of experimental data, $F_a/(F_a)_{\max}$, with calculated non-dimensional air bubble frequency using Equation (3). The calculated results at upper, lower, and lateral nappes were in qualitative agreement with experimental data, although there was some scatter. It is noted that the characteristic values of μ_0 and σ_0 differed among the upper, lower, and side nappes (Table 2). For the upper nappe, the air bubbles do not easily diffuse and instead transport to the inside of the water due to the buoyancy. This result in the air bubbles being drawn closer to the free surface ($\mu_0 = 0.210$). For the side nappe, the air bubbles can easily diffuse and transport to the interior of the water ($\mu_0 = 0.283$) because of the effect of large transverse turbulent diffusion [15]. For the lower nappe, the air bubbles are dragged by the buoyant force to the interior of the water ($\mu_0 = 0.326$).

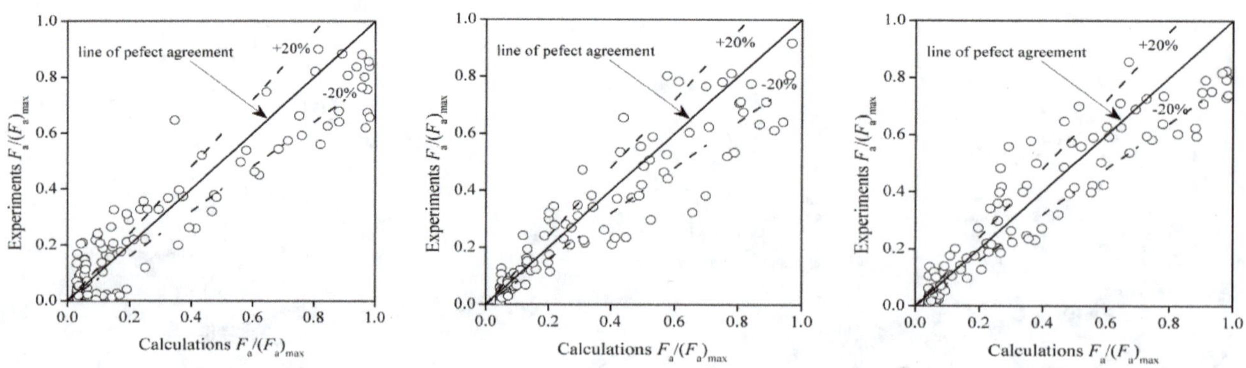

(a) (b) (c)

Figure 6. Comparison of Equation (3) with the experimental data. (**a**) upper nappe; (**b**) lower nappe; (**c**) side nappe.

Table 2. Values of μ_0 and σ_0.

Nappe	μ_0	σ_0
lower	0.326	0.419
upper	0.210	0.413
side	0.283	0.423

Air concentration distribution is just the external manifestation of air–water flows, whereas the internal factors include the count rate and size of the air bubbles. So, it is crucial to understand the features of the air bubble frequency and the air bubble chord length.

The air bubble frequency distributions at various positions along the jet obtained in this study and those reported by Chanson [12] and Toombes and Chanson [19] are shown in Figure 7. It is evident that the air bubble frequency initially increases and then decreases with the increase of air concentrations in the upper, lower, and side nappes. At each cross-section, the air bubble frequency profiles reach to an apex which corresponds to air concentrations of approximately 50%. The profiles tend to zero at extremely low and extremely high air concentrations. Overall, the distributions of the dimensionless air bubble frequency can be well fitted by the parabolic function:

$$\frac{F_a}{(F_a)_{\max}} = 4C(1-C) \tag{4}$$

Qualitatively, the calculated results of Equation (4) correspond with the results of Chanson [12] and Toombes and Chanson [19] (Figure 7). It is worth noting that the data [12] slightly deviate from the fully developed supercritical flows, suggesting that Equation (4) has broad applicability.

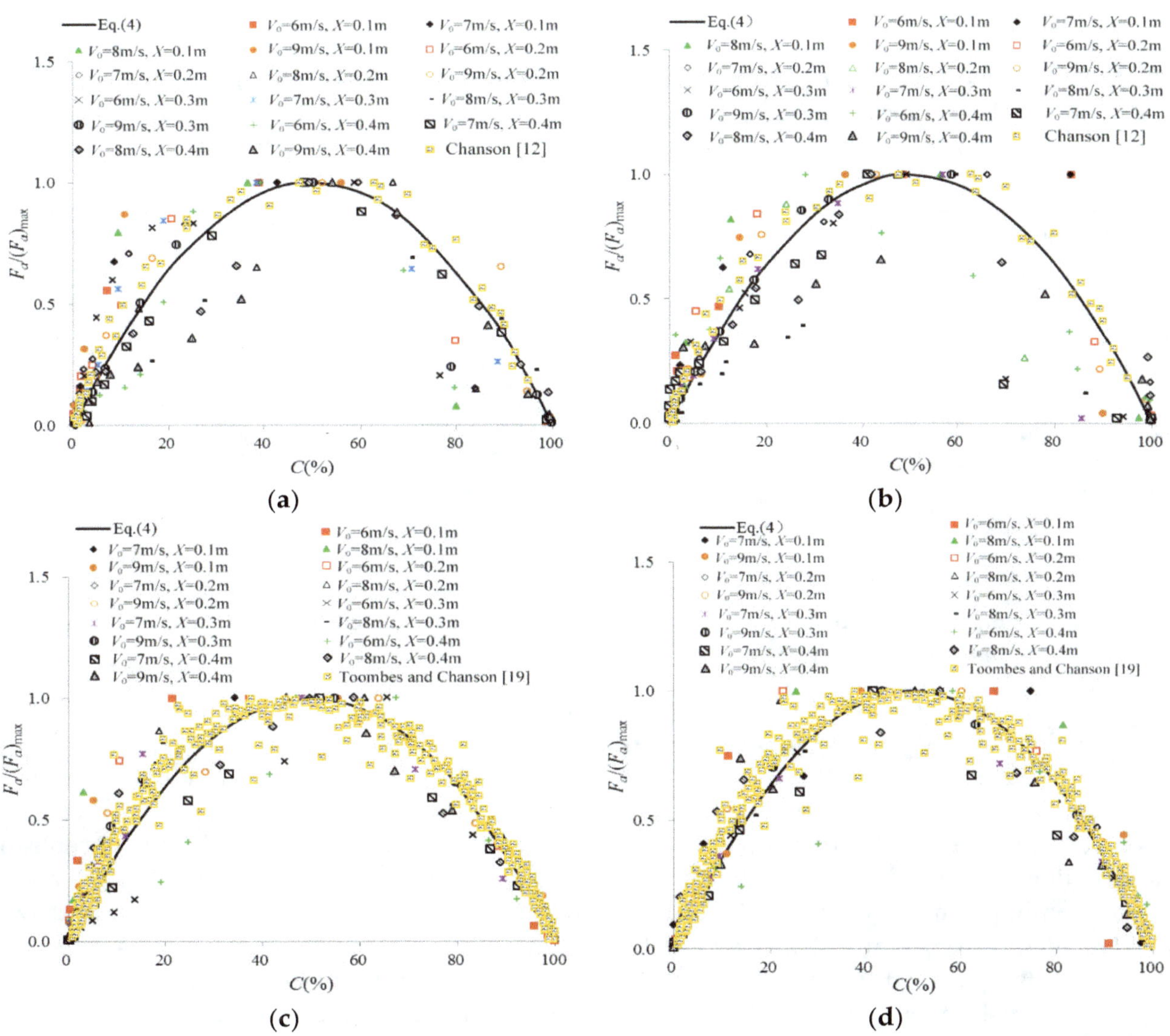

Figure 7. *Cont.*

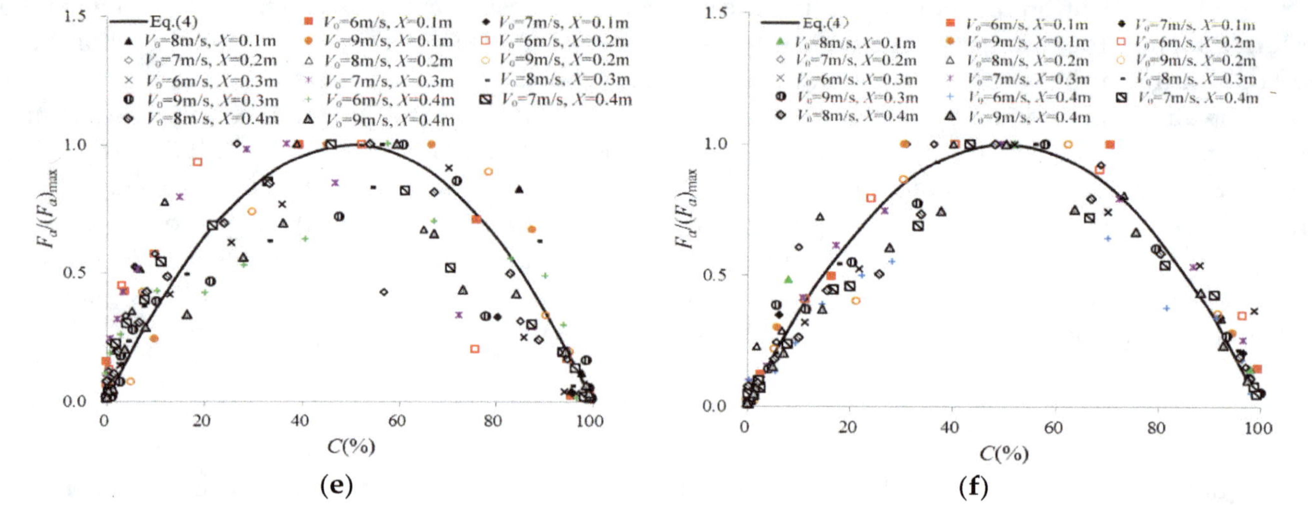

Figure 7. Dimensionless air bubble frequency as a function of the air concentration fitted with Equation (4). Upper nappe (**a**) $(h, b) = (4.5 \text{ cm}, 4.5 \text{ cm})$, $\theta = 25\%$. (**b**) $(h, b) = (6.5 \text{ cm}, 6.5 \text{ cm})$, $\theta = 10\%$; lower nappe (**c**) $(h, b) = (4.5 \text{ cm}, 4.5 \text{ cm})$, $\theta = 25\%$. (**d**) $(h, b) = (6.5 \text{ cm}, 6.5 \text{ cm})$, $\theta = 10\%$; side nappe (**e**) $(h, b) = (4.5 \text{ cm}, 4.5 \text{ cm})$, $\theta = 25\%$. (**f**) $(h, b) = (6.5 \text{ cm}, 6.5 \text{ cm})$, $\theta = 10\%$.

The majority of the data fall within the ±20% error lines as shown in Figure 8. The calculated non-dimensional air bubble frequency using Equation (4) are in reasonably good agreement with the experimental data at upper, lower, and lateral nappes.

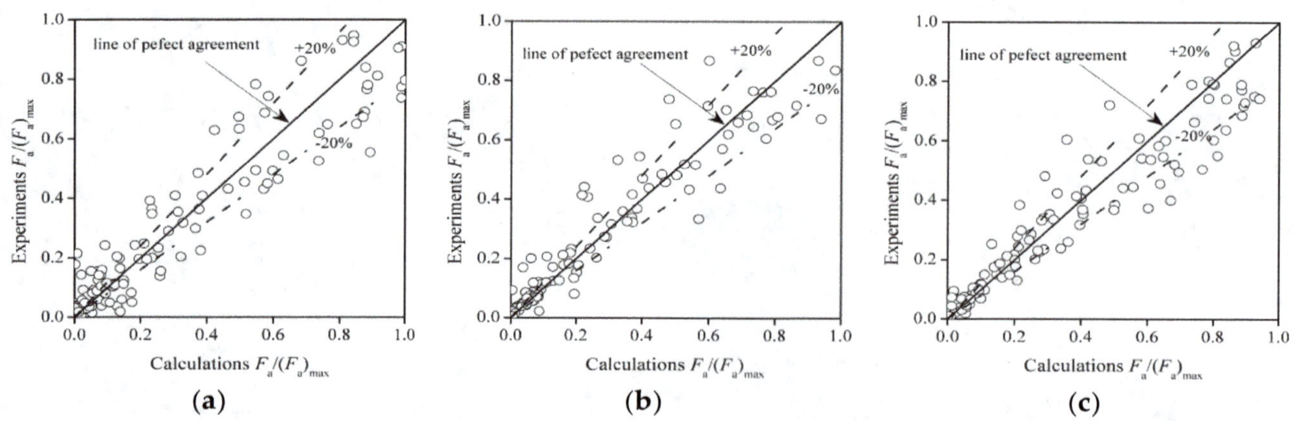

Figure 8. Comparison of experimental data with calculated values of dimensionless air bubble frequency. (**a**) upper nappe; (**b**) lower nappe; (**c**) side nappe.

The relationships between air concentration and the relative distance, which corresponds to the maximum air bubble frequency are presented in Figure 9 where a large fluctuation of air concentration ($C = 30\% \sim 60\%$) in the initial regime of aeration can be observed. The fluctuation may be due to the aeration layer located close to the pressure outlet, which may be thin and unstable. The air–water layer becomes stable with the development of the aeration further downstream. The air concentration corresponding to the maximum air bubble frequency gradually moves toward to 50% line. The result is consistent with the findings of Figure 7. Actually, the maximum air bubble frequency does not always coincide with $C = 50\%$ accurately. Factors such as the average size and length scales of discrete air and water elements may affect the local air concentration and flow conditions in air–water flows [13,31].

The relative location corresponding to the maximum air bubble frequency is shown in Figure 10. It can be seen that the relative location of the maximum air bubble frequency is 0.210, 0.326 and 0.283 times the thicknesses of the air–water layers in the upper, lower and side nappes, respectively. The results are consistent with the results of Figure 5. For the upper nappe, the air can easily be drawn into the water because the air–water external interface directly interacts with the atmosphere. However, the air bubbles cannot easily diffuse and transport to the inside of the jet because of the effect of buoyancy. For the lower nappe, the air bubbles can easily diffuse and move into the interior of the jet due to the local air rotation and eddy currents, which are driven by the air–water interfacial turbulence coupled with the positive effect of buoyancy.

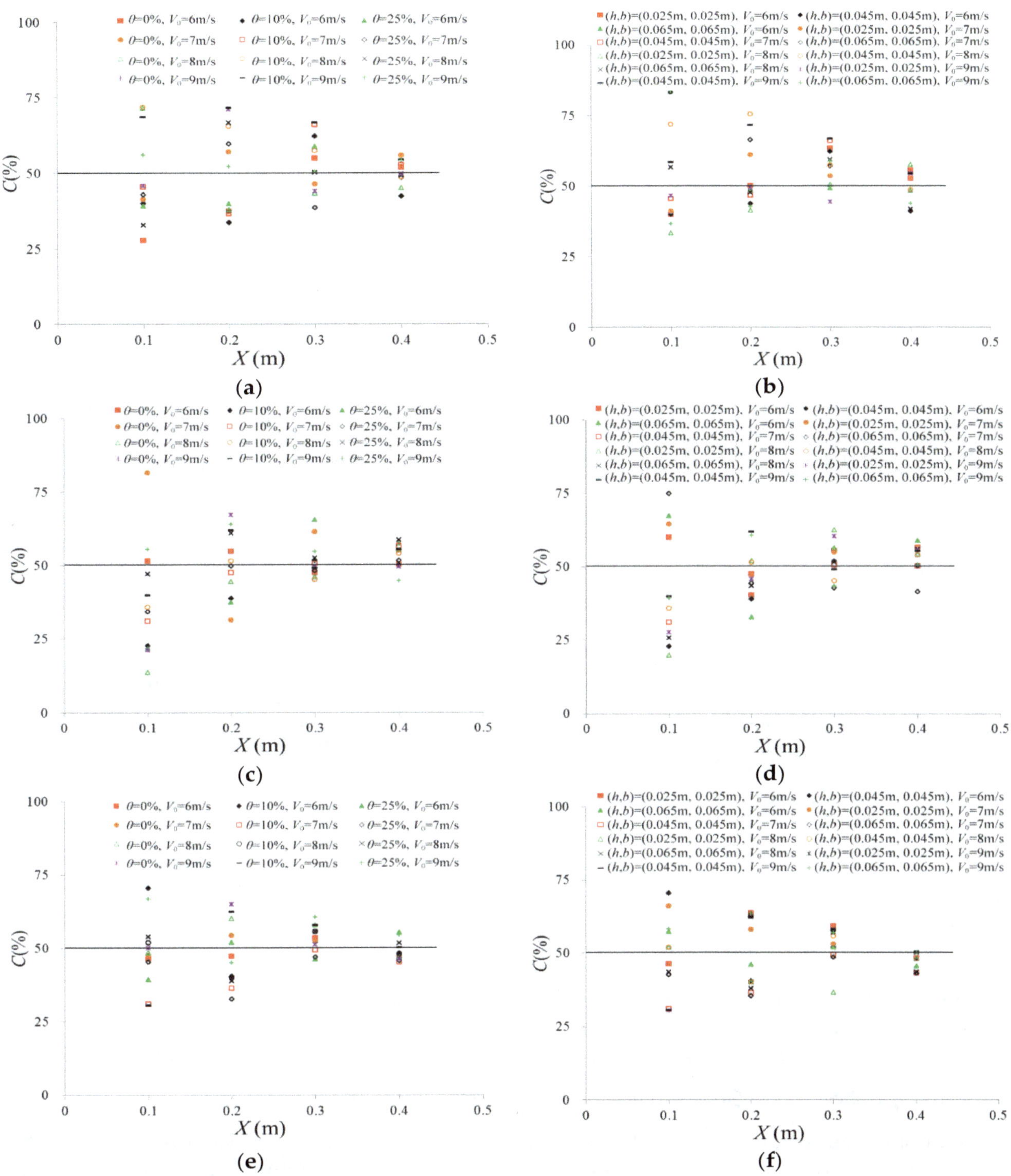

Figure 9. Air concentration corresponding to the maximum air bubble frequency. Upper nappe (**a**) $(h, b) = (4.5\ \text{cm}, 4.5\ \text{cm})$. (**b**) $\theta = 10\%$; lower nappe (**c**) $(h, b) = (4.5\ \text{cm}, 4.5\ \text{cm})$. (**d**) $\theta = 10\%$; side nappe (**e**) $(h, b) = (4.5\ \text{cm}, 4.5\ \text{cm})$. (**f**) $\theta = 10\%$.

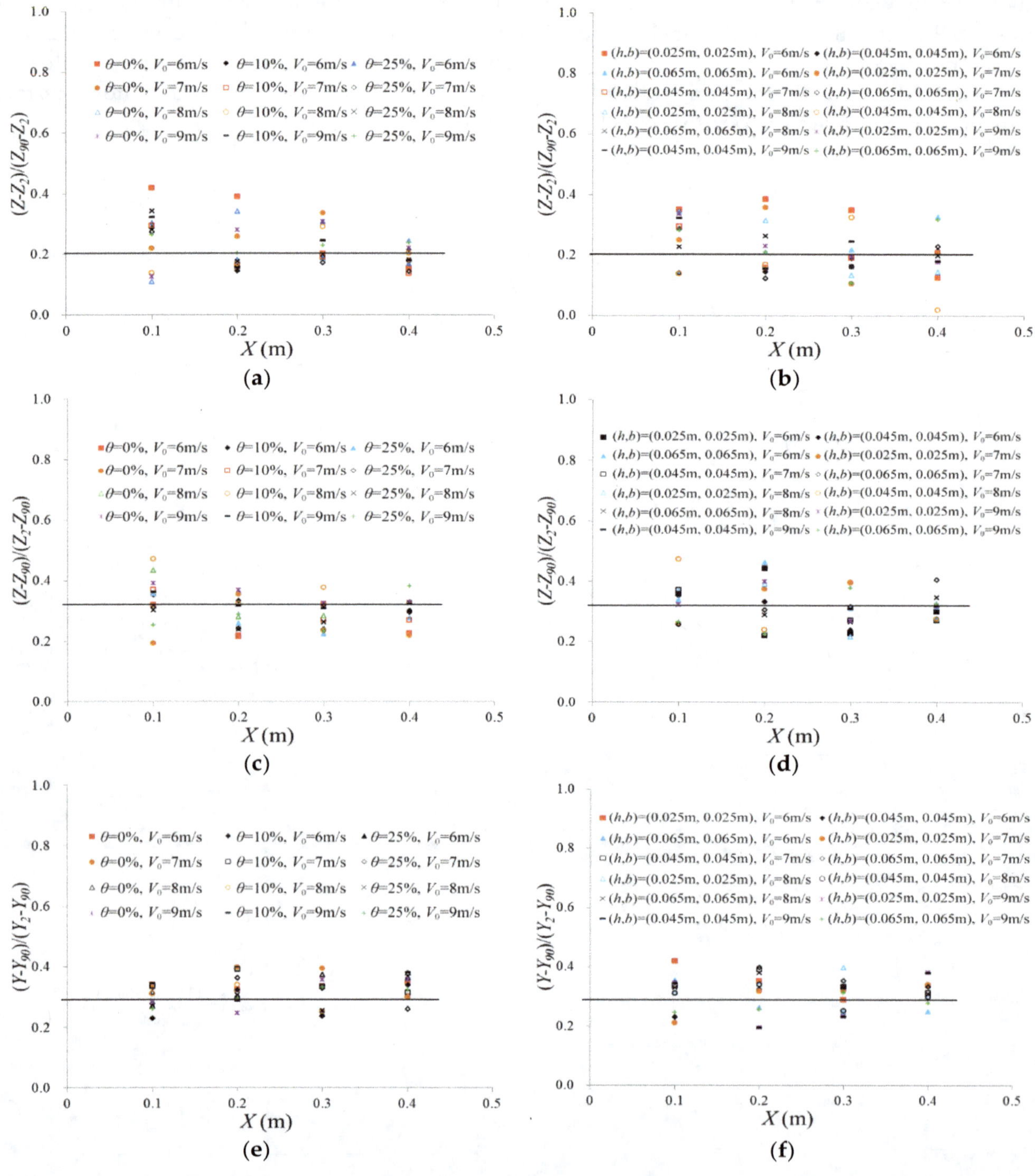

Figure 10. Relative location corresponding to the maximum dimensionless air bubble frequency. Upper nappe (**a**) (h, b) = (4.5 cm, 4.5 cm). (**b**) θ = 10%; lower nappe (**c**) (h, b) = (4.5 cm, 4.5 cm). (**d**) θ = 10%; side nappe (**e**) (h, b) = (4.5 cm, 4.5 cm). (**f**) θ = 10%.

3.2. Air Bubble Chord Length Distributions

Air bubble size is another parameter reflecting the characteristics of air–water flows. It is difficult to measure the size of air bubbles since their shapes vary greatly, are complex and highly changeable. This value can only be indirectly determined from the bubble chord length which is obtained using the double-tip conductivity probe. Here, the chord length $(ch_{ab})_{\text{mean}}$ is adopted to assess the size of the air bubbles, which is defined as follows:

$$(ch_{ab})_{\text{mean}} = \frac{\sum_{j=1}^{N} (ch)_j n_j}{\sum_{j=1}^{N} n_j} \tag{5}$$

where n_j is the count of air bubbles in the bubble chord length interval $\Delta(ch_{ab})_j$ ($\Delta(ch_{ab})_j$ = 0.1 mm in the experiments); and $(ch_{ab})_j$ is the average value of the chord length interval (e.g., the probability of chord length from 1.0 to 1.1 mm is represented by the label 1.05 mm).

The air bubble chord lengths $(ch_{ab})_{\text{mean}}$ are presented in Figure 11 at various positions in the upper, lower and side nappes. It is note that the air bubble chord length distributions obtained at the upper, lower, and side nappes have similar shapes. The air bubble chord length in the vicinity of the air–water interface is largest and then gradually decreases toward the inside of the water. A broad spectrum of air bubble chord lengths, i.e., from less than 0.1 mm to greater than 100 mm, is observed at each location and cross-section. At high air concentration (i.e., $C \geq 90\%$), chord length of many air bubbles can reach up to 100 mm or even more. These large values may be large air packets and air volumes surrounding the water structure (e.g., droplets). The reason is that the regime with high air concentration is always close to the air–water interfaces, which become uneven under the influence of turbulent forces. Air in the vicinity of the free surface that are trapped by the generated waves were treated as air bubbles by the conductivity probe. In addition, the shape of the air bubble is not completely spherical, in most cases, it is oval or banding, causing it to be measured to be much larger than it actually is. Indeed, it is nearly impossible to distinguish between air-bubbles and an un-enclosed bubble structure. The results indicate that the large air bubbles are constantly sheared and tore as they move towards the interior of the water. This results in air bubbles located deeper inside the body of water to be much smaller in size. Figure 9 also suggests that for $C < 90\%$, bubble chord lengths will most likely be no more than 20 mm and that bubbles close to the pressure outlet will have small sizes.

A positive correlation is observed between the air concentration and air bubble chord length among the upper, lower and side nappes (Figure 12). The distribution can be defined according to the function:

$$(ch_{ab})_{\text{mean}} = \frac{1}{a_1 + a_2 * C^{a_3}} \tag{6}$$

where a_1, a_2 and a_3 are empirical coefficients that may be related to the interaction. For the present experimental data, $a_1 = 0.212$, $a_2 = -0.006$ and $a_3 = 0.767$ were used. The coefficients of determination were set to be 0.85. 0.93 and 0.91 for the upper, lower and side nappes respectively.

Figure 13 presents a plot of experimental data, $(ch_{ab})_{\text{mean}}$, with calculated results using Equation (6). Since most of the values of $(ch_{ab})_{\text{mean}}$ calculated from Equation (6) fall around the line of perfect agreement in an area within $\pm 20\%$ error lines, although some discrepancy exists. The results exhibit agreement with the theoretical distribution curve of Equation (6) for all flow conditions at various positions.

As the relationships between the air concentration and air bubble frequency, as well as the air bubble chord length have been discussed above, the relationship between the air bubble count rate and bubble size is also considered (Figure 14). It is evident that the correlations are clearly non-symmetric, unimodal and positively skewed and may therefore be represented by a probability density function defined by the Gamma distribution. According to mathematical statistics, the gamma distribution is a continuous probability function that can be written as follows:

$$Ga(x) = \frac{1}{\Gamma(\alpha)} \beta^{-\alpha} x^{\alpha-1} e^{-\frac{x}{\beta}}, x > 0 \tag{7}$$

where α is the shape parameter, β is the scale parameter, and $\Gamma(x) = \int_0^\infty t^{x-1} e^{-t} dt$ is the gamma Function, and $\Gamma(x+1) = x\Gamma(x)$.

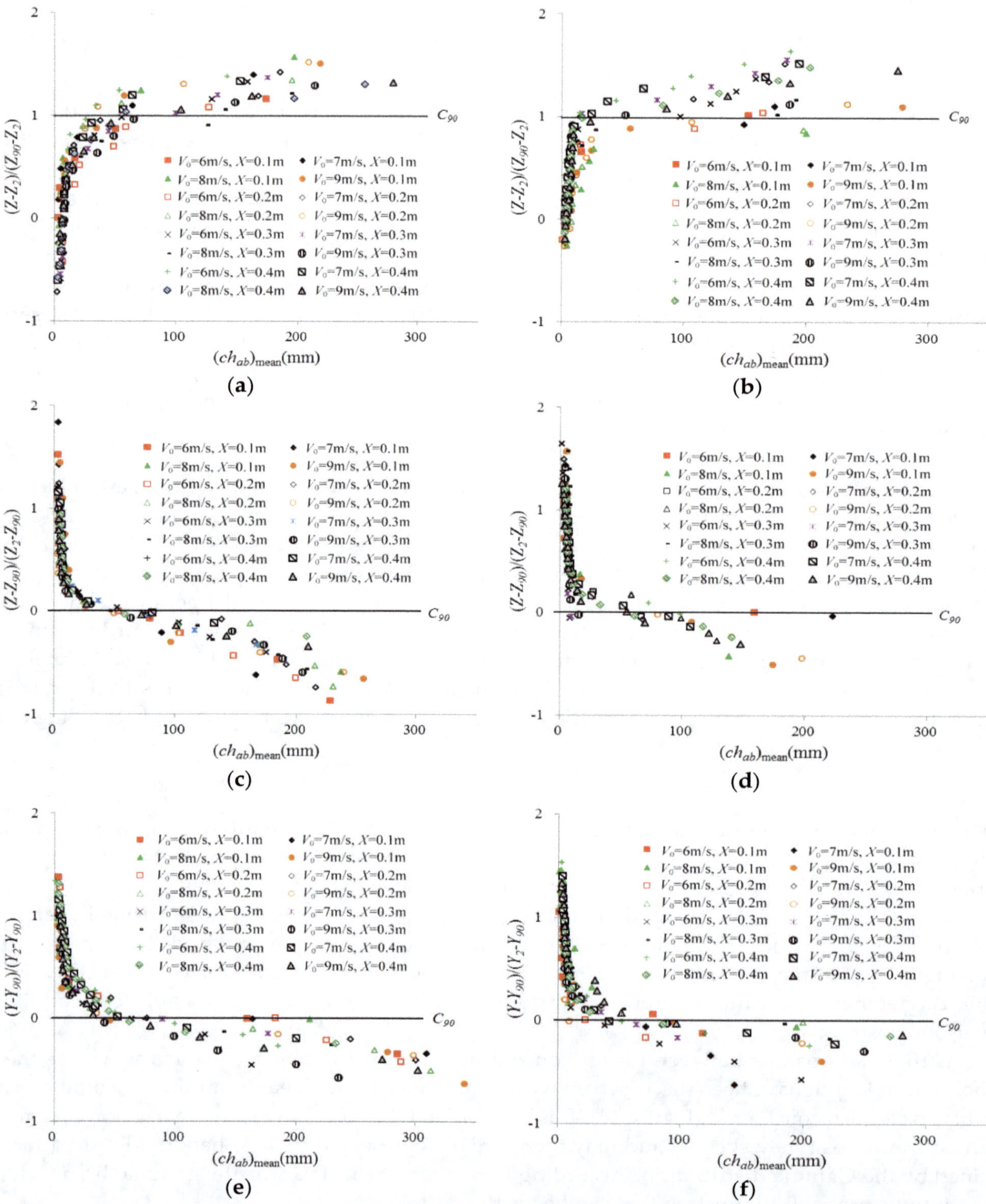

Figure 11. Air bubble chord length distributions. Upper nappe (**a**) $(h, b) = (4.5\text{ cm}, 4.5\text{ cm})$, $\theta = 0\%$. (**b**) $(h, b) = (2.5\text{ cm}, 2.5\text{ cm})$, $\theta = 10\%$; lower nappe (**c**) $(h, b) = (4.5\text{ cm}, 4.5\text{ cm})$, $\theta = 0\%$. (**d**) $(h, b) = (2.5\text{ cm}, 2.5\text{ cm})$, $\theta = 10\%$; side nappe (**e**) $(h, b) = (4.5\text{ cm}, 4.5\text{ cm})$, $\theta = 0\%$. (**f**) $(h, b) = (2.5\text{ cm}, 2.5\text{ cm})$, $\theta = 10\%$.

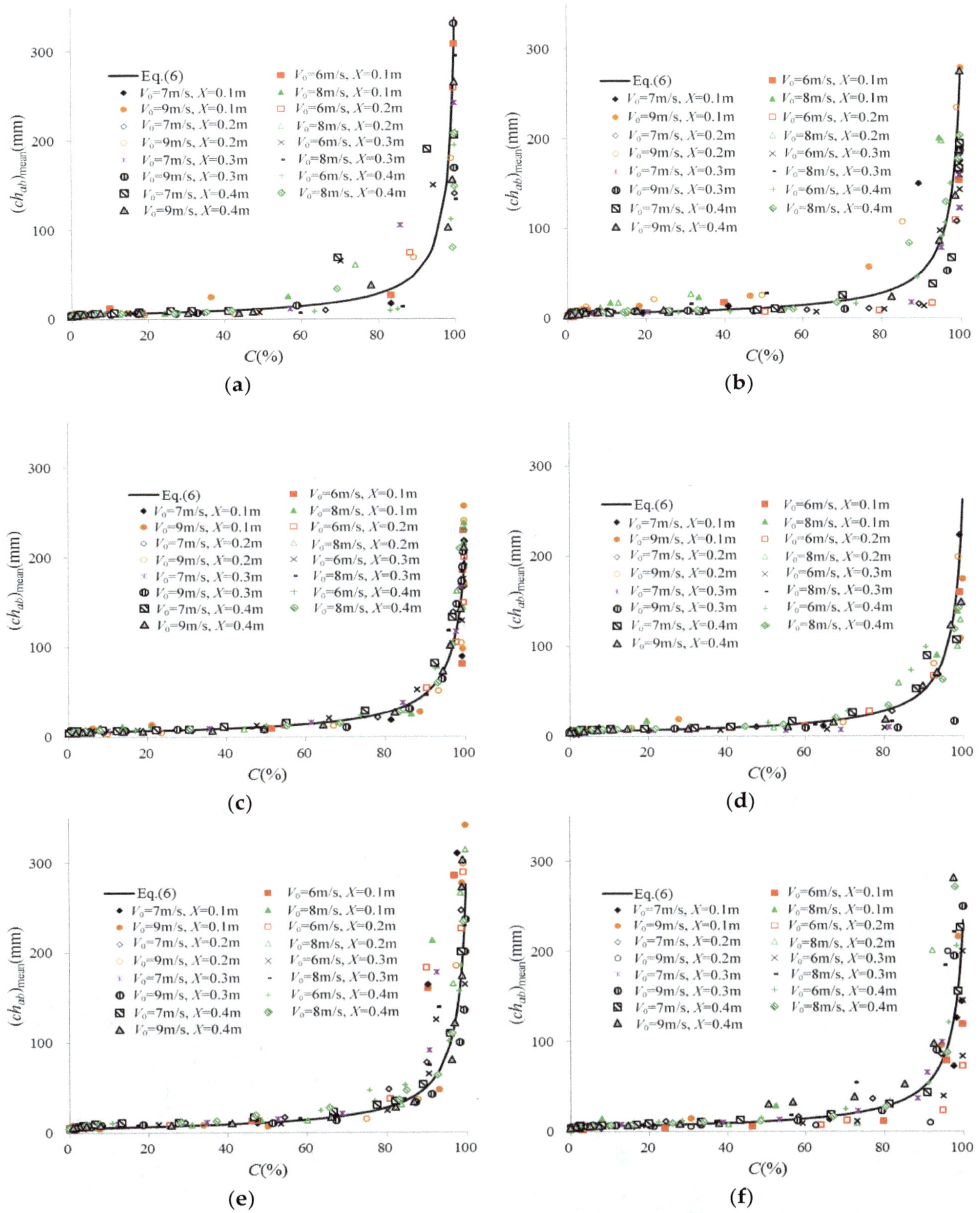

Figure 12. Measured air bubble chord length as a function of the air concentration. The solid lines are determined from Equation (6). Upper nappe (**a**) (h, b) = (4.5 cm, 4.5 cm), θ = 0%. (**b**) (h, b) = (2.5 cm, 2.5 cm), θ = 10%; lower nappe (**c**) (h, b) = (4.5 cm, 4.5 cm), θ = 0%. (**d**) (h, b) = (2.5 cm, 2.5 cm), θ = 10%; side nappe (**e**) (h, b) = (4.5 cm, 4.5 cm), θ = 0%. (**f**) (h, b) = (2.5 cm, 2.5 cm), θ = 10%.

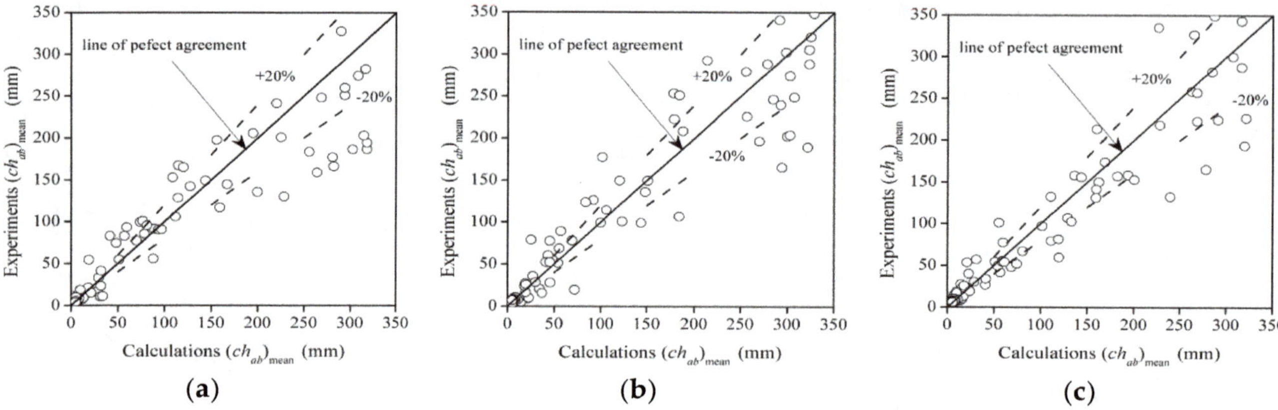

Figure 13. Comparison of experimental data with calculated values of air bubble chord length. (**a**) upper nappe; (**b**) lower nappe; (**c**) side nappe.

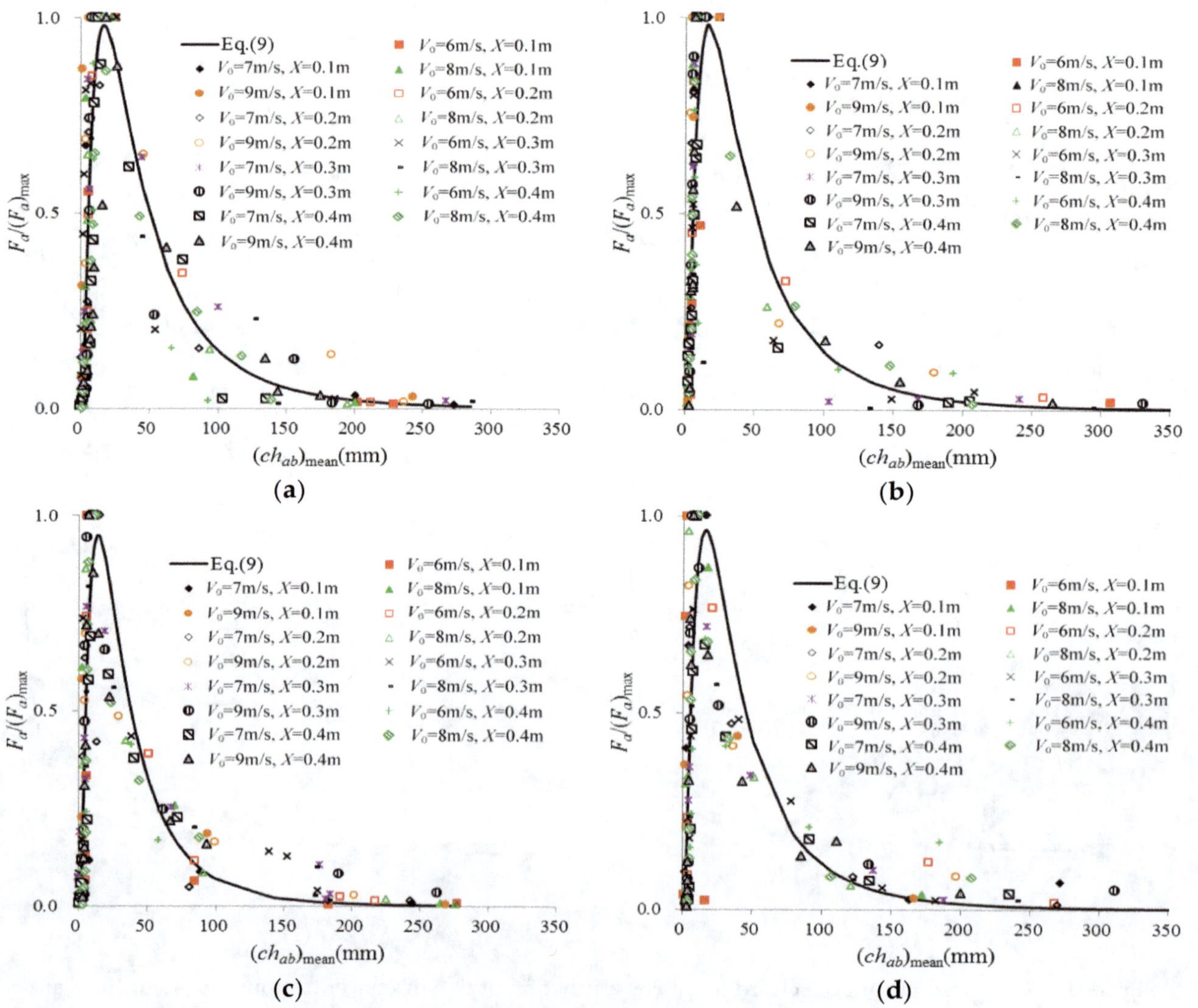

Figure 14. *Cont.*

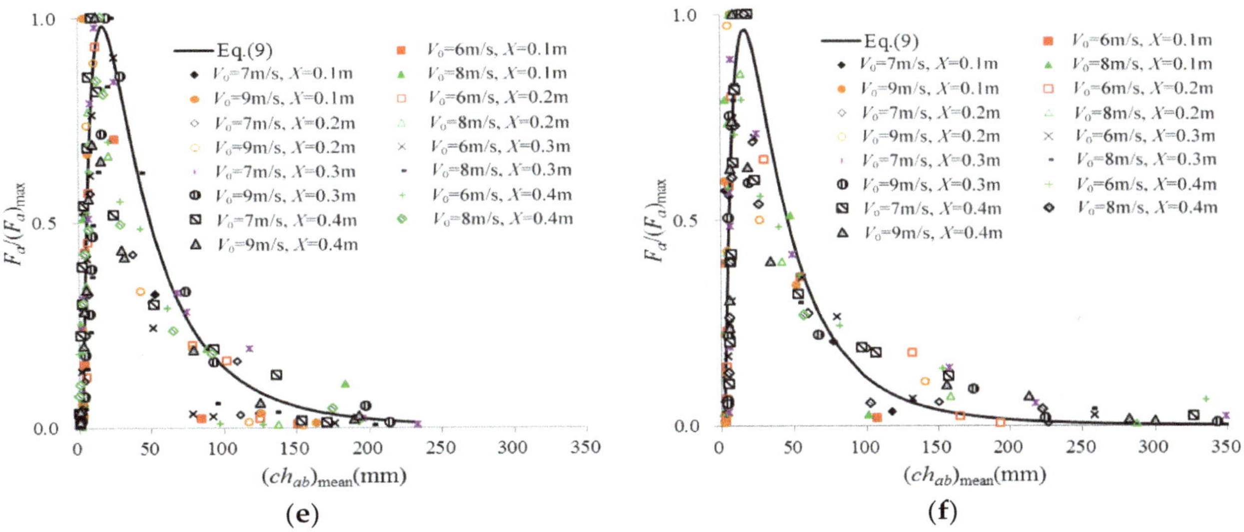

Figure 14. Dimensionless bubble frequency as a function of the bubble chord length compared with the theoretical curve derived from Equation (9). Upper nappe (**a**) (h, b) = (4.5 cm, 4.5 cm), θ = 10%. (**b**) (h, b) = (6.5 cm, 6.5 cm), θ = 10%; lower nappe (**c**) (h, b) = (4.5 cm, 4.5 cm), θ = 10%. (**d**) (h, b) = (6.5 cm, 6.5 cm), θ = 10%; side nappe (**e**) (h, b) = (4.5 cm, 4.5 cm), θ = 10%. (**f**) (h, b) = (6.5 cm, 6.5 cm), θ = 10%.

Equation (7) is a classic two-parameter gamma distribution. Considering the complexity of the air bubble frequency and chord length in aeration flows, the five-parameters form of the generalized gamma distribution (also known as the generalized gamma function with four parameters after translation) is used in this study. The probability density function is written as follows:

$$Ga(x) = \frac{1}{\Gamma(a_1)} a_2{}^{\eta_1 a_1} (x - a_3)^{\eta_2 a_1} e^{-a_4 * (x - a_3)^{a_5}} \qquad x > 0 \tag{8}$$

where η_1 and η_2 are the coefficients; a_1, a_2, a_3, a_4, and a_5 are the five parameters, of which a_1 and a_5 denote the shape parameters, a_2 and a_4 denote the scale parameters, and a_3 is the threshold parameter; and $\Gamma(\cdot)$ is the gamma function.

Based on the experimental data and the theoretical distribution curve of Equation (8), the relationship between the air bubble frequency and air bubble chord length can be fitted as follows:

$$\frac{F_a}{(F_a)_{max}} = 115[(ch_{ab})_{mean} - 1.5]^{3.8} * e^{-7.65[(ch_{ab})_{mean} - 1.5]^{0.25}} \tag{9}$$

Examples of a "modified" five-parameters gamma distribution that were used to fit the air bubble chord length distributions (Figure 14). The data is reasonably well represented by a "modified" gamma function distribution, although there is significant scatter.

Additionally, Figure 15 presents a plot of experimental data, $F_a/(F_a)_{max}$, with calculated non-dimensional air bubble frequency using Equation (9). Since most of the values of $F_a/(F_a)_{max}$ calculated from Equation (9) fall around the line of perfect agreement in an area within ±20% error lines, suggesting that Equation (9) can be considered adequate in calculating the air bubbles chord length distribution. The "modified" gamma function is skewed to the left with a peak that lies in the small chord size values. The mode was found to be between 0.5 and 30 mm.

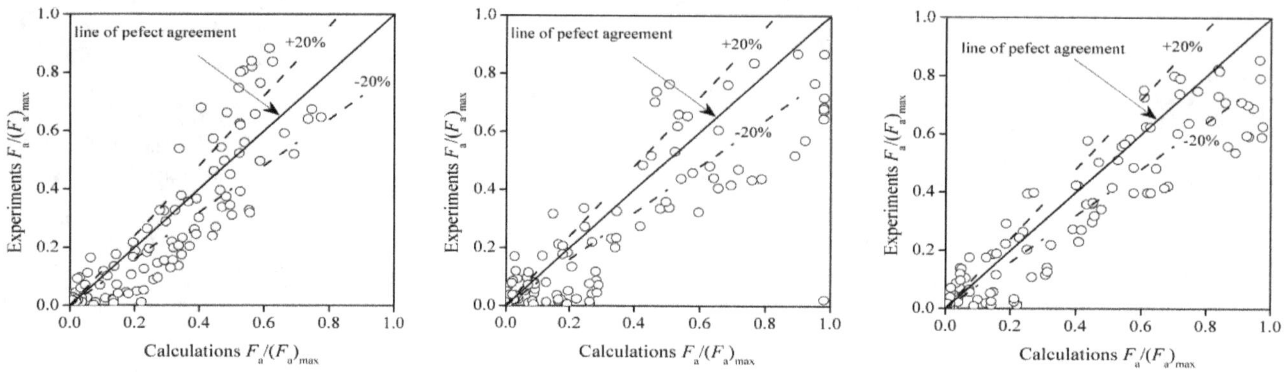

Figure 15. Comparison of experimental data with calculated values of dimensionless bubble frequency. (**a**) upper nappe; (**b**) lower nappe; (**c**) side nappe.

4. Discussion

Air concentration is a macro parameter that is used to describe air–water flows. However, it is still insufficient in providing insight that would lead to a deeper understanding of the said phenomena. The process of aeration defines the general developing process of air moving into a body of water, i.e., the air bubbles. As such, the micro characteristics of air bubbles such as its count rate and size are crucial parameters in the study of aeration flows.

A double-tip conductivity probe used in present study to detect the air–water interfaces at different fixed points downstream of a three-dimensional (vertical drop and lateral enlargement) aerator. Some simple expressions were developed to simulate the distributions of air bubble frequency and bubble chord sizes. The relationships between the air concentration, the air bubble frequency and the air bubble chord length were also established. The obtained air concentration distributions satisfy the analytical model of Chanson [12], Equation (4) for air bubble frequency distributions as observed in supercritical open channel flows and backward-facing step flows (Toombes and Chanson [19]). A "modified" gamma probability density function was found to provide a good fit to the air bubble chord length distributions measured from three-dimensional aerator flows. This is different from the log-normal probability density functions that Chanson (1997) used for supercritical open channel flows. The distributions of the air bubble chord length showed a positive correlation with the void fraction.

Physical modelling may provide some information on the flow motion if a suitable dynamic similarity is selected. Air–water flows depend on Froude, Reynolds, and Weber numbers. Strictly speaking, dynamic similarity of air concentration and air bubble dissipation for free-surface aeration on aerator flows on a reduced-scale model is not possible (Chanson [32]) because it is impossible to satisfy simultaneously Froude and Reynolds similarities. Hence, the experimental results cannot be directly extrapolated to prototypes unless working at full scale. Nevertheless, with the assumption of Froude similarity, the Reynolds number (Re) should be larger than 1×10^5 to minimize the scale effects (Chanson [32]; Heller [33]). This limit is considered in the present study (Re ~0.89–1.34 $\times 10^6$). Actually, results from scale experiments using Froude similarity are on the safe side for engineers with respect to air concentrations due to aeration tends to be underestimated in the smaller experiments.

Due to the complexity of bubble distribution in the air–water interfaces, the present results only serve as a preliminary investigation of the air bubble properties in the air–water mixtures downstream of a three-dimensional aerator. Hence, more detailed research is needed to study the air bubbles in aeration flows.

5. Conclusions

A number of experiments were performed under various configurations and approach flow velocities for a 3D aerator. The air concentration, air bubble frequency and bubble chord length were recorded in the developing shear layer of the upper, lower and side nappes downstream. The results are summarized as follows:

(1) The air bubble frequency distributions (yielding to the Gauss function) exhibit a unitary self-similarity along the air–water layer, presenting a trend which initially increases and then decreases from the air–water interface to the inside of the water. A quasi-parabolic relationship between the air bubble frequency and the air concentration is obtained for the upper, lower and side nappes.

(2) The air bubble frequency reaches to apex at approximately $C = 50\%$, and then decreases to zero as $C = 0\%$ and $C = 100\%$. The relative location at which the maximum air bubble frequency is observed is at 0.21, 0.326 and 0.283 times of the thickness of the air–water layers for the upper, lower and side nappes, respectively.

(3) The air bubble chord length decreases gradually from the free interface to the interior of the water. When the air bubble chord size is plotted against the air concentration, the resulting distribution follows a power-law function. The relationship between the air bubble frequency and bubble chord size exhibits a "modified" gamma function for the upper, lower and side nappes.

Author Contributions: S.L. and J.Z. proposed the research ideas and methods of the manuscript and writing, X.C. put forward the revise suggestion to the paper. J.C. is responsible for data collection and creating the figures.

Notations

The following symbols are used in this paper:

$a_{1,2,3,4,5}$	empirical coefficients
b	the width of sudden fall-expansion aerator
B	width of the pressure outlet
C	air concentration
$(ch_{ab})_{\text{mean}}$	the air bubble chord length size
$(ch_{ab})_j$	the average value of air bubble chord length
d_i	air bubble chord length
i	air bubble number
F_a	air bubble frequency
$(F_a)_{\max}$	the maximum bubble frequency
F_r	Froude Number
g	gravitational acceleration
h	the height of sudden fall-expansion aerator
H	height of the pressure outlet
q_{w}	the unit-width discharge
Re	Reynolds number
n_i	the number of detected bubbles
n_j	the count of air bubbles
N_a	the number of air-structures
R_i	a ratio of the bubble measured in the whole measurement time
t	time
V_0	the cross-sectional mean velocity at the pressure outlet
X, Y, Z	the horizontal, transverse and vertical coordinates of the lower nappe profile

θ	the slope with respect to the horizontal downstream bottom plane
Y_2, Z_2	characteristic air–water flow heights where the air concentration $C = 2\%$;
Y_{90}, Z_{90}	characteristic air–water flow heights where the air concentration $C = 90\%$
μ_0	characteristic depth corresponding to the air bubble frequency
σ_0	degree of deviation between the relative depth and its mean value
$Ga(x)$	the probability density function of the gamma function
α	the shape parameter
β	the scale parameter
$\Gamma(\cdot)$	the gamma function

References

1. Cassidy, J.; Elder, R. *Spillways of High Dams, Developments in Hydraulic Engineering—2*; Elsevier: New York, NY, USA, 1984.
2. Kells, J.A.; Smith, C.D. Reduction of cavitation on spillways by induced air entrainment. *Can. J. Civ. Eng.* **1991**, *18*, 358–377. [CrossRef]
3. Hager, W.H.; Pfister, M. Historical advance of chute aerators. In Proceedings of the 33rd IAHR Congress, Vancouver, BC, Canada, 9–14 August 2009; pp. 5827–5834.
4. Pfister, M.; Lucas, J.; Hager, W.H. Chute aerators: Pre aerated approach flow. *J. Hydraul. Eng.* **2011**, *137*, 1452–1461. [CrossRef]
5. Hager, W.H.; Boes, R.M. Hydraulic structures: A positive outlook into the future. *J. Hydraul. Res.* **2014**, *52*, 299–310. [CrossRef]
6. Low, H.S. *Model Studies of Clyde Dam Spillway Aerators*; Research Report, No. 86-6; Department of Civil Engineering, University of Canterbury: Christchurch, New Zealand, 1986.
7. Zhang, J.M.; Wang, Y.R.; Yang, Q.; Xu, W.L. Scale effects of incipient cavitation for high-speed flows. *Proc. Inst. Civ. Eng. Water Manag.* **2013**, *166*, 141–145. [CrossRef]
8. Chanson, H. A study of Air Entrainment and Aeration Devices on a Spillway Model. Ph.D. Thesis, Department of Civil Engineering, University of Canterbury, Christchurch, New Zealand, 1988.
9. Chanson, H. *Air Bubble Entrainment in Free-Surface Turbulent Flows: Experimental Investigations*; Research Report No. CH46/95; Department of Civil Engineering, University of Queensland: Brisbane, Australia, 1995.
10. Chanson, H.; Brattberg, T. *Experimental Investigations of Air Bubble Entrainment in Developing Shear Layer*; Research Report, No. CH48/97; Department of Civil Engineering, The University of Queensland: Brisbane, Australia, 1997.
11. Brattberg, T.; Chanson, H.; Toombes, L. Experimental investigations of free-surface aeration in the developing flow of two-dimensional water jets. *J. Fluids Eng.* **1998**, *120*, 738–744. [CrossRef]
12. Chanson, H. Air bubble entrainment in open channels: Flow structure and bubble size distributions. *Int. J. Multiphase Flow* **1997**, *23*, 193–203. [CrossRef]
13. Toombes, L. Experimental Study of Air-Water Flow Properties on Low-Gradient Stepped Cascades. Ph.D. Thesis, Department of Civil Engineering, University of Queensland, Brisbane, Australia, 2002.
14. Li, S.; Zhang, J.M.; Chen, X.Q.; Zhou Gordon, G.D. Air concentration and velocity downstream of an expanding chute aerator. *J. Hydraul. Res.* **2018**, *56*, 412–423. [CrossRef]
15. Li, S.; Zhang, J.M.; Chen, X.Q.; Chen, J.G.; Zhou Gordon, G.D. Cavity length downstream of a sudden fall-expansion aerator in chute. *Water Sci. Technol. Water Supply* **2018**, *18*, 2053–2062. [CrossRef]
16. Pfister, M.; Hager, W.H. Deflector-generated jets. *J. Hydraul. Res.* **2009**, *47*, 466–475. [CrossRef]
17. Kramer, K.; Hager, W.H. Air transport in chute flows. *Int. J. Multiphase Flow* **2005**, *31*, 1181–1197. [CrossRef]
18. Pfister, M.; Hager, W.H. Chute aerators. I: Air transport characteristics. *J. Hydraul. Eng.* **2010**, *136*, 352–359. [CrossRef]
19. Toombes, L.; Chanson, H. Interfacial aeration and bubble count rate distributions in a supercritical flow past a backward-facing step. *Int. J. Multiphase Flow* **2008**, *34*, 427–436. [CrossRef]
20. Zhang, J.M.; Chen, J.G.; Xu, W.L.; Wang, Y.R.; Li, G.J. Three-dimensional numerical simulation of aerated flows downstream sudden fall aerator expansion-in a tunnel. *J. Hydrodyn. Ser. B* **2011**, *23*, 71–80. [CrossRef]
21. Toombes, L.; Chanson, H. Free-surface aeration and momentum exchange at a bottom outlet. *J. Hydraul. Res.* **2007**, *45*, 100–110. [CrossRef]

22. Fuhrhop, H.; Schulz, H.E.; Wittenberg, H. Solution for spillway chute aeration through bottom aerators. *Int. J. CMEM* **2014**, *2*, 298–312. [CrossRef]
23. Bai, Z.L.; Peng, Y.; Zhang, J.M. Three-Dimensional Turbulence Simulation of Flow in a V-Shaped Stepped Spillway. *J. Hydraul. Eng.* **2017**, *143*, 06017011. [CrossRef]
24. Duarte, R. Influence of Air Entrainment on Rock Scour Development and Block Stability in Plunge Pools. Ph.D. Thesis, EPFL-LCH, Lausanne, Switzerland, 2014.
25. Peng, Y.; Mao, Y.F.; Wang, B.; Xie, B. Study on C-S and P-R EOS in pseudo-potential lattice Boltzmann model for two-phase flows. *Int. J. Mod. Phys. C* **2017**, *28*. [CrossRef]
26. Peng, Y.; Wang, B.; Mao, Y.F. Study on force schemes in pseudopotential lattice Boltzmann model for two-phase flows. *Math. Probl. Eng.* **2018**, *2018*, 6496379. [CrossRef]
27. Peng, Y.; Zhang, J.M.; Meng, J.P. Second order force scheme for lattice Boltzmann model of shallow water flows. *J. Hydraul. Res.* **2017**, *55*, 592–597. [CrossRef]
28. Chanson, H. Hydraulics of aerated flows: Qui pro quo? *J. Hydraul. Res.* **2013**, *51*, 223–243. [CrossRef]
29. Chen, X.P.; Shao, D.C. Measuring bubbles sizes in self-aerated flow. *Water Res. Hydropower Eng.* **2006**, *37*, 33–36.
30. Bai, R.; Liu, S.; Tian, Z.; Wang, W.; Zhang, F. Experimental Investigation of Air–Water Flow Properties of Offset Aerators. *J. Hydraul. Eng.* **2017**, *144*, 04017059. [CrossRef]
31. Gonzalez, C.A. An Experimental Study of Free-Surface Aeration on Embankment Stepped Chutes. Ph.D. Thesis, Department of Civil Engineering, The University of Queensland, Brisbane, Australia, 2005.
32. Chanson, H. Turbulent air-water flows in hydraulic structures: Dynamic similarity and scale effects. *Environ. Fluid Mech.* **2009**, *9*, 125–142. [CrossRef]
33. Heller, V. Scale effects in physical hydraulic engineering models. *J. Hydraul. Res.* **2011**, *49*, 293–306. [CrossRef]

4

An Analysis of the Factors Affecting Hyporheic Exchange based on Numerical Modeling

Jie Ren [1,*], Xiuping Wang [1], Yinjun Zhou [2,*], Bo Chen [1] and Lili Men [1]

1 State Key Laboratory of Eco−hydraulics in Northwest Arid Region of China, Xi'an University of Technology, Xi'an 710048, China; wxpyl1992@163.com (X.W.); chenbo7289@163.com (B.C.); li15029069869@163.com (L.M.)

2 Changjiang River Scientific Research Institute, Wuhan 430010, China

* Correspondence: renjie@xaut.edu.cn (J.R.); zhouyinjun1114@126.com (Y.Z.)

Abstract: The hyporheic zone is a transition zone for the exchange of matter and energy between surface water and subsurface water. The study of trends and sensitivities of bed hyporheic exchanges to the various influencing factors is of great significance. The surface−groundwater flow process was simulated using a multiphysics computational fluid dynamics (CFD) method and compared to previous flume experiments. Based on that, the single-factor effects of flow velocity (u), water depth (H), dune wave height (h), and bed substrate permeability (κ) on hyporheic exchange in the bed hyporheic zone were investigated. The sensitivity analysis of various factors (H, u, dune wavelength (L), h, bed substrate porosity (θ), κ, and the diffusion coefficient of solute molecules (D_m)) in the surface−subsurface water coupling model was done using orthogonal tests. The results indicated that u, h, and κ were positively related, whereas H was negatively related to hyporheic exchange. H and u showed large effects, whereas κ, D_m, and θ had moderate effects, and L and h showed small effects on hyporheic exchange. This study provides valuable references for the protection and recovery of river ecology.

Keywords: hyporheic exchange; surface−groundwater flow process; influencing factors; orthogonal tests; sensitivity analysis

---

1. Introduction

The hyporheic zone plays an important role as an interface between subsurface and surface water [1,2], and has important influences on exchanges of water, nutrients and heat [3–5], pollutant migration [6,7], and quality of surface and subsurface water [8–10]. In regards to water quality, hyporheic exchange controls the temperature pattern [11,12], induces diffusion of solutes on the bed [13], increases the residence time of solutes [14], accelerates circulation of nutrients [15,16], and increases opportunities for biological and geochemical processing by decelerating migrations of dissolved and suspended substances [8,17]. Hence, hyporheic exchange has a significant effect on rivers and subsurface water systems [18,19].

Hyporheic exchanges exhibit complicated temporal and spatial variations due to influences of water fluctuations [20], in-stream structures and channel flow rates [21–23], bedform morphology [24–26], sediment penetrability [22,27–29], and rainfall patterns [20]. Recently, hyporheic exchanges have been the focus of intensive research [14,30–36]. For instance, hyporheic exchanges were measured in streams using ion or dye tracers [37]. The dynamics of the coupled system for unidirectional flow in the water column and a triangular interface on dunes were investigated [38]. Based on that, sensitivity analysis was performed via multiple computational fluid dynamics (CFD) simulations and interactions between turbulent water-column flows, current topography-driven flows in underlying permeable sediments, and ambient subsurface water discharge from deep subsurface waters were investigated [39]. Sawyer et al. [40] proposed equations for bed pressure profiles and hyporheic exchange rates in the vicinity of a channel-spanning log that can be used to evaluate the impacts of the removal or introduction of large woody debris on hyporheic mixing. Schmadel et al. [41] developed a framework relating diel hydrologic fluctuations to hyporheic exchange in simple bedform morphologies and simulated subsurface water flows under time-varying boundary conditions using an aquifer bounded by a straight stream and hill slope. The lattice Boltzmann method has been used successfully to study some complex flows [42–45]. The lattice Boltzmann method was used by Peng et al. [43] to investigate solute transport in shallow water flows. This suggests that the present model has great potential to predict morphological change in shallow water flows.

Laboratory flume tests have also been employed to investigate hyporheic exchanges [46–49]. Tonina and Buffington [50] reported a series of recirculating flume tests to investigate hyporheic exchanges in pool-riffle channels spanning a broad range of discharge and bedform morphologies. Wu and Hunkeler [51] investigated flow processes in a conduit—sediment system using both a model resembling a siphon and a numerical model. Packman et al. [6] studied solute exchange in flat and dune-shaped beds using laboratory flume tests, where dye injections indicated that a combination of convective pore water flow and turbulent diffusion near the stream–subsurface interface is responsible for solute exchange on flat beds. Sawyer et al. [52] used laboratory flume tests and numerical simulations to quantify hyporheic fluids and heat exchanges induced by current interactions with channel-spanning logs. Fox et al. [53] used a novel laboratory flume system to investigate the effects of losing and gaining flow conditions on hyporheic exchange fluxes in a sandy streambed. Lu et al. [54] constructed a two-dimensional (2D) sand tank to study the influence of a clay lens with low permeability on the hyporheic zone under different surface flow conditions. Despite the field of hyporheic zone research being fairly mature, few studies have investigated the effects of individual factors on hyporheic exchange.

In this study, sensitivity analysis was applied to factors in hyporheic exchange via orthogonal tests and hyporheic exchange was studied by flume tests [55]. The flume used had a length, width, and cross-sectional length of 2 m, 0.3 m, and 1.5 m, respectively. Seven identical triangular ripples with a wave height of 0.02 m, wavelength of 0.2 m, and distance from trough to crest of 15 cm were designed for the flume. A surface—subsurface flow coupling model was established and compared to previous flume experiments [55]. The effects of various factors, including water depth (H), flow velocity (u), dune wave height (h), bed substrate porosity (θ), bed substrate permeability (k), and coefficient of solute molecules (D_{m}) on hyporheic exchanges were investigated using orthogonal tests and sensitivity analysis. Additionally, numerical simulations of the injection of dye to the dune were developed for the surface—subsurface hyporheic exchange. The evaluation parameter selected was the time to equilibrium for solute concentrations at observation points.

2. Methodology

2.1. Governing Equations for Fluid Flow

The turbulent flow in the test section of the flume was simulated following the multiphysics computational fluid dynamic (CFD) approach of Cardenas and Wilson [38]. Turbulent flow is simulated by solving the Reynolds-averaged Navier—Stokes (RANS) equations with the k—ω turbulence closure model by Wilcox [56]. The pore water flow is simulated by solving a steady state groundwater flow model using COMSOL Multiphysics. These two sets of equations are coupled through the pressure distribution at the sediment—water interface [31]. Janseen et al. [55] used exactly the same combination of software. For an incompressible fluid, the steady state RANS equations are defined as:

$$\frac{\partial U_i}{\partial x_i} = 0 \tag{1}$$

$$\rho U_j \frac{\partial U_i}{\partial x_j} = -\frac{\partial P}{\partial x_i} + \frac{\partial}{\partial x_j}(2\mu S_{ij} - \rho \overline{u'_j u'_i}) \tag{2}$$

where ρ refers to the fluid density, μ refers to the dynamic viscosity (assumed standard for water), $U_{i\ or\ j}$ $(i, j = 1, 2$ where $i \neq j)$ refers to the time-averaged velocity, $u_i{}'$ $(i = 1, 2)$ refer to the fluctuations in the instantaneous velocity components in $x_{i\ or\ j}$ $(i, j = 1, 2,$ where $i \neq j)$ directions, and P refers to the time-averaged pressure. The strain rate tensor (S_{ij}) is defined as:

$$S_{ij} = \frac{1}{2}(\frac{\partial U_i}{\partial x_j} + \frac{\partial U_j}{\partial x_i}) \tag{3}$$

The Reynolds stresses, τ_{ij}, are related to the mean strain rates by:

$$\tau_{ij} = -\overline{u_j{}' u_i{}'} = v_t(2S_{ij}) - \frac{2}{3}\delta_{ij}k \tag{4}$$

where v_t refers to the kinematic eddy viscosity, δ_{ij} refers to the Kronecker delta, and k refers to the turbulent kinetic energy. The $k-\omega$ turbulence closure scheme [56] was employed due to its advantages in simulating separated flows with adverse pressure gradients, including flow over dunes where a pronounced eddy is present [57,58]. The eddy viscosity in this closure scheme is:

$$v_t = \frac{k}{\omega} \tag{5}$$

where the specific dissipation, ω, is defined as the ratio of the turbulence dissipation rate ε to k:

$$\omega = \frac{\varepsilon}{\beta^* k} \tag{6}$$

where β^* refers to the closure coefficient.

The steady-state migration equations for k and ω are:

$$\rho \frac{\partial (U_j k)}{\partial x_j} = \rho \tau_{ij} \frac{\partial U_i}{\partial x_j} - \beta^* \rho \omega k + \frac{\partial}{\partial x_j}\left[(\mu + \mu_t \sigma_k)\frac{\partial k}{\partial x_j}\right] \tag{7}$$

$$\rho \frac{\partial (U_j \omega)}{\partial x_j} = \alpha \frac{\rho \omega}{k} \tau_{ij} \frac{\partial U_i}{\partial x_j} - \beta \rho \omega^2 + \frac{\partial}{\partial x_j}\left[(\mu + \mu_t \sigma_\omega)\frac{\partial \omega}{\partial x_j}\right] \tag{8}$$

The standard closure coefficients for the $k-\omega$ scheme are from Wilcox [56]: $\alpha = 5/9$, $\beta = 3/40$, $\beta^* = 9/100$, $\sigma_k = \sigma_\omega = 0.5$.

The 2D pore water flow in sand was modeled by solving the steady-state subsurface water flow equation:

$$\nabla(-\frac{\kappa}{\mu}\nabla \times P) = 0 \tag{9}$$

where κ refer to isotropic permeability. The parenthetical term is the Darcy flux or Darcy velocity (Q). The solution for simulating turbulent flow for pressure along the interface of sediment and water is described as a Dirichlet boundary at the top of the porous domain, whereas simulations of other porous flows corresponding to the flume have no flow boundaries. The flow field divided by porosity was used as the input for particle tracking, whereas dispersion is neglected as this study focuses on the convective flow paths. The pore water flows were simulated using the finite element code COMSOL Multiphysics.

The solute migration model was established based on convective diffusion equations [59]:

$$\frac{\partial C}{\partial t} = D_{\mathrm{m}} \frac{\partial^2 C}{\partial x_i^2} - u_i \frac{\partial C}{\partial x_i} + \frac{\partial}{\partial x_i}\left(D_{ij}\frac{\partial C}{\partial x_i}\right) \tag{10}$$

where C is solute concentration, t is time, D_m is the molecular diffusion coefficient in porous media, and D_{ij} is the mechanical dispersion coefficient ensor. Index $i, j = 1, 2$.

The equation to obtain D_{ij} is [60]:

$$D_{ij} = \alpha_T U \delta_{ij} + (\alpha_L - \alpha_T)\frac{u_i u_j}{U} \tag{11}$$

where α_T and α_L are transverse and longitudinal dispersivities, U is the pore velocity magnitude, and δ_{ij} is the Kronecker delta function.

The Millington–Quirk model of the effective diffusion coefficient is:

$$\tau = \theta^{-\frac{1}{3}} \tag{12}$$

where θ refers to the porosity.

2.2. Calculation Model

The numerical calculation model proposed in the present study is based on the flume tests reported by Janssen et al. [55]. This model has a length of 1.5 m, including 0.05 m buffer segments at inlets and outlets and seven identical triangular dunes (crest curvature radius = 0.02 m, bed substrate height = 0.09 m, water depth H = 0.1 m). Figure 1 shows the 2D flume model in which arrows describe the flow directions.

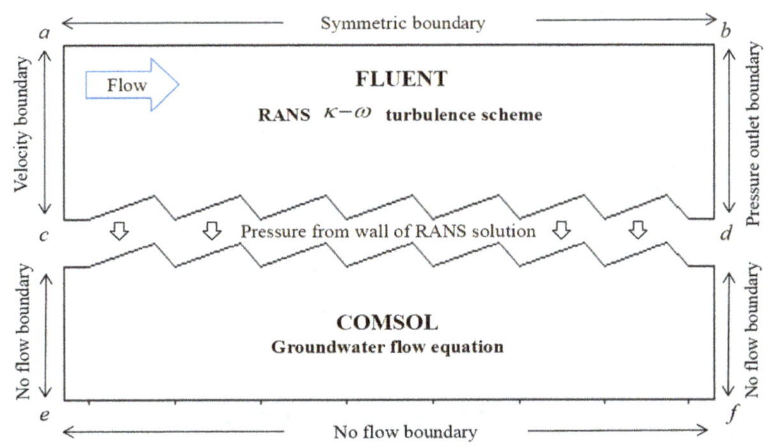

Figure 1. Two-dimensional flume model.

The present study assumed fluids to be steady and incompressible, bed substrates to be homogeneous, and isotropic and dune configurations to be free of any shifts. Fluent is a CFD software used for simulating and analyzing fluid flow and heat exchange problems in complex geometric regions. Using Fluent software to study the eddy current structure, flow separation, and water–sand interface pressure distribution results of surface water modeling under riverbed morphological disturbance has high precision. As shown in Figure 1, the overlying water was simulated by the CFD–Fluent; COMSOL Multiphysics is a software based on the finite element method. It is one of the softwares used for multiphysics coupling simulation by solving partial differential equations. It has good post-processing function. It is more accurate to use COMSOL software to describe the hyporheic exchange in a groundwater model. Therefore, we used the commercial finite element software COMSOL Multiphysics to solve the groundwater flow equations in the porous sediment. Pressure distributions at the interface of water and sand as determined by the CFD–Fluent were based on the coupling of surface and subsurface water.

The 6th dune was located in the middle of the flume, and the flow interference was lower. In addition, Janssen et al. [55] had measured the vertical velocity of the section above the water–sand interface of the 5th, 6th, and 7th sections. Based on this, the 6th dune was selected as the subject. With the height of the dune trough defined as "0", four feature observation points (A, B, C, and D) at −2 cm were selected, as shown in Figure 2. More specifically, observation points A and D were situated at the trough, observation point B was situated 8 cm to the right of observation point A (1/2 of the upstream face), and observation point C was situated at the crest. The dye injection into bed substrate were simulated as circular areas with observation points as the centers and radii of 0.5 cm to monitor the hyporheic exchange routes. In the present study, $CaCl_2$ solution was used as the dye and the time to equilibrium for concentrations at observation points was chosen as the evaluation parameter; the concentration variation was reflected by the dimensionless C/C_0 (C_0 = initial concentration).

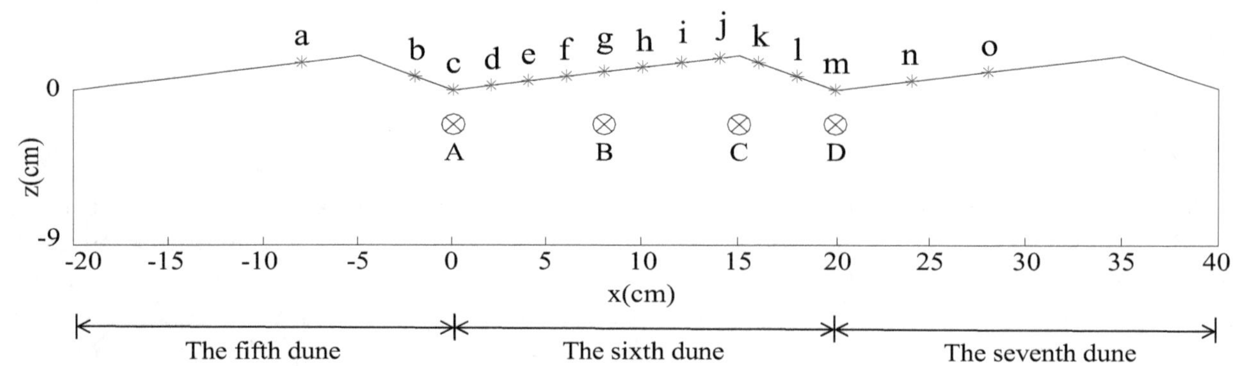

Figure 2. Layout of monitoring points.

Table 1 summarizes the parameters used in the proposed calculation model [55].

Table 1. Parameters used in the calculation model.

Fluid Density ρ (kg m^{-3})	Kinematic Eddy Viscosity ν_t (m^2 s^{-1})	Porosity θ (%)	Permeability κ (m^2)	Molecular Diffusion Coefficient D_m (m^2 s^{-1})
998.8	1.10×10^{-6}	40	1.50×10^{-11}	5.00×10^{-11}

2.3. Grid Division and Boundary Conditions

The overlying water grid was generated using the Gambit. The boundary layer had a thickness of 1 cm, meshes near wall boundaries had sizes of 0.2 mm, the growth rate was 1.06, and a total of 25 layers were generated. The boundary layer used had a thickness of 1.1 cm, and 440,000 meshes were generated for the overlying water. The bed substrate was divided into 10,000 meshes with the mass of 0.95. The constant velocity inlet boundary is referred to as *ac*, *bd* refers to the constant pressure outlet boundary, and *ab* refers to the free fluid surface as a symmetric boundary. In shallow water applications the top of the *ab* domain is actually a free surface, but in our simulations the water depth was large enough to replace the free surface with the symmetry condition [55]. The variable *cd* refers to the interface of water and sand as a sliding-free wall boundary. Although recent studies have emphasized the effects of wall permeability on the mass and momentum exchange at the interface of water and sand, it is a valid approximation for interfacial flow with lower porosity and permeability porous media such as considered in this study [31]. Flux free boundaries are referred to as *ce*, *ef*, and *df*.

2.4. Model Evaluation

Root mean square error (RMSE), coefficient of determination (R^2), and the relative error (Re) were used to quantitatively evaluate the accuracy of the calculation model:

$$\text{RMSE} = \sqrt{\sum_{i=1}^{n} (O_i - S_i)^2 / n} \tag{13}$$

$$R^2 = 1 - \sum_{i=1}^{n} (O_i - S_i)^2 / \sum_{i=1}^{n} (O_i - \overline{O})^2 \tag{14}$$

$$\text{Re} = \sqrt{\sum_{i=1}^{n} (O_i - S_i)^2 / \sum_{i=1}^{n} O_i^2} \tag{15}$$

where O_i, S_i, n, and $\overline{O}$ refer to the measured value, simulated value, sample size, and average value, respectively.

The consistency between measured and simulated values was measured using RMSE to verify the model. The RMSE is a non-negative value and a low RMSE indicates good consistency between measured and simulated values [61]. R^2 is the coefficient of determination of the linear regression equation (y = x) between measured and simulated values and a large R^2 indicates good consistency between measured and simulated values [62]. Re is the relative error between measured and simulated values and a low Re indicates good consistency between measured and simulated values [63].

2.5. Orthogonal Tests

Orthogonal testing involves the use of orthogonal list-based multi-factor tests and result analysis. Representative points with a homogeneous distribution and good comparability were selected for testing based on the principle of orthogonality. The orthogonal test is used to assess result trends with reduced testing cycles. The sensitivity of each factor on the evaluation index was determined by range analysis of the results of orthogonal tests.

Suppose M and N refer to different influencing factors, t refers to the factor level, M_i refers to the ith ($i = 1, 2, \ldots, t$) level of factor M, and X_{ij} refers to the ith ($I = 1, 2, \ldots, n$) level of factor j ($j = M, N$). F_s ($s = 1, 2, \ldots, n$) refers to the testing result at X_{ij}. The statistical parameters were calculated by:

$$K_{ij} = \frac{1}{n}\sum_{s=1}^{n} F_s - \overline{F} \tag{16}$$

where K_{ij}, n, F_s, and $\overline{F}$ refer to the average of factor j at the ith level, the testing cycles of factor j at the ith value, the value of the evaluation index in the sth test, and the average of the evaluation index, respectively.

The range (R_j) is the evaluation parameter for range analysis of factor sensitivity and can be calculated by:

$$R_j = \text{Max}\{K_{1j}, K_{2j}, \cdots\} - \text{Min}\{K_{1j}, K_{2j}, \cdots\} \tag{17}$$

A large R_j suggests that variation of a specific factor has a significant effect on the evaluation parameter, indicating high sensitivity of hyporheic exchange to this factor and vice versa.

Based on the 2D dune-shaped surface—subsurface coupling mathematical model, u, H, h, L, κ, θ, and D_m were selected as the factors affecting hyporheic exchange. Three levels (−20%, average, +20%) were designed for each factor. Table 2 summarizes the calculation parameters and average level of each factor.

Table 2. Orthogonal tests of sensitivities of affecting factors and their levels in the model.

Factor Level	u (m s^{-1})	H (m)	h (m)	L (m)	K (m^2)	Θ (%)	D_m (m^2 s^{-1})
1	0.056	0.080	0.016	0.160	1.2×10^{-11}	32	4.0×10^{-11}
2	0.070	0.100	0.020	0.200	1.5×10^{-11}	40	5.0×10^{-11}
3	0.084	0.120	0.024	0.240	1.8×10^{-11}	48	6.0×10^{-11}

3. Results and Discussion

3.1. Model Validation

The model was calibrated using data obtained at monitoring points c–m on the 6th dune under low flow conditions (2.1 L s^{-1}) and the calibrated model was further validated by using data obtained at monitoring points a, b, n, and o on the 5th and the 7th dunes, as shown in Figure 2. The average flow velocity was 0.07 m s^{-1}. Vertical simulated and measured values of velocity vectors at the monitoring points were obtained by the particle image velocimetry (PIV) technique by unifying model size and boundary conditions, as shown in Figure 3. Table 3 shows the results of the evaluation of the model simulation accuracy. A close correlation between observed and measured values was found, with RMSEs at monitoring points c–m on the 6th dune ranging between 0.0025–0.0055 m s^{-1}, significantly lower than the average flow velocity; R^2 at monitoring point k was 0.8935, whereas those at the other monitoring points exceeded 0.9; Re was 3.26–7.13%. Monitoring points a and b on the 5th dune and monitoring points n and o on the 7th dune were then investigated. The RMSEs obtained were 0.0033–0.0063 m s^{-1}, significantly lower than the average flow velocity; the R^2 values of b, n, and o exceeded 0.95, whereas that of a was 0.88; Re was in a reasonable range (4.7–8.53%). In summary, the proposed model exhibited excellent simulation accuracy and can precisely describe the dynamic migration of bed solutes.

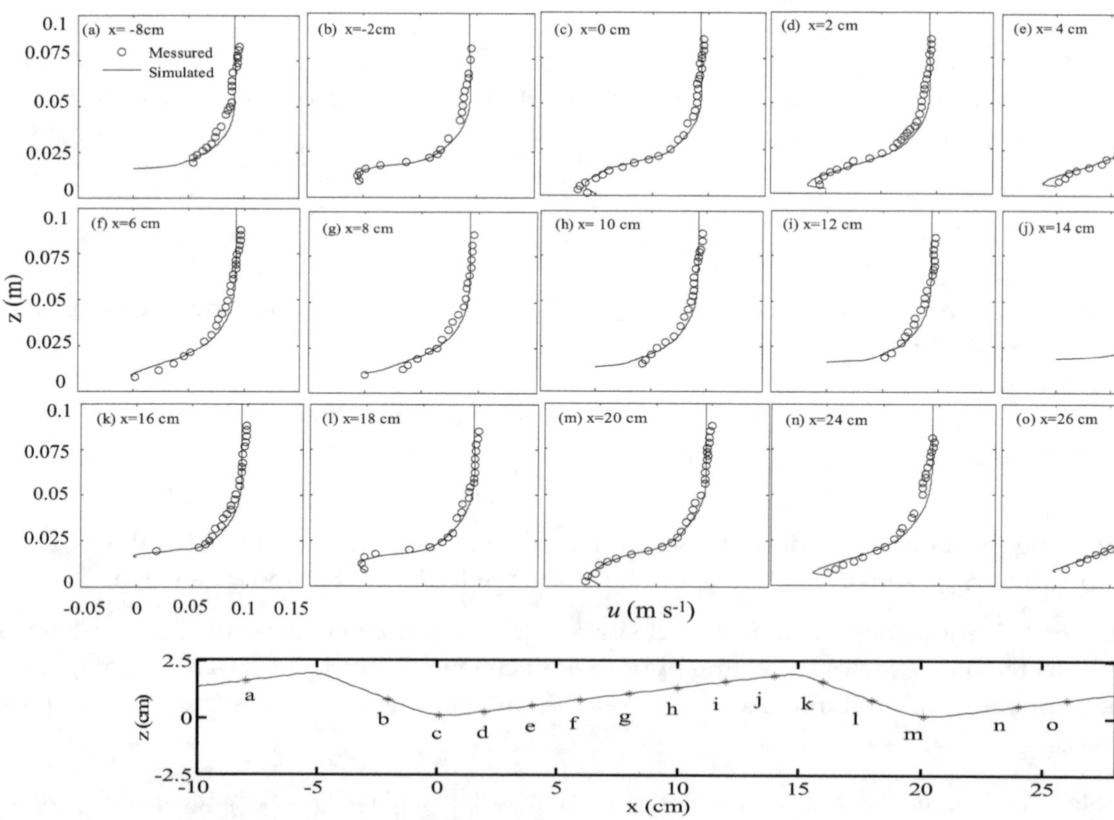

Figure 3. Comparison of vertical simulation results and experimental velocity vector results at different monitoring points (adapted from Janssen et al. [55]).

Table 3. Accuracy of vertical simulation of velocity vectors at different monitoring points.

Monitoring Points	RMSE (m s^{-1})	R^2	Re %
a	0.0046	0.8831	5.61
b	0.0033	0.9918	4.70
c	0.0044	0.9890	5.81
d	0.0047	0.9808	6.48
e	0.0054	0.9653	7.13
f	0.0055	0.9242	6.76
g	0.0050	0.9606	6.53
h	0.0053	0.9038	6.59
i	0.0044	0.8935	5.13
j	0.0034	0.9009	3.79
k	0.0042	0.9480	4.93
l	0.0041	0.9856	5.25
m	0.0025	0.9957	3.26
n	0.0063	0.9553	8.53
o	0.0039	0.976	5.05

3.2. *Effects of Flow Velocity on Hyporheic Exchange*

As flow velocity is a key factor affecting hyporheic exchange on a bed downstream of a dam, hyporheic exchanges at u = 0.056 m s^{-1}, 0.070 m s^{-1}, and 0.084 m s^{-1} were investigated. Figure 4 shows the cloud chart of concentrations. Depth and range of hyporheic exchange of solutes per unit time showed a positive relationship with u owing to interactions of surface water and subsurface water; seepage depth was 3.8 cm, 4.5 cm, and 5.5 cm and at u = 0.056 m s^{-1}, 0.070 m s^{-1}, and 0.084 m s^{-1}, respectively. In other words, the flow velocity was positively related to hyporheic exchange.

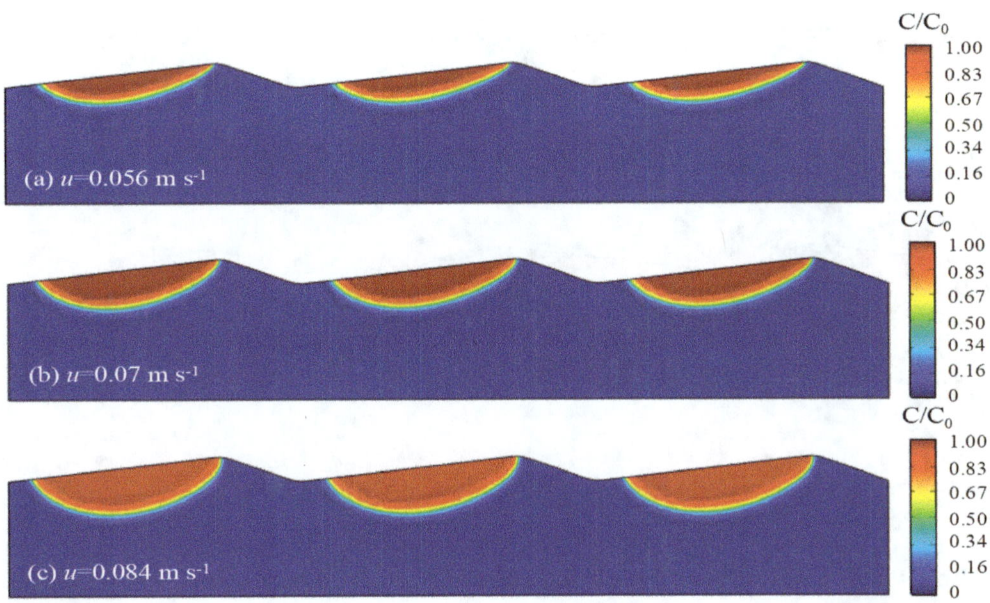

Figure 4. Distribution of concentration.

Table 4 shows the velocity effect on fluxes in hyporheic exchange, with flow velocity showing a positive relationship with water and solute fluxes on the interface of water and sand. In the vertical direction, opposite directions in water flux were observed between the upstream and downstream faces, demonstrating the hyporheic exchange process.

Table 4. Hyporheic exchange flux at different flow velocities.

	Vertical Water Flux on the Upstream Face ($m^2\ s^{-1}$)	Vertical Water Flux on the Downstream Face ($m^2\ s^{-1}$)	Overall Water Flux at the Interface ($m^2\ s^{-1}$)	Overall Solute Flux at the Interface ($mol\ m^{-1}\ s^{-1}$)
$u = 0.056\ m\ s^{-1}$	-1.42×10^{-8}	5.20×10^{-9}	8.12×10^{-8}	2.49×10^{-7}
$u = 0.070\ m\ s^{-1}$	-2.25×10^{-8}	8.28×10^{-9}	1.29×10^{-7}	4.26×10^{-7}
$u = 0.084\ m\ s^{-1}$	-3.28×10^{-8}	1.21×10^{-8}	1.88×10^{-7}	6.51×10^{-7}

Figure 5 shows stress distributions at u = 0.056 m s^{-1}, 0.070 m s^{-1}, and 0.084 m s^{-1}. Stress distributions were consistent across different flow velocities. The maximum, minimum, and negative pressure zones were observed at midstream of the upstream face, at the crest, and the downstream face, respectively. In addition, the pressure increased continuously with u.

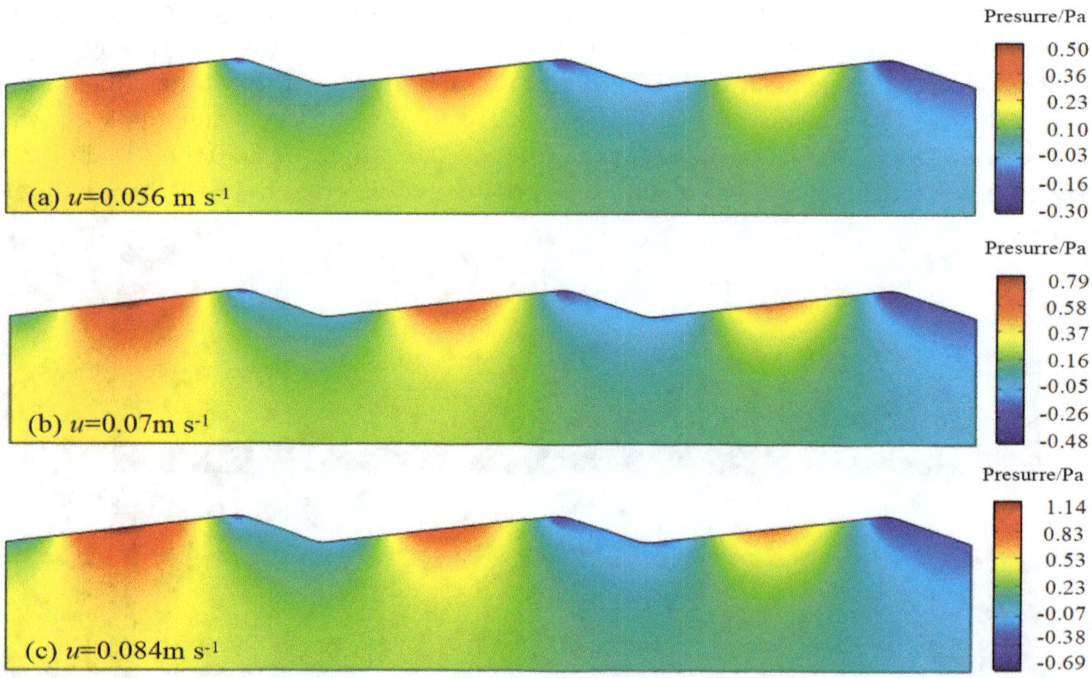

Figure 5. Distribution of the pressure field.

Figure 6 shows concentrations at the observation points over time. The curve gradient reflects the hyporheic exchange rate. As shown in Figure 6, u had negative and positive relationships with the time to equilibrium for concentrations at observation points (t) and the curve gradient, respectively, indicating an increasing hyporheic exchange rate. A minimum t of observation point B was evident at the upstream slope, indicating a maximum hyporheic exchange rate at this point. This can be attributed to the fact that maximum pressure was observed at midstream of the upstream face, which has a greater impact on observation point B.

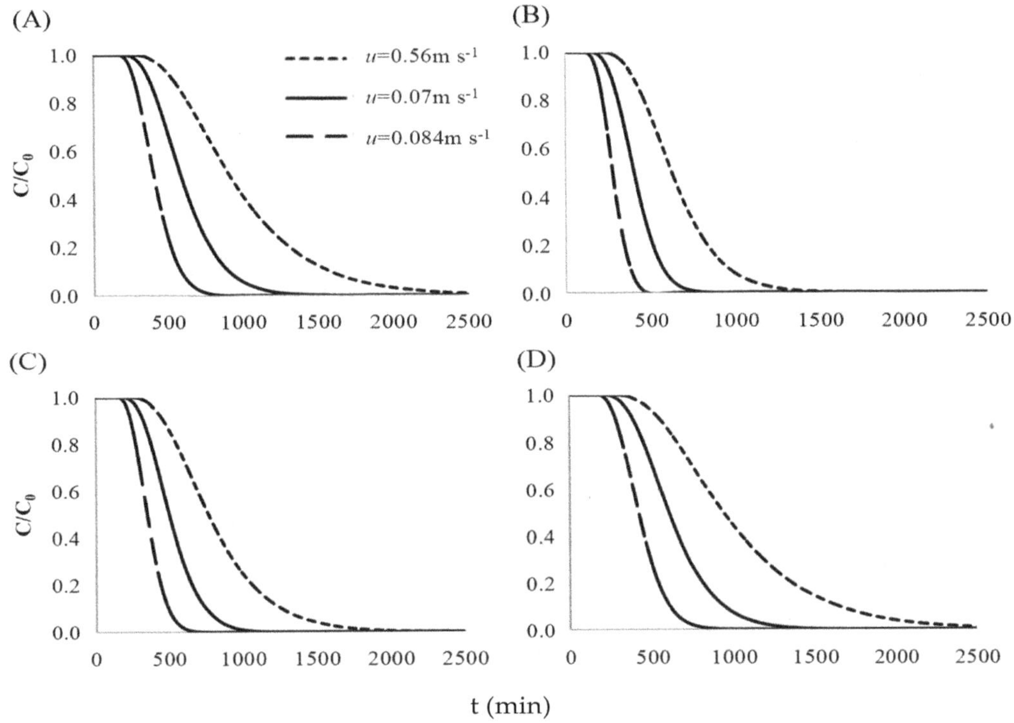

Figure 6. Concentrations at different monitoring points over time: (**A**) Observation point A, (**B**) Observation point B, (**C**) Observation point C, (**D**) observation point D.

3.3. *Effects of Water Depth on Hyporheic Exchange*

Hyporheic exchanges at $H = 0.08$ m, 0.1 m, and 0.12 m were investigated as H has a significant effect on hyporheic exchange. Figure 7 shows the cloud chart of concentrations. H evidently showed negative relationships with hyporheic exchange and the migration depth of solute per unit time, resulting in a reduced affected area. More specifically, the seepage depth was 5.1 cm, 4.5 cm, and 4.3 cm at $H = 0.08$ m, 0.1 m, and 0.12 m, respectively. In other words, H was negatively related to hyporheic exchange.

Table 5 summarizes hyporheic exchange fluxes as a function of H. As observed, hyporheic exchange flux at the interface of water and sand showed a negative relationship with H. In the vertical direction, fluxes on the upstream and downstream faces were different and in opposite directions, indicating the presence of hyporheic exchange.

Table 5. Hyporheic exchange flux at different water depths.

	Vertical Water Flux on the Upstream Face ($m^2\ s^{-1}$)	**Vertical Water Flux on the Downstream Face ($m^2\ s^{-1}$)**	**Overall Water Flux at the Interface ($m^2\ s^{-1}$)**	**Overall Solute Flux at the Interface ($mol\ m^{-1}\ s^{-1}$)**
$H = 0.08$ m	-2.86×10^{-8}	1.70×10^{-8}	1.62×10^{-7}	5.50×10^{-7}
$H = 0.10$ m	-2.25×10^{-8}	8.28×10^{-9}	1.29×10^{-7}	4.26×10^{-7}
$H = 0.12$ m	-2.05×10^{-8}	7.45×10^{-9}	1.19×10^{-7}	3.86×10^{-7}

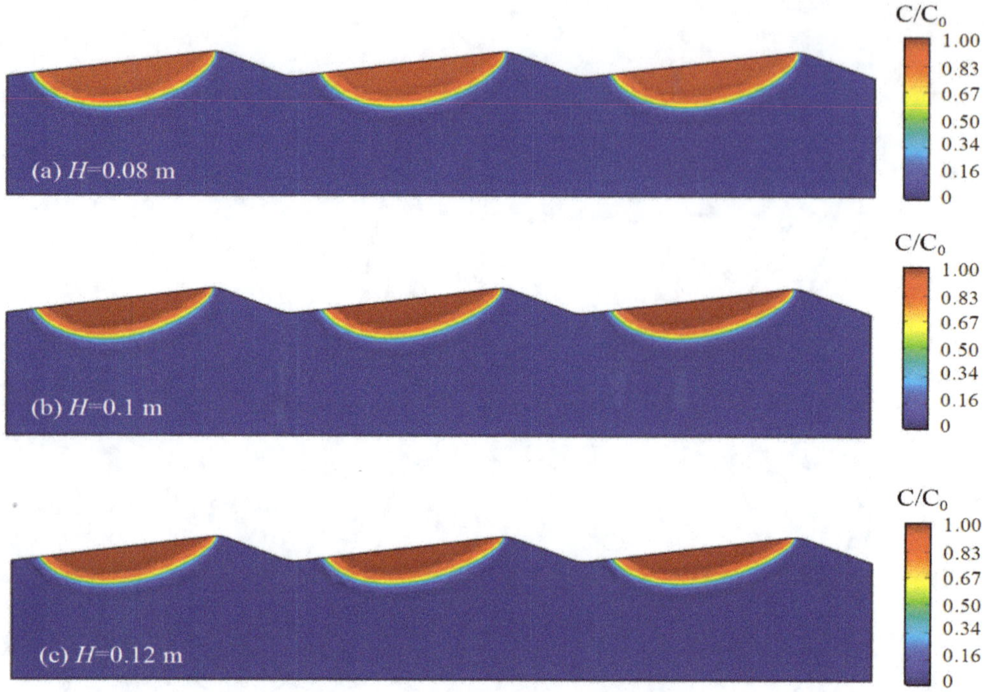

Figure 7. Distribution of concentrations.

Figure 8 shows that the stress distributions as a function of H were highly consistent with those as a function of u. In other words, the maximum and minimum pressures were observed at midstream of the upstream face and the crest, whereas the pressure values were different. In addition, the pressure decreased as H increased.

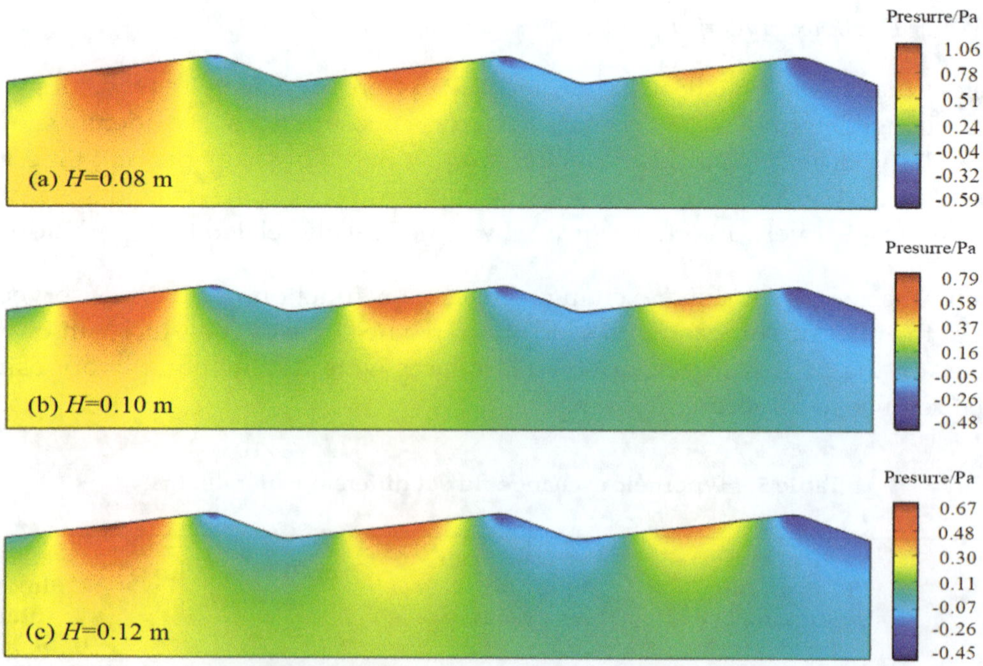

Figure 8. Distribution of the pressure field.

Figure 9 shows the solute concentrations at observation points A, B, C, and D over time as a function of H. A positive relationship between H and t was evident, indicating that hyporheic exchange was negatively related to H. The curve gradient, and thus the hyporheic exchange rate, was maximum at observation point B, followed by that at observation point C, whereas the hyporheic exchange rates of observation points A and D at the trough were similar and relatively low.

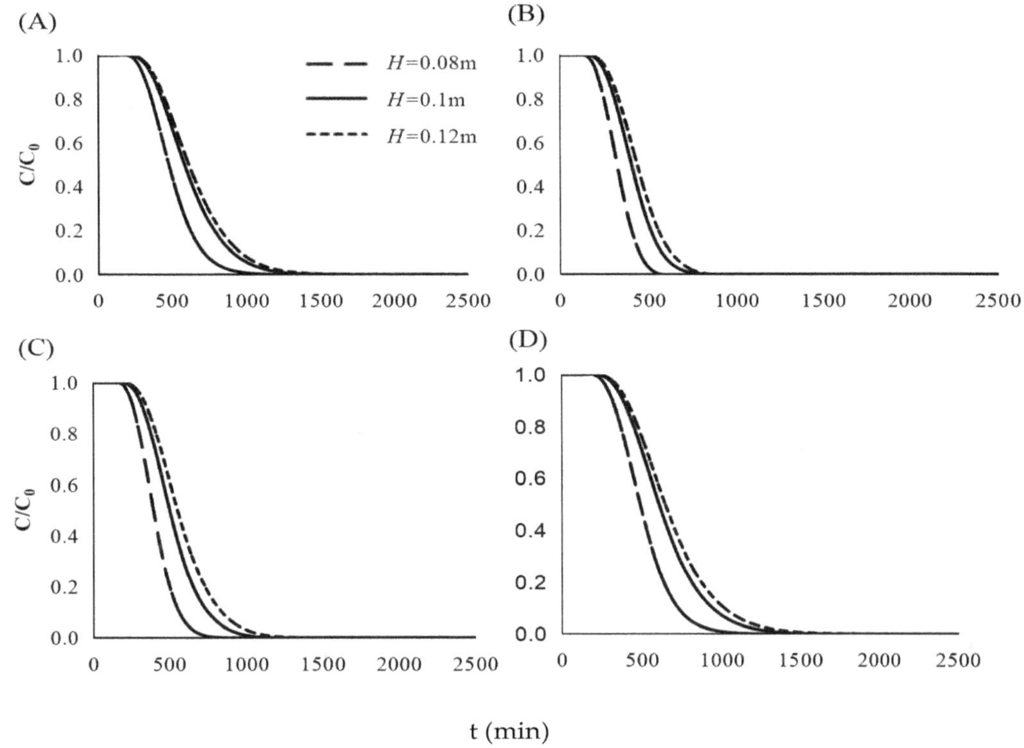

Figure 9. Concentrations at different monitoring points over time: (**A**) Observation point A, (**B**) Observation point B, (**C**) Observation point C, (**D**) observation point D.

3.4. *Effects of Dune Wave Height on Hyporheic Exchange*

As irregular bedform morphology has significant effects on hyporheic exchange, hyporheic exchanges at $h = 0.016$ m, 0.02 m, and 0.024 m were investigated. Figure 10 shows the cloud chart of concentrations. h showed positive relationships with surface roughness, depth, and range of solutes. In other words, hyporheic exchange increased with h; seepage depth was 4.1 cm, 4.5 cm, and 5.2 m at $h = 0.016$ m, 0.02 m, and 0.024 cm, respectively.

Table 6 shows hyporheic exchange flux as a function of h, where it is evident that bedform morphology roughness showed a positive relationship with the overall water and solute fluxes in hyporheic exchange at the sediment—water interfaces.

Table 6. Hyporheic exchange flux at different dune wave heights.

	Vertical Water Flux on the Upstream Face ($m^2\ s^{-1}$)	**Vertical Water Flux on the Downstream Face ($m^2\ s^{-1}$)**	**Overall Water Flux at the Interface ($m^2\ s^{-1}$)**	**Overall Solute Flux at the Interface ($mol\ m^{-1}\ s^{-1}$)**
$h = 0.16$ m	-1.62×10^{-8}	8.18×10^{-9}	1.03×10^{-7}	3.20×10^{-7}
$h = 0.20$ m	-2.25×10^{-8}	8.28×10^{-9}	1.29×10^{-7}	4.26×10^{-7}
$h = 0.24$ m	-2.93×10^{-8}	8.63×10^{-9}	1.58×10^{-7}	5.53×10^{-7}

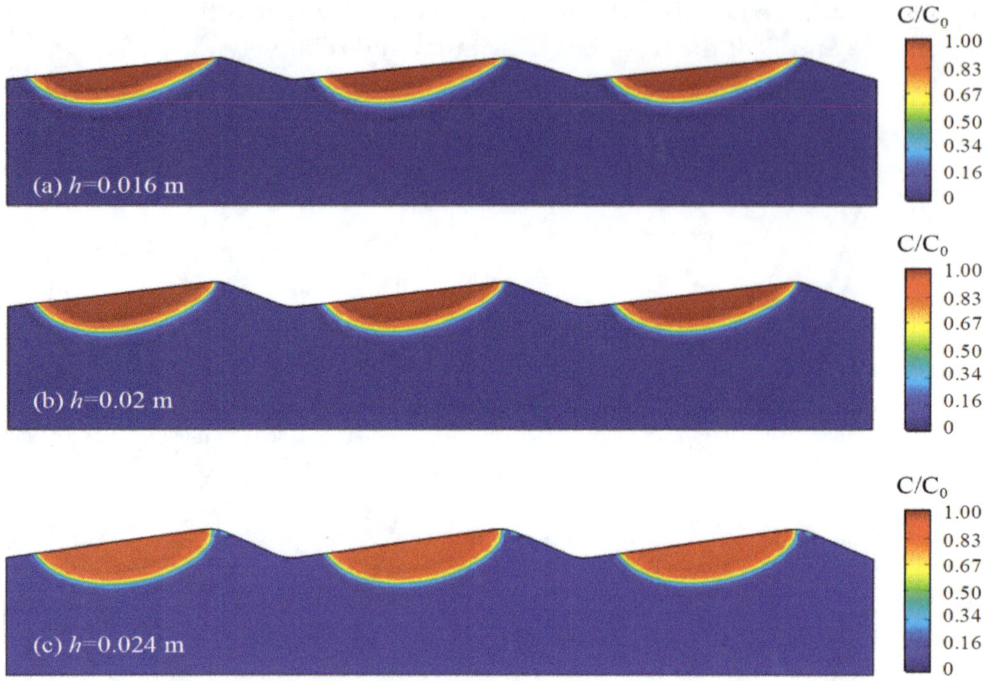

Figure 10. Distribution of concentration.

Figure 11 shows that the stress distributions as a function of h were highly consistent with those as a function of u and H, although the stress values were different. Indeed, the maximum pressure at h = 0.024 m was twice that at h = 0.016 m, whereas pressures in the negative pressure zone were homogeneous.

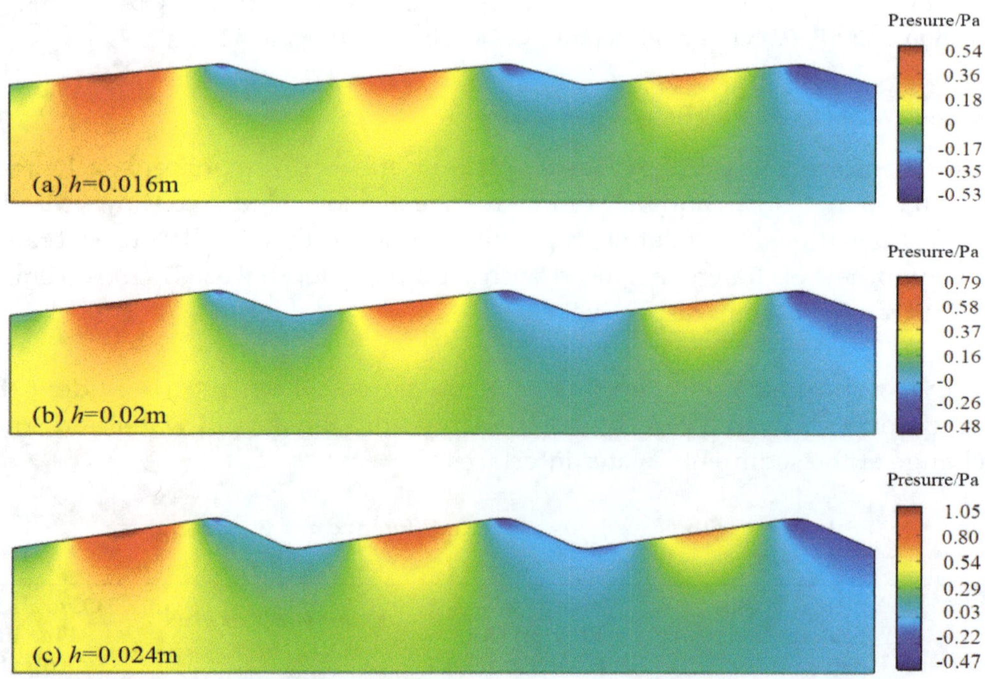

Figure 11. Distribution of pressure field.

Figure 12 shows the solute concentrations at observation points A, B, C, and D over time as a function of h. h showed a negative relationship with t at observation points A, B, C, and D, resulting in enhanced and accelerated hyporheic exchange. Hence, h was positively related to hyporheic exchange.

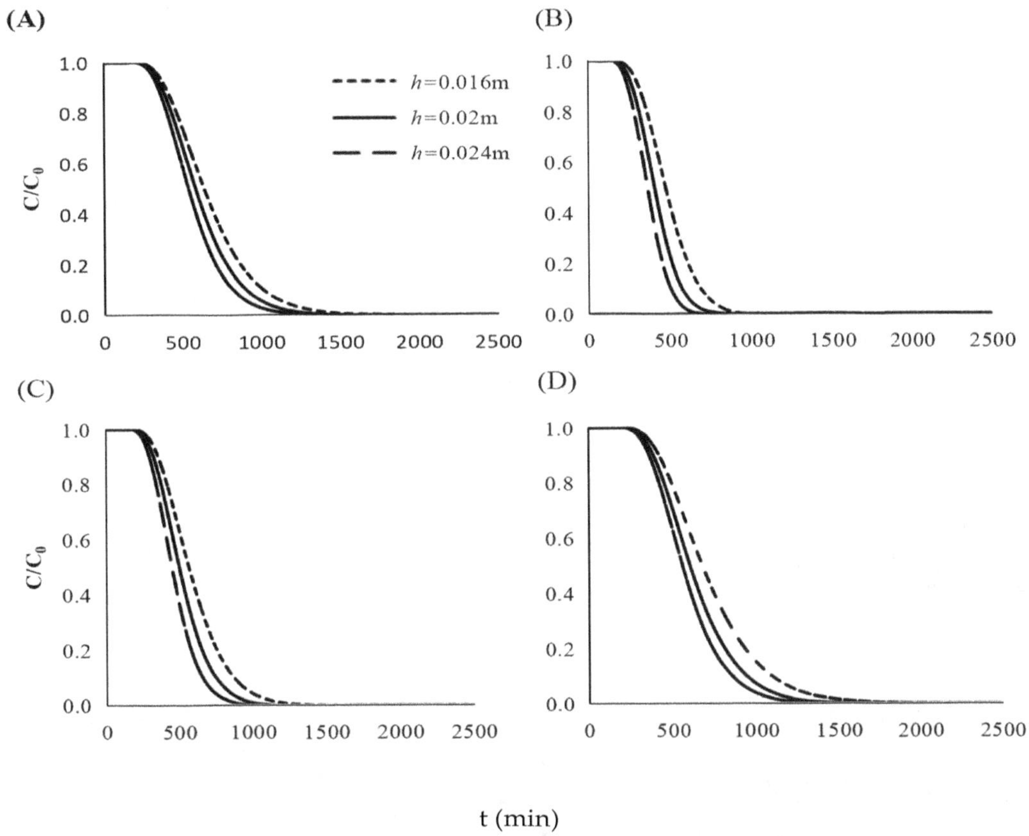

Figure 12. Concentrations at different monitoring points as a function of time: (**A**) Observation point A, (**B**) Observation point B, (**C**) Observation point C, (**D**) observation point D.

3.5. *Effects of Bed Substrate Permeability on Hyporheic Exchange*

As the structure of bed sediments has a significant effect on hyporheic exchange and the permeability of bed substrate has a determining effect on the hyporheic exchange depth, hyporheic exchanges at $\kappa = 1.2 \times 10^{-11}$ m^2, 1.5×10^{-11} m^2, and 1.8×10^{-11} m^2 were investigated. Figure 13 shows the cloud chart of concentrations in these cases. Seepage depth was 4.2 cm, 4.5 cm, and 5 cm at $\kappa = 1.2 \times 10^{-11}$ m^2, 1.5×10^{-11} m^2, and 1.8×10^{-11} m^2, respectively. In other words, the depth and range of solute exchange increased with κ, indicating that hyporheic exchange was positively related to κ.

Table 7 shows hyporheic exchange flux as a function of κ. Bed substrates with high permeabilities accelerated hyporheic exchange by accelerating the penetration of fluids and solutes through the interface of water and sand.

Table 7. Hyporheic exchange flux at different bedform permeabilities.

	Vertical Water Flux on the Upstream Face (m^2 s^{-1})	Vertical Water Flux on the Downstream Face (m^2 s^{-1})	Overall Water Flux at the Interface (m^2 s^{-1})	Overall Solute Flux at the Interface (mol m^{-1} s^{-1})
$\kappa = 1.2 \times 10^{-11}$ m^2	-1.80×10^{-8}	6.63×10^{-9}	1.03×10^{-7}	3.30×10^{-7}
$\kappa = 1.5 \times 10^{-11}$ m^2	-2.25×10^{-8}	8.28×10^{-9}	1.29×10^{-7}	4.26×10^{-7}
$\kappa = 1.8 \times 10^{-11}$ m^2	-2.70×10^{-8}	9.94×10^{-9}	1.55×10^{-7}	5.23×10^{-7}

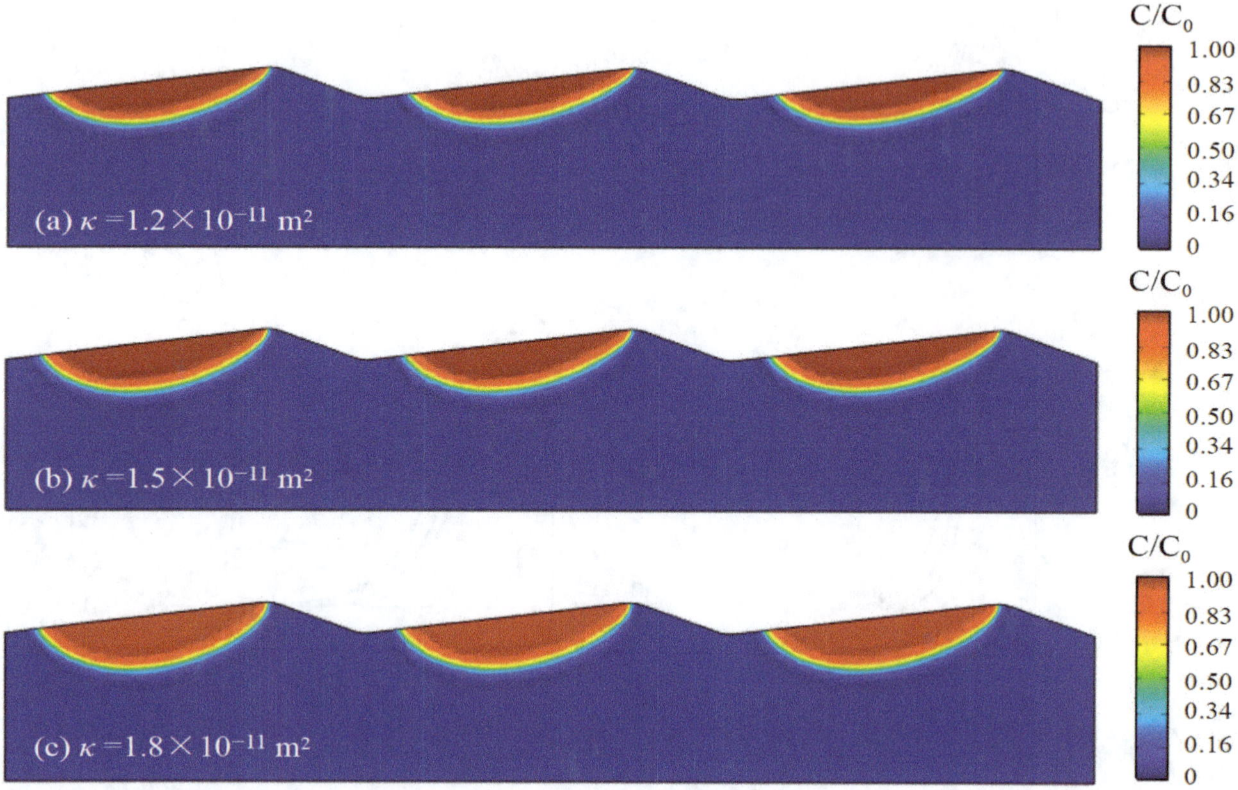

Figure 13. Distribution of concentration.

Figure 14 shows stress distributions as a function of κ, with no variations in stress as u, H, and h remained constant. Figure 15 shows the solute concentrations at observation points A, B, C, and D over time. A negative relationship was evident between κ and t, demonstrating that increasing κ accelerates hyporheic exchange. In addition, stress distributions as a function of κ were highly consistent with those as a function of u, H, and h. More specifically, the order of the curve gradient from largest to smallest was that at observation point B (maximum hyporheic exchange rate), followed by that at observation point C at the crest, and then that at observation points A and D at the trough.

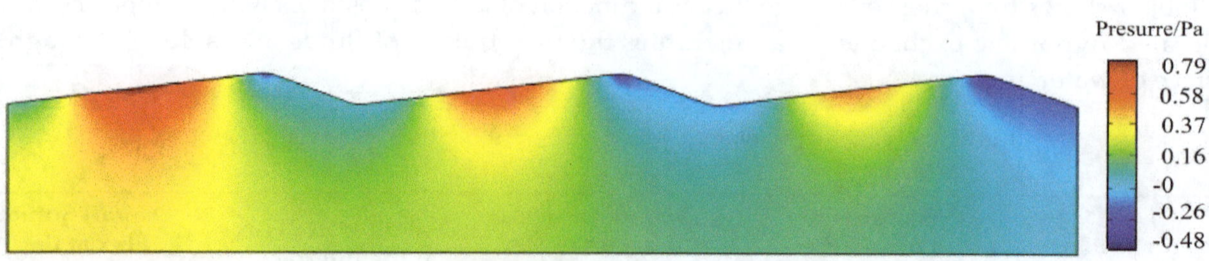

Figure 14. Distribution of pressure fields on bedforms with different permeabilities.

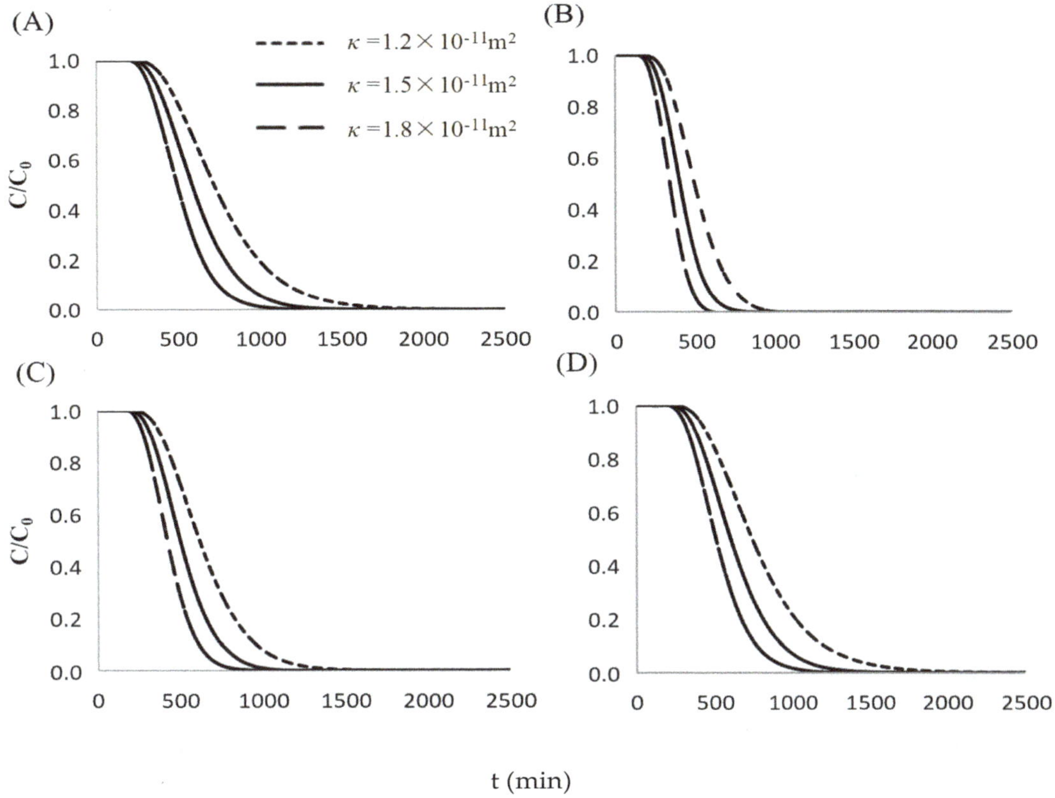

Figure 15. Concentrations at different monitoring points as a function of time: (**A**) Observation point A, (**B**) Observation point B, (**C**) Observation point C, (**D**) observation point D.

3.6. Sensitivity Analysis

To investigate the process and intensity of hyporheic exchange, 18 orthogonal tests were designed. Observation points I and III at the trough and II at the crest on the 6th dune (see Figure 16) were selected and injections of dye ($CaCl_2$ solution in this case) into the bed substrate were simulated by circles with observation points as centers and radii of 0.5 cm to monitor the hyporheic exchange routes. t at observation points I, II, and III (t_I, t_{II}, and t_{III}, respectively) were employed as the evaluation indices. As shown in Figure 16, the minimum, negative, and maximum pressure zones were observed at the crest, the downstream face, and trough and the upstream face, respectively.

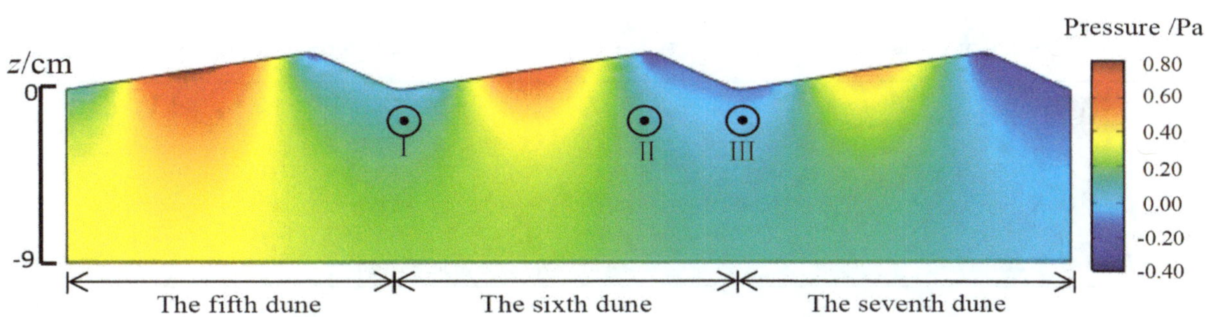

Figure 16. Cloud charts of pressure distributions at the 5th, 6th and 7th dunes of observation points I, II, and III.

Table 8 summarizes the schemes and results of the L_{18} (2×3^7) orthogonal tests of factors affecting hyporheic exchange. Tables 9–11 show the results of range analysis of factors affecting hyporheic exchange at observation points I, II, and III, respectively, based on t_I, t_{II}, and t_{III}, respectively.

Table 8. Schemes and results of the L_{18} (2×3^7) orthogonal tests of factors affecting hyporheic exchange.

Test	Empty	u (m s^{-1})	H (m)	h (m)	L (m)	κ (m^2)	θ (%)	D_m (m^2 s^{-1})	t_I (min)	t_{II} (min)	t_{III} (min)
1	1	0.056	0.08	0.016	0.16	1.20×10^{-11}	32	4.00×10^{-11}	3183	2155	3411
2	1	0.056	0.1	0.02	0.2	1.50×10^{-11}	40	5.00×10^{-11}	3800	2776	4027
3	1	0.056	0.12	0.024	0.24	1.80×10^{-11}	48	6.00×10^{-11}	4537	3485	4949
4	1	0.07	0.08	0.016	0.2	1.50×10^{-11}	48	6.00×10^{-11}	2390	1852	2621
5	1	0.07	0.1	0.02	0.24	1.80×10^{-11}	32	4.00×10^{-11}	884	714	952
6	1	0.07	0.12	0.024	0.16	1.20×10^{-11}	40	5.00×10^{-11}	3493	2590	3718
7	1	0.084	0.08	0.02	0.16	1.80×10^{-11}	40	6.00×10^{-11}	758	557	791
8	1	0.084	0.1	0.024	0.2	1.20×10^{-11}	48	4.00×10^{-11}	1661	1132	1727
9	1	0.084	0.12	0.016	0.24	1.50×10^{-11}	32	5.00×10^{-11}	885	852	986
10	2	0.056	0.08	0.024	0.24	1.50×10^{-11}	40	4.00×10^{-11}	2013	1347	2080
11	2	0.056	0.1	0.016	0.16	1.80×10^{-11}	48	5.00×10^{-11}	3760	2819	3995
12	2	0.056	0.12	0.02	0.2	1.20×10^{-11}	32	6.00×10^{-11}	4702	3585	5169
13	2	0.07	0.08	0.02	0.24	1.20×10^{-11}	48	5.00×10^{-11}	2463	1774	2463
14	2	0.07	0.1	0.024	0.16	1.50×10^{-11}	32	6.00×10^{-11}	1553	1083	1620
15	2	0.07	0.12	0.016	0.2	1.80×10^{-11}	40	4.00×10^{-11}	1624	1350	1761
16	2	0.084	0.08	0.024	0.2	1.80×10^{-11}	32	5.00×10^{-11}	449	321	449
17	2	0.084	0.1	0.016	0.24	1.20×10^{-11}	40	6.00×10^{-11}	1646	1510	1782
18	2	0.084	0.12	0.02	0.16	1.50×10^{-11}	48	4.00×10^{-11}	1680	1271	1816

Table 9. Results of range analysis of factors affecting hyporheic exchange at observation point I.

Factors	u	H	h	L	κ	θ	D_m
K_1	1361.3	−428.5	−56.5	100.0	553.5	−361.8	−463.7
K_2	−236.7	−87.2	76.7	133.2	−251.0	−82.2	170.5
K_3	−1124.7	515.7	−20.2	−233.2	−302.5	444.0	293.2
R_j	2486.0	944.2	133.2	366.3	856.0	805.8	756.8
Susceptibility	$u > H > \kappa > \theta > D_m > L > h$						

Table 10. Results of range analysis of factors affecting hyporheic exchange at observation point II.

Factors	u	H	h	L	κ	θ	D_m
K_1	962.7	−397.5	24.5	14.0	392.5	−280.2	−403.7
K_2	−171.3	−59.5	47.7	104.2	−201.7	−43.5	123.5
K_3	−791.3	457.0	−72.2	−118.2	−190.8	323.7	280.2
R_j	1754.0	854.5	119.8	222.3	594.2	603.8	683.8
Susceptibility	$u > H > D_m > \theta > \kappa > L > h$						

Table 11. Results of range analysis of factors affecting hyporheic exchange at observation point III.

Factors	u	H	h	L	κ	θ	D_m
K_1	1476.4	−492.9	−36.1	96.4	582.9	−364.2	−504.2
K_2	−272.9	−111.6	74.3	163.6	−270.4	−102.2	144.3
K_3	−1203.6	604.4	−38.2	−260.1	−312.6	466.4	359.9
R_j	2680.0	1097.3	112.5	423.7	895.5	830.7	864.2
Susceptibility	$u > H > D_m > \theta > \kappa > L > h$						

Figure 5 shows a histogram of range analysis of t for $CaCl_2$ concentration at the observation points. As shown in Figure 17, u and H (especially κ) have dominant effects on hyporheic exchange; k, D_m, and θ also have significant effects on hyporheic exchange, whereas L and h ($L > h$) have relatively low effects. These results are consistent with those of previous studies [46,64]. For instance, Wörman et al. [64] reported a dominant effect of surface water flow velocity on hyporheic exchange and a significant effect of H.

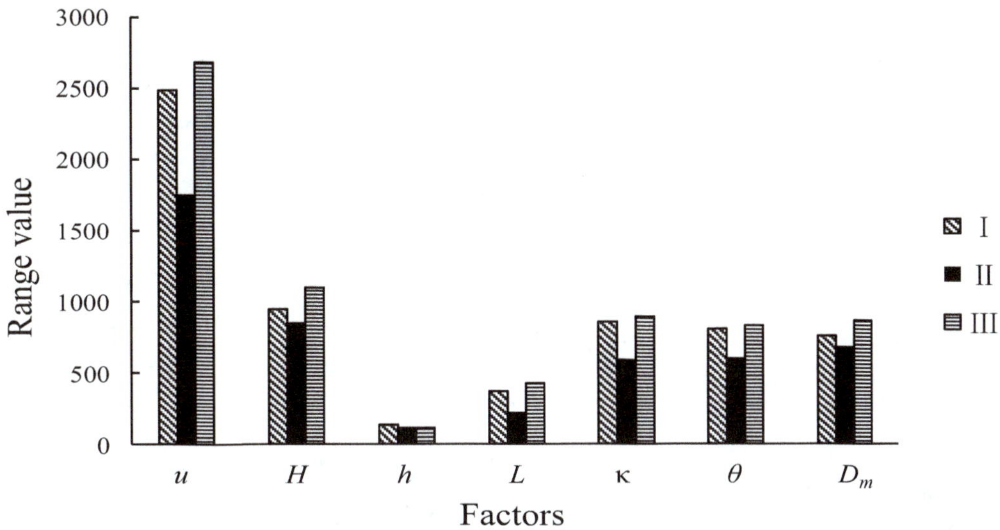

Figure 17. Ranges of factors affecting hyporheic exchange at observation points I, II, and III.

3.7. Migration Routes of Solutes

Cloud charts of stress distribution at the 6th dune and solute migration routes at different moments in different schemes were obtained based on the sensitivities of factors affecting the hyporheic exchange. Figure 18 shows the cloud chart of stress distribution at the 6th dune in Test 16 (red arrows denote the pore seepage field) and Figure 19 shows solute migration routes at 0 min, 60 min, 180 min, 360 min, 540 min, and 720 min. The maximum pressure was observed at half of the upstream face, whereas the downstream face and the trough were negative pressure zones. As the pressure difference causes exchange of surface water and subsurface water, the pore seepage field was divided into two parts. On the upstream face, reverse flows were observed at locations with heights lower than that corresponding to the maximum pressure, whereas normal flows were observed at other locations. Meanwhile, solute fields at the three observation points approached the surface water until disappearance, as shown in Figure 19.

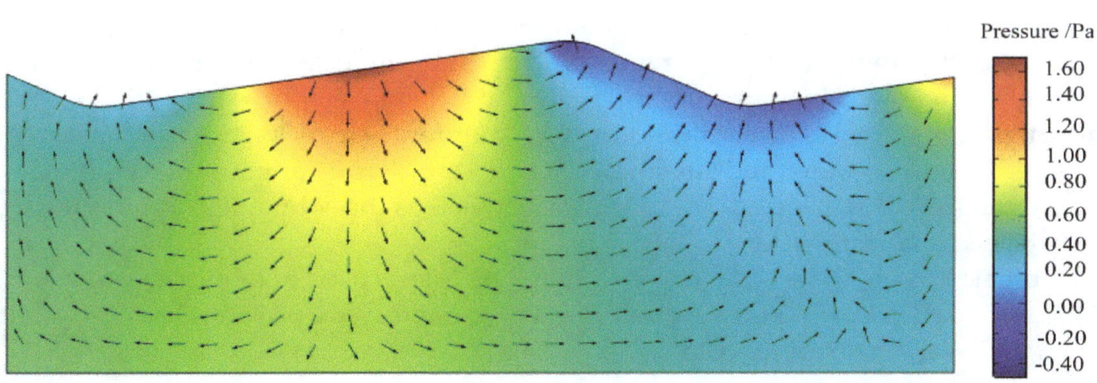

Figure 18. Cloud chart of pressures and distribution of seepage field at the 6th dune in Test 16.

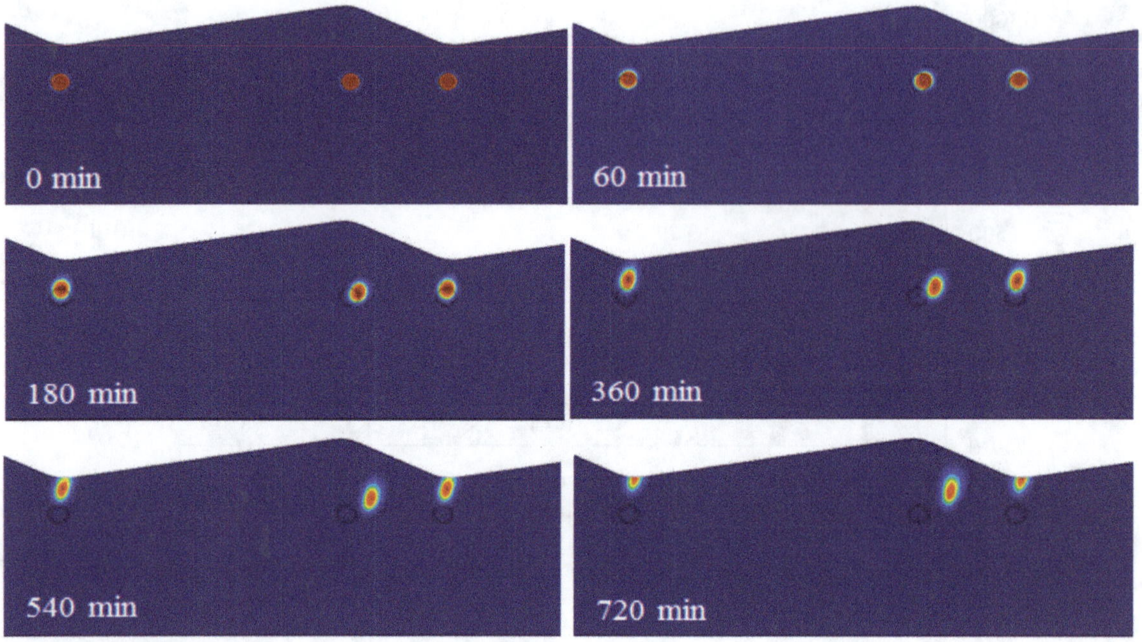

Figure 19. Migration routes of solutes at observation points I, II, and III at the 6th dune in Test 16.

4. Conclusions

A 2D dune-shaped surface–subsurface coupling mathematical model based on the RANS equation, the $k-\omega$ turbulence model, and a steady state groundwater flow model was proposed. Janseen et al. [55] used exactly the same combination of the multiphysics computational fluid dynamic (CFD) approach and COMSOL Multiphysics to solve these equations. The coupling mathematical model is verified by previous flume experiments [55], and numerical model simulations were verified against observations using the RMSE, R^2, and Re as evaluation indices. The maximum RMSE obtained was 0.0055 m s^{-1}, significantly lower than the average flow velocity; $R^2 > 0.9$ in all cases, indicating good fitting effectiveness; Re was 3.26–8.53%. In summary, simulation results were highly consistent with experimental results; therefore, it is argued that the proposed calculation model is reliable.

The single-factor effect of u, H, h, and κ on hyporheic exchange were investigated by numerical simulations. The results indicated that u, h, and κ were positively related to hyporheic exchange, whereas H was negatively related to hyporheic exchange.

Sensitivity analysis of parameters in the surface–subsurface coupling model and the solute migration model was conducted using orthogonal tests based on the simulation of dye injection into the dune with t used as the evaluation parameter. The results indicated that the sensitivities of u and H were the highest, followed by those of κ, D_m, θ, L, and h ($L > h$).

Solute migration routes at different moments in different schemes were then obtained. Owing to the exchange of surface water and subsurface water, solute fields at monitoring points A, C, and D approached the surface water until disappearance, whereas the solute field at monitoring point B approached the bed substrate.

The hyporheic exchange was shown to be affected by interactions of multiple factors and u and H exhibited the largest effects on hyporheic exchange. The present study provides knowledge vital to the protection and recovery of riverine ecology.

Author Contributions: J.R. and X.W. jointly analyzed the data and wrote the paper. Y.Z., B.C., L.M. provided critical feedback on the manuscript.

References

1. Angermann, L.; Lewandowski, J.; Fleckenstein, J.H.; Nutzmann, G. A 3D analysis algorithm to improve interpretation of heat pulse sensor results for the determination of small−scale flow directions and velocities in the hyporheic zone. *J. Hydrol.* **2012**, *475*, 1–11. [CrossRef]
2. Boano, F.; Harvey, J.W.; Marion, A.; Packman, A.I.; Revelli, R.; Ridolfi, L.; Worman, A. Hyporheic flow and transport processes: Mechanisms, models, and biogeochemical implications. *Rev. Geophys.* **2014**, *52*, 603–679. [CrossRef]
3. Bencala, K.E. Hyporheic zone hydrological processes. *Hydrol. Process.* **2000**, *14*, 2797–2798. [CrossRef]
4. Boano, F.; Poggi, D.; Revelli, R.; Ridolfi, L. Gravity-driven water exchange between streams and hyporheic zones. *Geophys. Res.* **2009**, *36*, 146–158. [CrossRef]
5. Zheng, L.; Cardenas, M.B.; Wang, L. Temperature effects on nitrogen cycling and nitrate removal−production efficiency in bed form−induced hyporheic zones. *J. Geophys. Res.* **2016**, *121*, 1086–1103. [CrossRef]
6. Packman, A.I.; Salehin, M.; Zaramella, M. Hyporheic Exchange with Gravel Beds: Basic Hydrodynamic Interactions and Bedform−Induced Advective Flows. *J. Hydraul. Eng.* **2004**, *130*, 647–656. [CrossRef]
7. Pinay, G.; Okeefe, T.C.; Edwards, R.T.; Naiman, R.J. Nitrate removal in the hyporheic zone of a salmon river in Alaska. *River Res. Appl.* **2009**, *25*, 367–375. [CrossRef]
8. Jones, J.B.; Mulholland, P.J. *Streams and Ground Waters*; Academic: San Diego, CA, USA, 2000.
9. Boulton, A.J.; Hancock, P.J. Rivers as groundwater−dependent ecosystems: A review of degrees of dependency, riverine processes and management implications. *Aust. J. Bot.* **2006**, *54*, 133–144. [CrossRef]
10. Fleckenstein, J.H.; Krause, S.; Hannah, D.M.; Boano, F. Groundwater−surface water interactions: New methods and models to improve understanding of processes and dynamics. *Adv. Water Resour.* **2010**, *33*, 1291–1295. [CrossRef]
11. Swanson, T.E.; Cardenas, M.B. Diel heat transport within the hyporheic zone of a pool−riffle−pool sequence of a losing stream and evaluation of models for fluid flux estimation using heat. *Limnol. Oceanogr.* **2010**, *55*, 1741–1754. [CrossRef]
12. Norman, F.A.; Cardenas, M.B. Heat transport in hyporheic zones due to bedforms: An experimental study. *Water Resour. Res.* **2014**, *50*, 3568–3582. [CrossRef]
13. Marion, A.; Zaramella, M. Diffusive behavior of bedform−induced hyporheic exchange in rivers. *J. Environ. Eng.* **2005**, *131*, 1260–1266. [CrossRef]
14. Lautz, L.K.; Siegel, D.I.; Bauer, R.L. Impact of debris dams on hyporheic interaction along a semi-arid stream. *Hydrol. Process.* **2006**, *20*, 183–196. [CrossRef]
15. Bardini, L.; Boano, F.; Cardenas, M.B.; Revelli, R.; Ridolfi, L. Nutrient cycling in bedform induced hyporheic zones. *Geochim. Cosmochim. Acta* **2012**, *84*, 47–61. [CrossRef]
16. Zarnetske, J.P.; Haggerty, R.; Wondzell, S.M.; Baker, M.A. Dynamics of nitrate production and removal as a function of residence time in the hyporheic zone. *J. Geophys. Res.* **2011**, *116*, 1–12. [CrossRef]
17. Battin, T.J.; Kaplan, L.A.; Findlay, S.E.; Hopkinson, C.S.; Marti, E.; Packman, A.I.; Newbold, J.D.; Sabater, F. Biophysical controls on organic carbon fluxes in fluvial networks. *Nat. Geosci.* **2008**, *1*, 95–100. [CrossRef]
18. Boano, F.; Revelli, R.; Ridolfi, L. Effect of streamflow stochasticity on bedform−driven hyporheic exchange. *Adv. Water Resour.* **2010**, *33*, 1367–1374. [CrossRef]
19. Karwan, D.L.; Saiers, J.E. Hyporheic exchange and streambed filtration of suspended particles. *Water Resour. Res.* **2012**, *48*. [CrossRef]
20. Liu, D.; Zhao, J.; Chen, X.; Li, Y.; Weiyan, S.; Feng, M. Dynamic processes of hyporheic exchange and temperature distribution in the riparian zone in response to dam−induced water fluctuations. *Geosci. J.* **2018**, *22*, 465–475. [CrossRef]
21. Hester, E.T.; Doyle, M.W. In-stream geomorphic structures as drivers of hyporheic exchange. *Water Resour. Res.* **2008**, *44*. [CrossRef]
22. Menichino, G.T.; Hester, E.T. Hydraulic and thermal effects of in−stream structure−induced hyporheic exchange across a range of hydraulic conductivities. *Water Resour. Res.* **2014**, *50*, 4643–4661. [CrossRef]
23. Rana, S.M.M.; Scott, D.T.; Hester, E.T. Effects of in−stream structures and channel flow rate variation on transient storage. *J. Hydrol.* **2017**, *548*, 157–169. [CrossRef]

24. Worman, A.; Packman, A.I.; Marklund, L.; Harvey, J.W.; Stone, S.H. Exact three-dimensional spectral solution to surface-groundwater interactions with arbitrary surface topography. *Geophys. Res. Lett.* **2006**, *33*. [CrossRef]
25. Cardenas, M.B.; Wilson, J.L. Hydrodynamics of coupled flow above and below a sediment–water interface with triangular bedforms. *Adv. Water Resour.* **2007**, *30*, 301–313. [CrossRef]
26. Caruso, A.; Ridolfi, L.; Boano, F. Impact of watershed topography on hyporheic exchange. *Adv. Water Resour.* **2016**, *94*, 400–411. [CrossRef]
27. Fox, A.; Laube, G.; Schmidt, C.; Fleckenstein, J.H.; Arnon, S. The effect of losing and gaining flow conditions on hyporheic exchange in heterogeneous streambeds. *Water Resour. Res.* **2016**, *52*, 7460–7477. [CrossRef]
28. Pryshlak, T.T.; Sawyer, A.H.; Stonedahl, S.H.; Soltanian, M.R. Multiscale hyporheic exchange through strongly heterogeneous sediments. *Water Resour. Res.* **2015**, *51*, 9127–9140. [CrossRef]
29. Su, X.; Shu, L.; Lu, C. Impact of a low-permeability lens on dune–induced hyporheic exchange. *Hydrolog. Sci. J.* **2018**, *63*, 818–835. [CrossRef]
30. Bhaskar, A.S.; Harvey, J.W.; Henry, E.J. Resolving hyporheic and groundwater components of streambed water flux using heat as a tracer. *Water Resour. Res.* **2012**, *48*. [CrossRef]
31. Chen, X.; Cardenas, M.B.; Chen, L. Three–dimensional versus two dimensional bed form–induced hyporheic exchange. *Water Resour. Res.* **2015**, *51*, 2923–2936. [CrossRef]
32. Endreny, T.; Lautz, L.; Siegel, D. Hyporheic flow path response to hydraulic jumps at river steps: Hydrostatic model simulations. *Water Resour. Res.* **2011**, *47*, 1198–1204. [CrossRef]
33. Laattoe, T.; Werner, A.D.; Post, V.E.A. Spatial periodicity in bed form–scale solute and thermal transport models of the hyporheic zone. *Water Resour. Res.* **2014**, *50*, 7886–7899. [CrossRef]
34. Langston, G.; Hayashi, M.; Roy, J.W. Quantifying groundwater–surface water interactions in a proglacial moraine using heat and solute tracers. *Water Resour. Res.* **2013**, *49*, 5411–5426. [CrossRef]
35. Stonedahl, S.H.; Harvey, J.W.; Wörman, A.; Salehin, M.; Packman, A.I. A multiscale model for integrating hyporheic exchange from ripples to meanders. *Water Resour. Res.* **2010**, *46*, 308–316. [CrossRef]
36. Trauth, N.; Schmidt, C.; Maier, U.; Vieweg, M.; Fleckenstein, J.H. Coupled 3–D stream flow and hyporheic flow model under varying stream and ambient groundwater flow conditions in a pool–riffle system. *Water Resour. Res.* **2013**, *49*, 5834–5850. [CrossRef]
37. Harvey, J.W.; Wagner, B.J. Quantifying hydrologic interactions between streams and their subsurface hyporheic zones. In *Streams Ground Waters*; Academic Press: Cambridge, MA, USA, 2000; pp. 3–44.
38. Cardenas, M.B.; Wilson, J.L. Dunes, turbulent eddies, and interfacial exchange with permeable sediments. *Water Resour. Res.* **2007**, *430*, 199–212. [CrossRef]
39. Cardenas, M.B.; Wilson, J.L. Exchange across a sediment-water interface with ambient groundwater discharge. *J. Hydrol.* **2007**, *346*, 69–80. [CrossRef]
40. Sawyer, A.H.; Cardenas, M.B.; Buttles, J. Hyporheic exchange due to channel–spanning logs. *Water Resour. Res.* **2011**, *47*, W08502. [CrossRef]
41. Schmadel, N.M.; Ward, A.S.; Lowry, C.S.; Malzone, J.M. Hyporheic exchange controlled by dynamic hydrologic boundary conditions. *Geophys. Res. Lett.* **2016**, *43*, 4408–4417. [CrossRef]
42. Peng, Y.; Zhou, J.G.; Burrows, R. Modelling the free surface flow in rectangular shallow basins by lattice Boltzmann method. *J. Hydraul. Eng.* **2011**, *137*, 1680–1685. [CrossRef]
43. Peng, Y.; Zhou, J.G.; Burrows, R. Modelling solute transport in shallow water with the lattice boltzmann method. *Comput. Fluids* **2011**, *50*, 181–188. [CrossRef]
44. Peng, Y.; Zhang, J.M.; Zhou, J.G. Lattice Boltzmann Model Using Two–Relaxation–Time for Shallow Water Equations. *J. Hydraul. Eng.* **2016**, *142*, 06015017. [CrossRef]
45. Peng, Y.; Zhang, J.M.; Meng, J.P. Second order force scheme for lattice Boltzmann model of shallow water flows. *J. Hydraul. Res.* **2017**, *55*, 592–597. [CrossRef]
46. Elliott, A.H.; Brooks, N.H. Transfer of nonsorbing solutes to a streambed with bed forms: Laboratory experiments. *Water Resour. Res.* **1997**, *33*, 137–151. [CrossRef]
47. Ju, L.; Zhang, J.J.; Chen, C.; Wu, L.S.; Zeng, L.Z. Water flux characterization through hydraulic head and temperature data assimilation: Numerical modeling and sandbox experiments. *J. Hydrol.* **2018**, *558*, 104–114. [CrossRef]
48. Stonedahl, S.H.; Roche, K.R.; Stonedahl, F.; Packman, A.I. Visualizing hyporheic flow through bedforms using dye experiments and simulation. *J. Vis. Exp.* **2015**, *105*. [CrossRef]

49. Zhou, T.; Endreny, T.A. Reshaping of the hyporheic zone beneath river restoration structures: Flume and hydrodynamic experiments. *Water Resour. Res.* **2013**, *49*, 5009–5020. [CrossRef]
50. Tonina, D.; Buffington, J.M. Hyporheic exchange in gravel bed rivers with pool−riffle morphology: Laboratory experiments and three−dimensional modeling. *Water Resour. Res.* **2007**, *43*, 208–214. [CrossRef]
51. Wu, Y.X.; Hunkeler, D. Hyporheic exchange in a karst conduit and sediment system—A laboratory analog study. *J. Hydrol.* **2013**, *501*, 125–132. [CrossRef]
52. Sawyer, A.H.; Cardenas, M.B.; Buttles, J. Hyporheic temperature dynamics and heat exchange near channel−spanning logs. *Water Resour. Res.* **2012**, *48*. [CrossRef]
53. Fox, A.; Boano, F.; Arnon, S. Impact of losing and gaining streamflow conditions on hyporheic exchange fluxes induced by dune−shaped bed forms. *Water Resour. Res.* **2014**, *50*, 1895–1907. [CrossRef]
54. Lu, C.; Zhuang, W.; Wang, S.; Zhu, X.; Li, H. Experimental study on hyporheic flow varied by the clay lens and stream flow. *Environ. Earth Sci.* **2018**, *77*, 482. [CrossRef]
55. Janssen, F.; Cardenas, M.B.; Sawyer, A.H.; Dammrich, T.; Krietsch, J.; Beer, D. A comparative experimental and multiphysics computational fluid dynamics study of coupled surface-subsurface flow in bed forms. *Water Resour. Res.* **2012**, *48*, 8514. [CrossRef]
56. Wilcox, D.C. *Turbulence Modeling for CFD*; DCW Industries, Inc.: La Canada, CA, USA, 1998; 540p.
57. Yoon, J.Y.; Patel, V.C. Numerical model of turbulent flow oversand dune. *J. Hydraul. Eng.* **1996**, *122*, 10–18. [CrossRef]
58. Cardenas, M.B.; Wilson, J.L. Comment on "Flow resistanceand bed form geometry in a wide alluvial channel" by Shu-Qing Yang, Soon-KeatTan, and Siow-Yong Lim. *Water Resour. Res.* **2006**, *42*. [CrossRef]
59. Cardenas, M.B.; Wilson, J.L.; Haggerty, R. Residence time of bedform−driven hyporheic exchange. *Adv. Water Resour.* **2008**, *31*, 1382–1386. [CrossRef]
60. de Marsily, G. *Quantitative Hydrogeology: Groundwater Hydrology for Engineers*; Academic Press: Orlando, FL, USA, 1986.
61. Mentaschi, L.; Besio, G.; Cassola, F.; Mazzino, A. Problems in RMSE−based wave model validations. *Ocean Model.* **2013**, *72*, 53–58. [CrossRef]
62. Quinino, R.C.; Reis, E.A.; Bessegato, L.F. Using the coefficient of determination R^2 to test the significance of multiple linear regression. *Teach. Stat.* **2013**, *35*, 84–88. [CrossRef]
63. Suñé, V.; Carrasco, J.A. Efficient implementations of the randomization method with control of the relative error. *Comput. Oper. Res.* **2005**, *32*, 1089–1114. [CrossRef]
64. Wörman, A.; Packman, A.I.; Johansson, H.; Jonsson, K. Effect of flow-induced exchange in hyporheic zones on longitudinal transport of solutes in streams and rivers. *Water Resour. Res.* **2002**, *38*, 2-1–2-15. [CrossRef]

5

Numerical Study of the Collapse of Multiple Bubbles and the Energy Conversion during Bubble Collapse

Jing Zhang [1,2,3], Lingxin Zhang [1,2,3,*] and Jian Deng [1,2,3]

1 State Key Laboratory of Fluid Power and Mechatronic Systems, Zhejiang University, Hangzhou 310027, China; zj121314@yeah.net (J.Z.); zjudengjian@zju.edu.cn (J.D.)
2 Key Laboratory of Soft Machines and Smart Devices of Zhejiang Province, Zhejiang University, Hangzhou 310027, China
3 Department of Mechanics, Zhejiang University, Hangzhou 310027, China
* Correspondence: zhanglingxin@zju.edu.cn.

Abstract: This paper investigates numerically the collapses of both a single cavitation bubble and a cluster consisting of 8 bubbles, concerning mainly on the conversions between different forms of energy. Direct numerical simulation (DNS) with volume of fluid (VOF) method is applied, considering the detailed resolution of the vapor-liquid interfaces. First, for a single bubble near a solid wall, we find that the peak value of the wave energy, or equivalently the energy conversion rate decreases when the distance between the bubble and the wall is reduced. However, for the collapses of multiple bubbles, this relationship between the bubble-wall distance and the conversion rate reverses, implying a distinct physical mechanism. The evolutions of individual bubbles during the collapses of multiple bubbles are examined. We observe that when the bubbles are placed far away from the solid wall, the jetting flows induced by all bubbles point towards the cluster centre, while the focal point shifts towards the solid wall when the cluster is very close to the wall. We note that it is very challenging to consider thermal and acoustic damping mechanisms in the current numerical methods, which might be significant contributions to the energy budget, and we leave it open to the future studies.

Keywords: cavitation bubble, multiple bubble collapse, pressure wave energy, energy conversion rate

---

1. Introduction

As a natural phenomenon, cavitation occurs frequently in a variety of engineering applications. As the local pressure in liquid is lower than its saturated vapor pressure, gas nucleus can develop into cavitation bubbles. When the local pressure is recovered or driven by high pressure waves, the cavitation bubbles collapse, leading to a series of physical scenarios such as micro jetting flows, shock wave emission, heat release etc. In hydraulic systems, cavitation can cause material damage, loss of power, and induce noises. On the other hand, bubble collapse can also be utilized as a tool in the fields of medical treatments and mining, such as drug delivery, ultrasound operation and cavity jet exploits [1]. Numerous experimental and numerical studies have been performed to study the bubble dynamics and the collapses of cavitation bubbles, in order to control or reinforce the effects of cavitation bubble collapse. However, the dynamics of multiple cavitation bubbles and their collapses, particularly the underlying behavior of energy conversion have not been well understood.

Historically, studies on bubble dynamics can date back to 1917, when Reyleigh proposed a theoretical spherical-bubble model [2]. This model was for an ideal spherical bubble, neglecting the effects of viscosity, surface tension, compressibility and mass transfer. It was followed by a series of modifications and perfections [3–8].

In the past few decades, a large amount of works have been undertaken to understand the characteristics of bubble collapse [9–24]. Lauterborn recorded the process of a single bubble collapsing near a rigid wall, in which the shock wave emission was captured by a schlieren system [25–27]. Moreover, the particle image velocimetry (PIV) technique was adopted to monitor the velocity fields by lauterborn & Vogel [28], who captured the micro jetting flows as the bubble collapses asymmetrically. They revealed the relationship between the noises induced by the bubble collapse and the dimensionless distance γ ($\gamma = L/R_{max}$, where L is the distance between the bubble centre and the rigid wall, and R_{max} is the maximum bubble radius). Gonzalez-Avila measured the pressure emission of laser-induced bubble collapse [29], in which the pressure amplitude was found to be up to $1k$ Bar when the bubble was close to the rigid boundary [30]. Merouan [31] measured the temperature variations during the growth and collapse of a single cavitation bubble. More recently, Fortes et al. proposed an analytic model to examine the pressure wave emission during bubble collapse, with the conversion rate of the wave energy found to be strongly influenced by gas content in the bubble [32].

However, not many efforts have been made to study the pressure wave emission from the collapses of multiple bubbles due to the difficulties in experimental setup and the consideration of compressibility of liquid in numerics [33–36]. Generally, a bubble cluster collapses inward, generating a high level impulsive pressure wave, and the symmetries for individual bubbles are lost due to the presence of the rigid wall and their nearby bubbles [37,38]. It has been shown that the shock wave emission focuses inward during the bubble cloud collapse, leading to a very high peak pressure, which can exceed that of the collapse of a single bubble with the equivalent volume [39]. Tiwari et al. [40] simulated the expansion and subsequent collapse of a hemispherical cluster of 50 bubbles adjacent to a plane rigid wall, with the detailed compressible-fluid mechanics of bubble-bubble interactions considered. They found that the peak pressure was associated with the centremost bubble, which caused a corresponding peak pressure on the nearby wall. Despite these advances in the few existing researches, the detailed bubble-scale dynamics of a cluster collapse remain poorly understood.

Here, we investigate the characteristics of the collapses of multiple bubbles by solving directly the Navier-Stokes equations considering the compressiblility of the fluid, with particular concerns on the energy conversion. The volume of fluid method is involved to capture the interfaces between the liquid and the cavitation bubbles. The simulation results are organized into two parts: in the first part, a single bubble is studied as a reference to our further multiple-bubble studies. In the second part, we focus the collapses of multiple bubbles, presenting the unique dynamic behaviors in pressure wave propagation and energy conversion, in contrast to the single bubble situation.

2. Numerical Methods

The numerical work is performed using a finite-volume based code. Assuming that the liquid is compressible and the cavitation bubbles are filled with pure vapour, and considering viscosity and surface tension, the governing equations are formularized as the following

$$\frac{d\rho}{dt} + \rho\nabla\vec{U} = 0, \tag{1}$$

$$\rho\frac{\partial\vec{U}}{\partial t} + \rho\vec{U}\cdot\nabla\vec{U} = \rho\vec{g} - \nabla p + 2\nabla\cdot\left(\mu\overline{\overline{D}}\right) - \frac{2}{3}\nabla\left(\mu\nabla\cdot\vec{U}\right) + \sigma\kappa\vec{N}, \tag{2}$$

$$\frac{d\alpha}{dt} = \frac{\partial\alpha}{\partial t} + \vec{U}\cdot\nabla\alpha = 0, \tag{3}$$

where k is surface curvature, σ is the surface tension coefficient, $\overline{\overline{D}}$ is the strain rate tensor, $\vec{N}$ is the unit normal vector of the interface, α is the volume fraction of the liquid, and ρ and μ are the density and the viscosity respectively of the mixture fluid, which are obtained by weighting of the volume fractions:

$$\rho = \alpha\rho_1 + (1-\alpha)\rho_2, \tag{4}$$

$$\mu = \alpha\mu_1 + (1-\alpha)\mu_2, \tag{5}$$

where subscripts 1 and 2 denote liquid phase (water) and gas phase (vapour), respectively. The densities of each phase are calculated by

$$\rho_1 = \rho_{10} + \psi_1 p, \tag{6}$$

$$\rho_2 = \rho_{20} + \psi_2 p, \tag{7}$$

where ψ_1 and ψ_2 are two constants associated with the respective compressibilities of water and vapour.

In our simulations, the space discretizations are second-order upwind for the convection terms and central differences for the Laplacian terms, respectively. The time discretization is first-order implicit Euler. The pressure-velocity coupling is obtained using the Pressure Implicit Split Operator (PISO) scheme. The preconditioned conjugate gradient (PCG) method is used to treat the pressure equation and the preconditioned biconjugate gradient (PBiCG) method is used for the velocity equations.

The computational domain is a cylinder with a radius of 25 mm, and a height of 50 mm. The grid distributions on some specific sections are shown in Figure 1. The bottom of the cylindrical flow domain is set as a rigid wall, the others are far field boundaries. The initial pressure of the liquid phase is 101,325 Pa, and the pressure of vapor is 3154 Pa. The initial flow is quiescent, and the initial bubbles are spherical, with the radius of 2 mm. The geometrical setup of the multiple-bubble case is shown in Figure 2. The dimensionless wall distance is defined as $\gamma = L/R_{max}$. After carefully carrying out self-consistency tests, we find that the current grid with a cell number of 25,581,192, and an initial maximum time step size of $dt = 1 \times 10^{-7}$ s, are sufficient to assure satisfactory independence of the results with respect to both mesh and time discretizations. We note that the time step is adjustable during the simulations to meet the requirement of local courant number $Co = 0.35$.

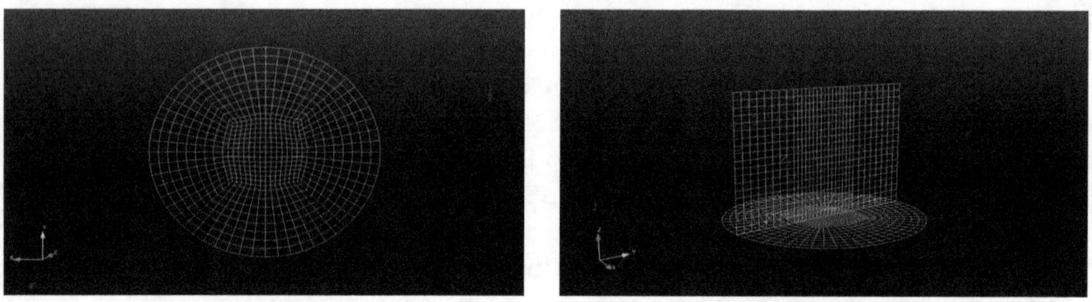

(**a**) Top view. (**b**) Side view.

Figure 1. (**a**) Top view and (**b**) side view of the grid distributions within the computational domain. Note that the grid in the central area of the cylindrical domain is refined where the bubbles are placed, and only a coarse grid with 10% of the cells is displayed to diagram the grid distributions.

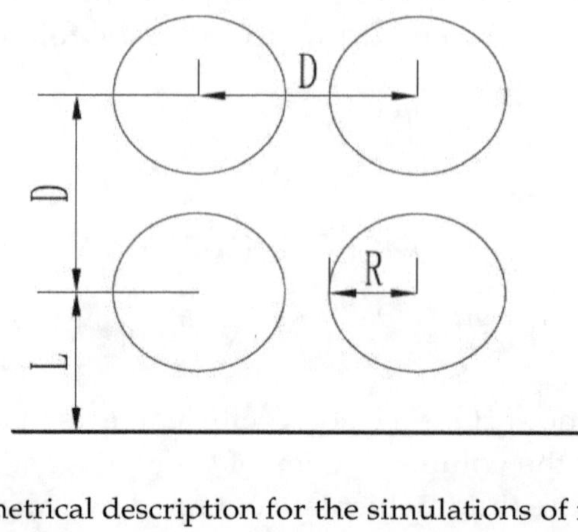

Figure 2. Geometrical description for the simulations of multiple bubbles.

To further validate the spatial resolutions, we carry out simulations with different numbers of cells along the diameter of a single bubble. Figure 3 presents the time histories of the dimensionless radius R^*, where $R^* = R/R_{max}$, and the time is non-dimensionalized according to $T^* = t/t_c$. Here, R_{max} is the maximum or the initial radius of the bubble, and t_c is the collapsing time predicated from the Reyleigh-plesset equation. It is observed that the result converges at 34 cells, which accords well with the theoretical solution. We show in Figure 4a the evolutions of bubble shapes during collapse simulated based on the current resolution of 34 cells. The numerical results are consistent with the previous experimental observation [41].

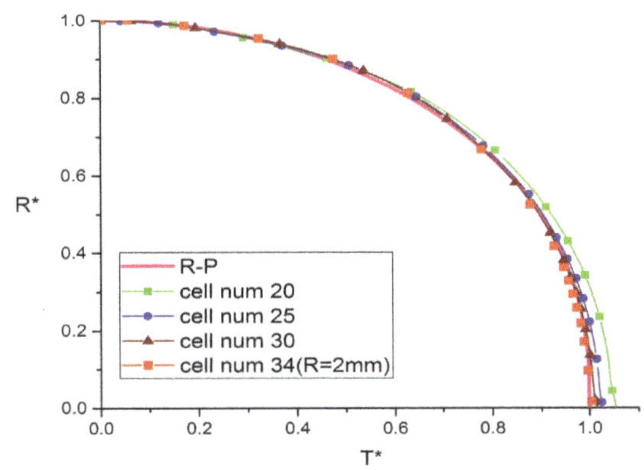

Figure 3. Time histories of radius for a single bubble resolved by different number of cells distributed along the diameter. The reference curve is the theoretical solution of the Rayleigh-Plesset equation.

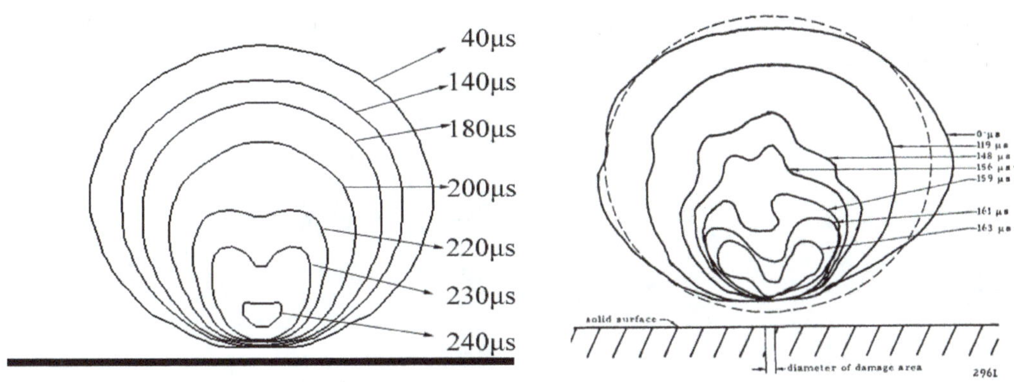

(a) (b)

Figure 4. Variations of the bubble shapes during collapse obtained by (**a**) the present numerical simulations and (**b**) previous experiments [41].

In the following sections, according to [42], the potential energy of a bubble is defined as

$$E_{pot} = \frac{4}{3}\pi R^3 \Delta p, \tag{8}$$

where R is the bubble radius and Δp is the pressure difference between two sides of the bubble interface, then the maximum potential energy of a cavitation bubble is defined as $E_{pot-max}$. The total kinetic energy of the flow domain is the integral:

$$E_k = \int \frac{1}{2}\rho U^2 dV. \tag{9}$$

We also define the wave energy emitted by the bubble collapse, following an acoustic approach [28,43], with the following form:

$$E_{wave} = \int \frac{\Delta P^2}{(\rho c)^2} dV, \tag{10}$$

where c is the sound speed in water, $c = 1500$ m/s in the current simulations. The conversion rate of the wave generation is defined as

$$\eta = \frac{E_{wave}}{E_{pot-max}} \tag{11}$$

3. Result

3.1. Collapse of a Single Bubble

In this section, we study a single cavitation bubble collapsing near a solid wall. To understand the presence of the solid wall on the non-spherical deformation, and consequently the energy conversion, we carry out a series of simulations by varying the distance between the bubble and the solid wall. The dimensionless distance γ ranges from 1.5 to 4.0.

In Table 1, we present the simulation results. First, we observe that the energy conversion rate reaches its maximum of 29.29% at $\gamma = 12.5$, which is not difficult to understand since the $\gamma = 12.5$ case corresponds to a distance very far from the solid wall, which we believe that the wall effect can be reasonably neglected. The values of maximum kinetic energy for all cases are around 4.0 mJ, with slight variations among different distances. We note that the energies and the conversions at $\gamma = 2.5$ and $\gamma = 4.0$ are very close, exhibiting nontrivial bounding effects of the solid wall.

Table 1. Energies and their conversion rates for various distances for the collapse of a single bubble (note the different units for the two energies).

γ	E_k (mJ)	E_{wave} (μJ)	η (%)
1.5	3.94	352.2	10.73
1.65	3.90	418.1	12.74
1.85	3.97	543.8	16.57
2.0	3.94	564.2	17.19
2.5	4.08	608.8	18.55
3.0	3.95	579.3	17.65
3.5	4.01	630.1	19.20
4.0	4.10	620.6	18.91
12.5	4.36	961.3	29.29

In Figure 5a,b, we present the time histories of different forms of energy at $\gamma = 12.5$ and $\gamma = 1.5$ respectively, to make a direct comparison between a small and a large distances. In Figure 5a, we observe that the potential energy of the bubble decreases during the process of collapse as the bubble shrinks, while the kinetic energy grows from zero, representing the accelerated process of the bubble collapse. As the bubble collapses to a singular point at 199 μs, the kinetic energy reaches its maximum value, while the potential energy drops to nearly zero value. During this whole period, the total energy is approximately equal to the initial bubble potential energy. In other words, the total mechanical energy remains approximately constant, or is conserved. During this process of bubble shrinkage, the potential energy is lost to the re-entrant jetting towards the bubble centre.

After the bubble vanishes, around 200 μs, a pressure wave is emitted and travels outwards, resulting in an intensive wave energy observed in Figure 5a. However, since we consider only limited forms of energy in the simulations and neglect thermal and acoustic damping mechanisms, which might be significant, the total energy is not conserved anymore. We understand that it is very challenging to quantify in detail any switch-over of these dominant damping mechanisms in such violent bubble collapses. We leave it open to the future studies. To help understand this process more intuitively, we present in Figure 6 the flow fields for three time instants demonstrating the alternative dominance of different energy forms. In Figure 6a, at the early stage, the bubble potential energy dominates, while it is converted to kinetic energy when the bubble starts to collapse (see Figure 6b), and in Figure 6c the pressure wave is emitted with only the wave energy prominent in the energy budget.

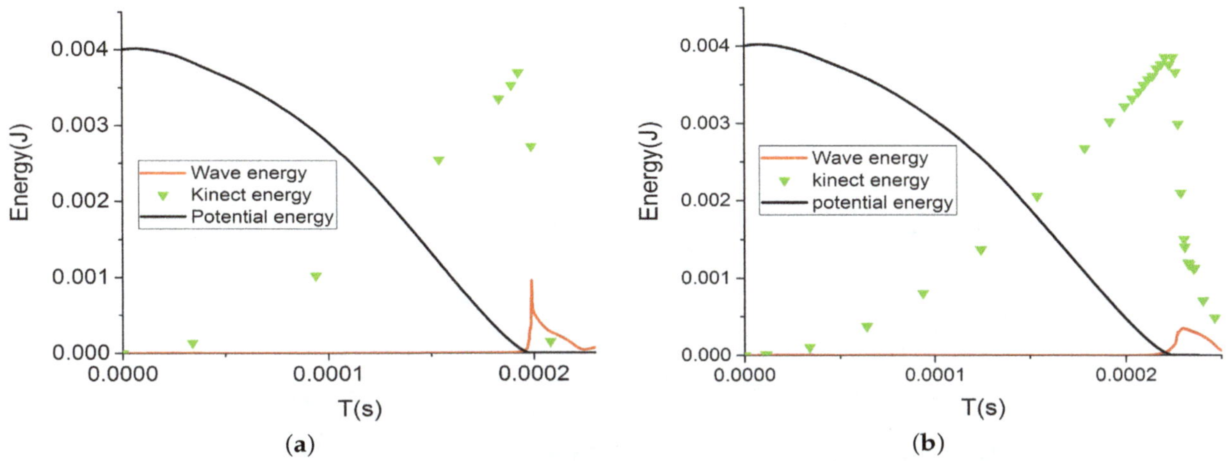

Figure 5. Time histories of different forms of energy in the flow domain during the collapse of a single bubble at (**a**) $\gamma = 12.5$ and (**b**) $\gamma = 1.5$. For both cases, $R_{max} = 2$ mm and ΔP = 101,325 Pa.

For comparison, we present the time histories of different forms of energy at $\gamma = 1.5$ in Figure 5b, in which case the initial bubble is very close to the solid wall. Apparently, the changing trend of each form of energy is roughly the same with its large distance counterpart (see Figure 5a). They still differ in some aspects: first, the wave energy appears later than the large distance case due to the confinement of the solid wall, which delays the collapse of the bubble. Second, during the whole process, even after the kinetic energy drops, it is always higher than the wave energy, implying a different physical mechanism. We conjecture that the presence of a rigid boundary can lead to asymmetric collapse, which requires more time. In the process of asymmetric collapse, high speed micro jetting flows are induced towards the rigid boundary, the liquid can thus be pushed away along the wall instead of collapsing towards a singular point. Therefore, it is easy to understand that the kinetic energy dominates even after the bubble collapse.

To get a comprehensive understanding of the distance effects, in Figure 7 we present the variations of wave energy with time for various distances corresponding to Table 1. When the bubble is located far away from the rigid boundary, e.g., γ = 12.5, the wave energy reaches the maximum value of 0.95 mJ at $t = 1.85$ ms within a very short period of 0.1 ms, and with a very sharp peak. As the distance is reduced, the peak value of wave energy decreases accordingly, and the sharp peak is replaced by a broad distribution of the high wave energy. In specific, the peak value of wave energy for $\gamma = 1.5$ is around 0.3 mJ, only one-third of that for $\gamma = 12.5$. It is interesting to find that there are secondary peaks in the wave energy distributions at γ = 2.5–4.0. We believe that these secondary peaks are induced by the pressure waves reflected by the solid boundary, which are difficult to be distinguished as we further reduce the distance. The comparison between small and large distances suggests that the near-wall collapses generate less wave energy due to the non-spherical characteristics, but with a longer duration in the computational domain.

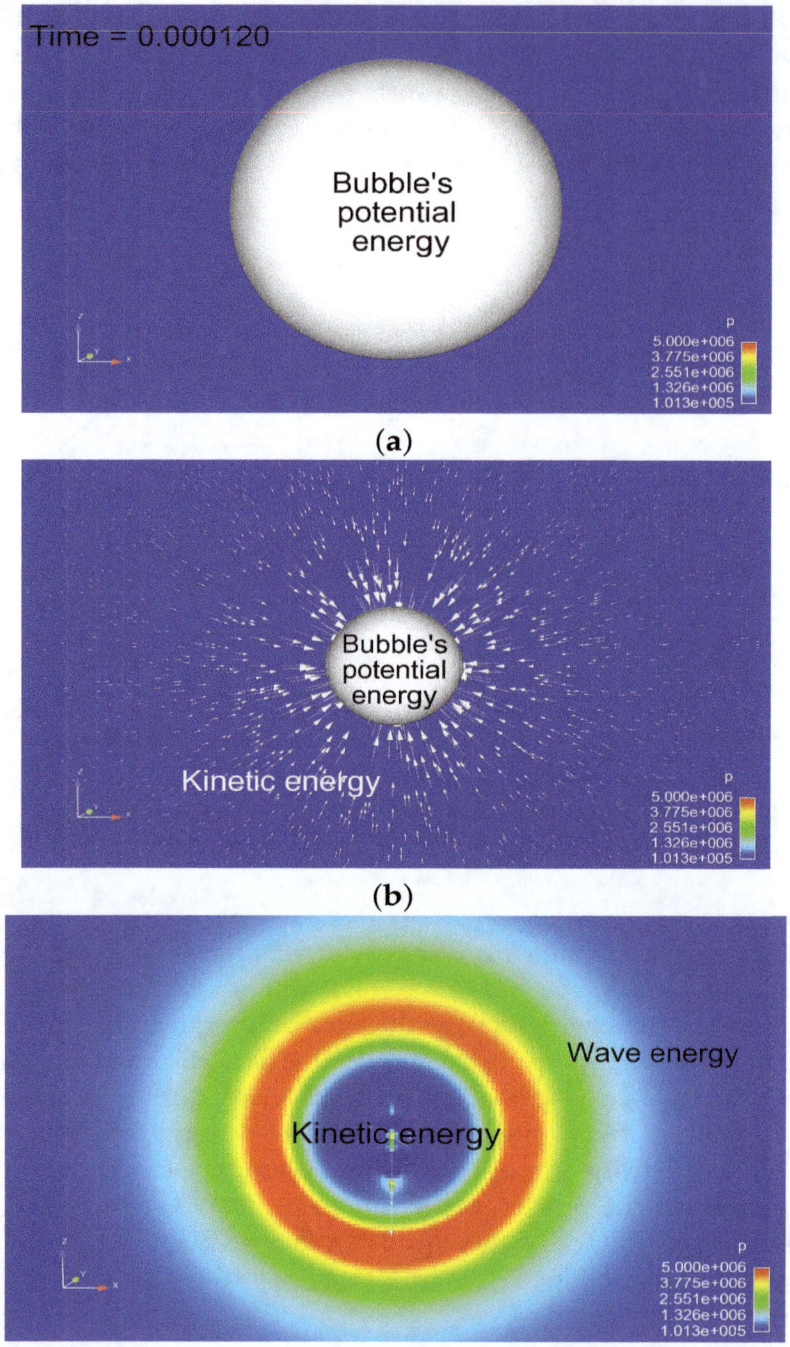

(a) (b) (c)

Figure 6. Flow fields with pressure contours for three time instants at $\gamma = 12.5$, demonstrating different dominant forms of energy: (**a**) potential energy, (**b**) kinetic energy and (**c**) wave energy.

Figure 8 shows the variations of kinetic energy with time for various distances. Again, their overall trends are the same, as we have discussed in Figure 5, and they are consistent with the wave energy variations shown in Figure 7. There are also some slight differences among various distances. In short, the wall effects on kinetic energy can be concluded as: first, the drop of kinetic energy is delayed as the bubble is placed closer to the wall. Second, the sharp peak is replaced by broader distributions of the kinetic energy as the bubble stays closer to the wall. Both conclusions are consistent with the wave energy variations, as shown in Figure 7.

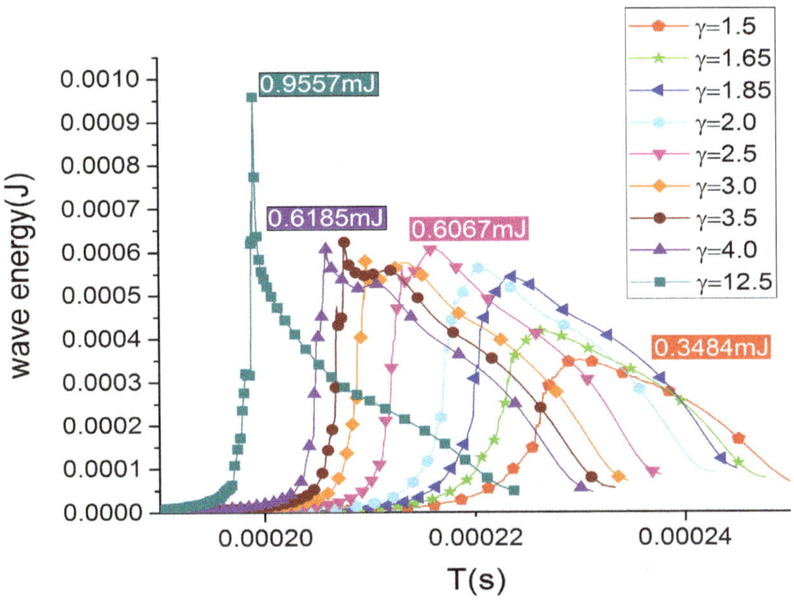

Figure 7. Wave energy variations in the flow fields at different values of γ.

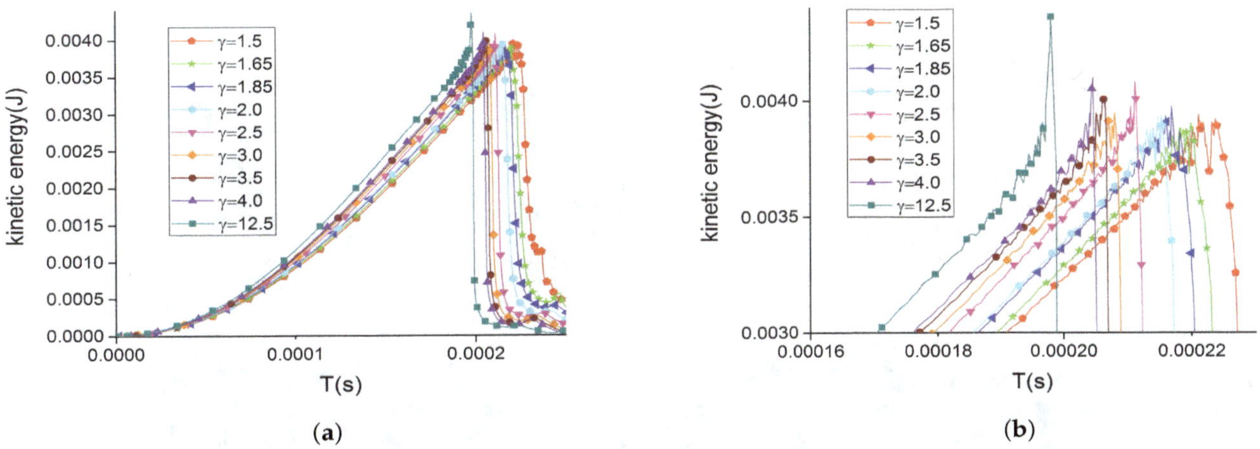

Figure 8. (**a**) Kinetic energy variations in the flow fields at different values of γ, and the enlarged views around the peak regions are shown in (**b**).

3.2. Collapses of Multiple Bubbles

The asymmetric bubble collapse can be brought by placing a solid wall, as we have studied in the last section for a single bubble, it can also be caused by the presence of nearby bubbles. In this section, we study the collapses of multiple bubbles, or more specifically, 8 bubbles, which are placed in a cubic region with two layers, and 4 bubbles on each layer.

In Table 2, we present the peak values of both the kinetic and wave energies, as well as their conversion rates. The value of E_k slightly grows with the distance γ, while η decreases as γ increases. In specific, the energy conversion rate at γ = 1.5 is approximately twice of that at $\gamma = 11.25$.

To examine the dynamic evolutions of individual bubbles, in Figures 9 and 10, we present the instantaneous shapes of the bubbles in three subsequent time instants. As shown in Figures 9a and 10a for the initial bubble configurations, the two cases represent respectively the small (γ = 1.5) and large (γ = 11.5) distances of the bubbles away from the solid wall. For the small distance, as shown in Figure 9, the wall effect leads to asymmetric collapses of the bubble cluster, with the bubbles in the upper layer collapsing faster than that in the lower layer, due to the solid wall below the lower layer. While at $\gamma = 12.5$, as shown in Figure 10, the asymmetry is brought only by the presence of nearby bubbles, therefore, the collapses and the evolutions of bubbles are symmetric with respect to the symmetry plane between the two layers, which is a straightforward demonstration that the solid wall does not affect the bubble collapse markedly when they are spaced sufficiently far away. The two cases differ in the focal point. For the large distance case, the jetting flows from all bubbles point to the centre of the bubble cluster, while for the small distance case, the focal point shifts towards the solid wall.

Table 2. Energies and their conversion rates for multiple bubbles.

γ	E_k (mJ)	E_{wave} (mJ)	η (%)
1.5	27.96	2.97	11.31
2.0	28.65	2.96	11.27
3.75	29.91	2.09	7.96
6.25	29.28	2.03	7.73
11.25	30.04	1.40	5.33

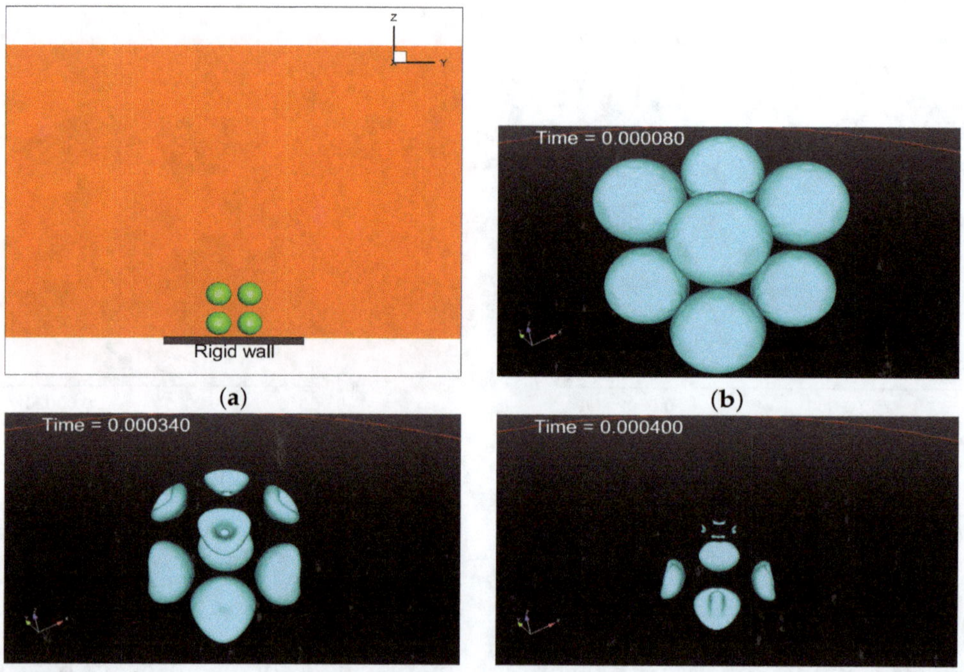

(a) (b) (c) (d)

Figure 9. Deformations of multiple bubbles at $\gamma = 1.5$ for (**a**) the initial shapes (side view), (**b**) T = 80 μs, (**c**) T = 340 μs and (**d**) T = 400 μs.

Figure 11a presents the potential energy variations of multiple bubbles. The drops of potential energy slow down at the small distances, which is apparent, because the collapses of bubbles have been delayed due to the confinement of the solid wall. Comparing to the single bubble, for the multiple bubbles, the solid wall plays a similar role, as the bubbles are very close to the wall. However, the trend of wave energy conversion is quite different with that of the single bubble, which might be raised by the strongly nonlinear interactions between the bubbles. Similar comparisons can be made in the kinetic energy variations, as shown in Figures 8b and 11b.

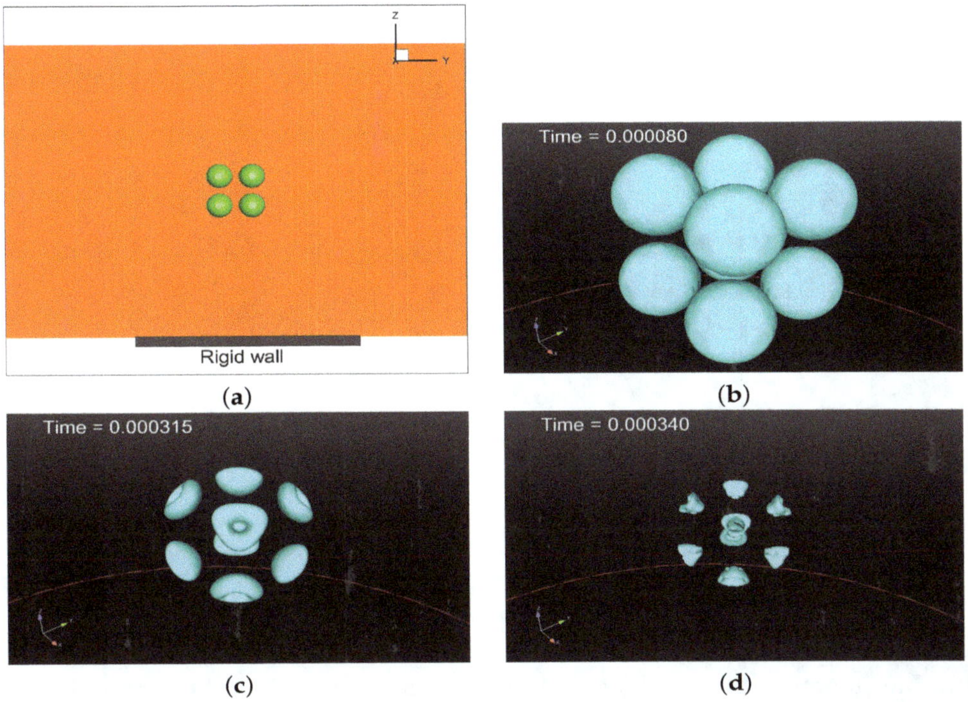

Figure 10. Deformations of multiple bubbles at γ = 11.5 for (**a**) the initial shapes (side view), (**b**) T = 80 μs, (**c**) T = 315 μs and (**d**) T = 340 μs.

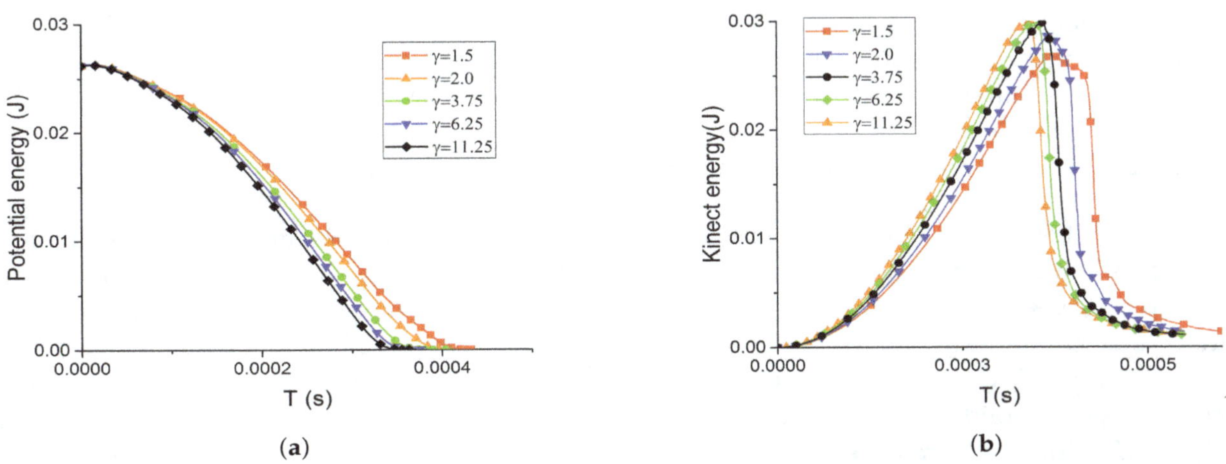

Figure 11. (**a**) Potential energy and (**b**) kinetic energy variations in the flow fields for the collapses of multiple bubbles.

The most remarkable difference between the single bubble and the multiple bubbles lies in the wave energy. Figure 12 presents the wave energy variations in the simulations of multiple bubbles. In contrast to that of the single bubble (see Figure 7), here, the ascending and descending branches of the wave energy variations are nearly symmetric with respect to the peak location. The peak values of wave energy for the small distance cases are higher than that of the large distance cases, resulting in the same behaviour for the energy conversion rate. This relationship between the peak wave energy is opposite to that of the single bubble.

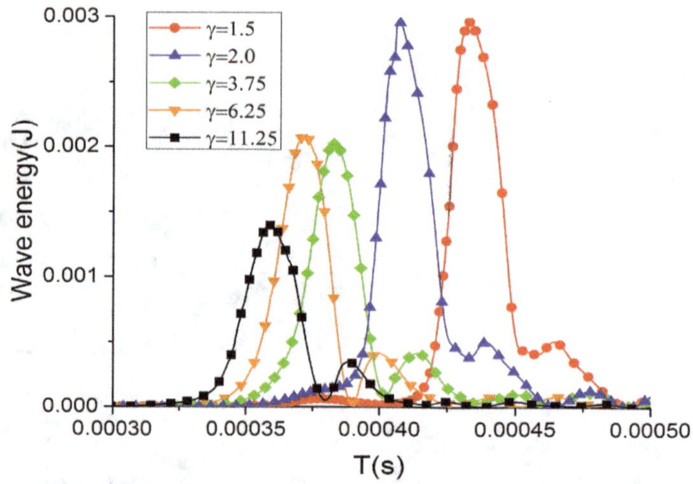

Figure 12. Variations of wave energy in the flow fields during the collapsed of multiple bubbles.

4. Conclusions

In the present work, we carry out numerical simulations to study the collapses for both a single bubble and a bubble cluster with 8 bubbles. Different forms of energy are evaluated to understand the underlying physical mechanisms of bubble collapse.

The first part concentrates on a single bubble collapsing near a solid wall. The intuitive physical rule tells that the energy should be conserved, which has indeed been observed during the first stage of collapse. As the bubble shrinks, the potential energy determined by the vapour volume converts into kinetic energy. Further, as the bubble vanishes, a part of kinetic energy starts to convert into other energy forms such as wave energy, or is lost due to the thermal and acoustic damping mechanisms, which have unfortunately not been considered in the current simulations. We report that a rigid boundary can affect the process of bubble collapse by deforming the bubble into non-spherical shapes, delaying the collapse, and consequently decreasing the conversion rate.

For a cluster consisting of 8 bubbles, their collapses are more complicated than the single bubble case, because the individual bubbles can be affected by both the solid wall and the surrounding bubbles, and both can break the symmetry of their geometric configurations. By examining the evolutions of individual bubbles during the collapse, we observe that when the bubbles are far away from the solid wall, the jetting flows from all bubbles point towards the cluster centre, while the focal point shifts towards the solid wall as the cluster is located very close to the wall. Moreover, we find that the collapse differentiates from the single bubble case mainly in the relationship between the bubble-wall distance and the wave energy, or equivalently the energy conversion rate.

Author Contributions: Conceptualization, J.Z. and L.Z.; Methodology, L.Z. and J.Z.; Investigation, J.Z.; Data Curation, J.Z.; Writing-Original Draft Preparation, J.Z. and L.Z.; Writing-Review and Editing, J.D.; Visualization, J.Z.; Funding Acquisition, L.Z. and J.D.

References

1. Ohl, S.W.; Klaseboer, E.; Khoo, B.C. Bubbles with shock waves and ultrasound: A review. *Interface Focus* **2015**, *5*, 20150019. [CrossRef] [PubMed]
2. Lordrayleigh, O.M.F.R.S. VIII. On the pressure developed in a liquid during the collapse of a spherical cavity. *Philos. Mag.* **1917**, *34*, 94–98.
3. Plesset, M.S. The Dynamics of Cavitation Bubbles. *J. Appl. Mech.* **1949**, *16*, 277–282. [CrossRef]

4. Nigmatulin, R.; Khabeev, N. Dynamics of vapor-gas bubbles. *Fluid Dyn.* **1976**, *11*, 867–871. [CrossRef]
5. Nigmatulin, R.; Khabeev, N. Heat exchange between a gas bubble and a liquid. *Fluid Dyn.* **1974**, *9*, 759–764. [CrossRef]
6. Ida, M.; Naoe, T.; Futakawa, M. Suppression of cavitation inception by gas bubble injection: A numerical study focusing on bubble-bubble interaction. *Phys. Rev. E* **2007**, *76*, 046309. [CrossRef] [PubMed]
7. Fuster, D.; Colonius, T. Modelling bubble clusters in compressible liquids. *J. Fluid Mech.* **2011**, *688*, 352–389. [CrossRef]
8. Chahine, G.L. Numerical simulation of bubble flow interactions. *J. Hydrodyn. Ser. B* **2009**, *21*, 316–332. [CrossRef]
9. Alloncle, A.; Dufresne, D.; Autric, M. Visualisation of laser-induced vapor bubbles and pressure waves. In *Bubble Dynamics and Interface Phenomena*; Springer: Dordrecht, The Netherlands, 1994; pp. 365–371.
10. Brujan, E.A.; Keen, G.S.; Vogel, A.; Blake, J.R. The final stage of the collapse of a cavitation bubble close to a rigid boundary. *Ultrason. Sonochem.* **2011**, *18*, 59–64. [CrossRef]
11. Dular, M.; Coutier-Delgosha, O. Thermodynamic effects during growth and collapse of a single cavitation bubble. *J. Fluid Mech.* **2013**, *736*, 44–66. [CrossRef]
12. Field, J.E. *Experimental Studies of Bubble Collapse*; Springer: Dordrecht, The Netherlands, 1994; pp. 17–31
13. Fujikawa, S.; Akamatsu, T. Effects of the non-equilibrium condensation of vapour on the pressure wave produced by the collapse of a bubble in a liquid. *J. Fluid Mech.* **1980**, *97*, 481–512. [CrossRef]
14. Goh, B.H.T.; Ohl, S.W.; Klaseboer, E.; Khoo, B.C. Jet orientation of a collapsing bubble near a solid wall with an attached air bubble. *Phys. Fluids* **2014**, *26*, 221–240. [CrossRef]
15. Isselin, J.C.; Alloncle, A.P.; Autric, M. On laser induced single bubble near a solid boundary: Contribution to the understanding of erosion phenomena. *J. Appl. Phys.* **1998**, *84*, 5766–5771. [CrossRef]
16. Kapahi, A.; Hsiao, C.T.; Chahine, G.L. *Shock-Induced Bubble Collapse versus Rayleigh Collapse*; IOP Publishing: Bristol, UK, 2015; p. 012128.
17. Philipp, A.; Lauterborn, W. Cavitation erosion by single laser-produced bubbles. *J. Fluid Mech.* **2000**, *361*, 75–116. [CrossRef]
18. Shaw, S.J.; Schiffers, W.P.; Gentry, T.P.; Emmony, D.C. The interaction of a laser-generated cavity with a solid boundary. *J. Acoust. Soc. Am.* **2000**, *107*, 3065. [CrossRef] [PubMed]
19. Shutler, N.D. A Photographic Study of the Dynamics and Damage Capabilities of Bubbles Collapsing Near Solid Boundaries. *J. Fluids Eng.* **1965**, *87*, 511. [CrossRef]
20. Smith, W.R.; Wang, Q.X. Viscous decay of nonlinear oscillations of a spherical bubble at large Reynolds number. *Phys. Fluids* **2017**, *29*, 082112. [CrossRef]
21. Sugimoto, Y.; Yamanishi, Y.; Sato, K.; Moriyama, M. Measurement of bubble behavior and impact on solid wall induced by fiber-holmium: YAG laser. *J. Flow Control Meas. Vis.* **2015**, *3*, 135–143. [CrossRef]
22. Supponen, O.; Kobel, P.; Obreschkow, D.; Farhat, M. The inner world of a collapsing bubble. *Phys. Fluids* **2015**, *27*, 94–98. [CrossRef]
23. Zhang, S.; Duncan, J.H.; Chahine, G.L. The final stage of the collapse of a cavitation bubble near a rigid wall. *J. Fluid Mech.* **2006**, *257*, 147–181. [CrossRef]
24. Wang, Z.; Pecha, R.; Gompf, B.; Eisenmenger, W. Single bubble sonoluminescence: Investigations of the emitted pressure wave with a fiber optic probe hydrophone. *Phys. Rev. E* **1999**, *59*, 1777. [CrossRef]
25. Lauterborn, W.; Vogel, A. *Shock Wave Emission by Laser Generated Bubbles*; Springer: Berlin, Heidelberg, 2013; pp. 67–103.
26. Lauterborn, W. Experimental investigation of cavitation-bubble collapse in the neighbourhood of a solid boundary. *J. Fluid Mech.* **1975**, *72*, 391–399. [CrossRef]
27. Lauterborn, W.; Kurz, T. Physics of bubble oscillations. *Rep. Prog. Phys.* **2010**, *73*, 106501. [CrossRef]
28. Vogel, A.; Lauterborn, W.; Timm, R. Optical and acoustic investigations of the dynamics of laser-produced cavitation bubbles near a solid boundary. *J. Fluid Mech.* **2006**, *206*, 299–338. [CrossRef]
29. Staudenraus, J.; Eisenmenger, W. Fibre-optic probe hydrophone for ultrasonic and shock-wave measurements in water. *Ultrasonics* **1993**, *31*, 267–273. [CrossRef]
30. Gonzalez Avila, S.R. The acoustic pressure generated by the non-spherical collapse of laser-induced cavitation bubbles near a rigid boundary. In Proceedings of the 10th International Symposium on Cavitation, Baltimore, MD, USA, 14–16 May 2018.

31. Merouan Hamdi, O.C.D. Measurements of the temperature variations during the growth and collapse of cavitation bubbles. In Proceedings of the 10th International Symposium on Cavitation, Baltimore, MD, USA, 14–16 May 2018.
32. Fortes-Patella, R.; Challier, G.; Reboud, J.L.; Archer, A. Energy Balance in Cavitation Erosion: From Bubble Collapse to Indentation of Material Surface. *J. Fluids Eng.* **2014**, *135*, 011303. [CrossRef]
33. Bui, T.T.; Ong, E.T.; Khoo, B.C.; Klaseboer, E.; Hung, K.C. A fast algorithm for modeling multiple bubbles dynamics. *J. Comput. Phys.* **2006**, *216*, 430–453. [CrossRef]
34. Han, B.; Köhler, K.; Jungnickel, K.; Mettin, R.; Lauterborn, W.; Vogel, A. Dynamics of laser-induced bubble pairs. *J. Fluid Mech.* **2015**, *771*, 706–742. [CrossRef]
35. Bremond, N.; Arora, M.; Ohl, C.D.; Lohse, D. Controlled multibubble surface cavitation. *Phys. Rev. Lett.* **2006**, *96*, 224501. [CrossRef]
36. Wang, Q. Multi-oscillations of a bubble in a compressible liquid near a rigid boundary. *J. Fluid Mech.* **2014**, *745*, 509–536. [CrossRef]
37. Chahine, G.L.; Duraiswami, R. Dynamical Interactions in a Multi-Bubble Cloud. *ASME J. Fluids Eng.* **1992**, *114*, 680–686. [CrossRef]
38. Tomita, Y.; Robinson, P.B.; Tong, R.P.; Blake, J.R. Growth and collapse of cavitation bubbles near a curved rigid boundary. *J. Fluid Mech.* **2002**, *466*, 259–283. [CrossRef]
39. Brujan, E.A.; Ikeda, T.; Matsumoto, Y. Shock wave emission from a cloud of bubbles. *Soft Matter* **2012**, *8*, 5777–5783. [CrossRef]
40. Tiwari, A.; Pantano, C.; Freund, J.B. Growth-and-collapse dynamics of small bubble clusters near a wall. *J. Fluid Mech.* **2015**, *775*, 1–23. [CrossRef]
41. Kling, C.L.; Hammitt, F.G. A photographic study of spark-induced cavitation bubble collapse. *J. Borderl. Stud.* **1970**, *94*, 75–90. [CrossRef]
42. Vogel, A.; Lauterborn, W. Time-resolved particle image velocimetry used in the investigation of cavitation bubble dynamics. *Appl. Opt.* **1988**, *27*, 1869–1876. [CrossRef]
43. Ward, B.; Emmony, D.C. The Energies and Pressures of Acoustic Transients Associated with Optical Cavitation in Water. *Opt. Acta Int. J. Opt.* **1990**, *37*, 803–811. [CrossRef]

6

Characteristics of Tidal Discharge and Phase Difference at a Tidal Channel Junction Investigated using the Fluvial Acoustic Tomography System

Mochammad Meddy Danial [1,2], Kiyosi Kawanisi [1,*] and Mohamad Basel Al Sawaf [1]

1 Department of Civil and Environmental Engineering, Graduate School of Engineering, Hiroshima University, 1-4-1 Kagamiyama, Higashi Hiroshima 739-8527, Japan; meddydanialstmt@gmail.com (M.M.D.); mbasel@hiroshima-u.ac.jp (M.B.A.S.)

2 Faculty of Engineering, Tanjungpura University, Pontianak 78124, Indonesia

* Correspondence: kiyosi@hiroshima-u.ac.jp.

Abstract: This study investigates the tidal discharge division and phase difference at branches connected to a channel junction. The tidal discharge at three branches (eastern, western, and northern branches) was continuously collected using the fluvial acoustic tomography system (FATS). The discharge asymmetry index was used to quantify the flow division between two seaward branches (eastern and western branches). The cross-wavelet method was applied to calculate the phase difference between the tidal discharge and water level. The discharge asymmetry index shows that the inequality of flow division is obviously prominent during the spring tide duration, where the eastern branch has the capability to deliver greater amounts of subtidal discharge, approximately 55–63%, compared with the western branch. However, the equality of flow division between the eastern and western channels can be observed clearly during the neap tide period. The wavelet analysis shows that the phase difference at the western branch is higher than at the eastern branch, because the geometry of the western branch is more convergent than that of the eastern branch. Accordingly, the amplitude of the tidal wave at the western branch is more magnified compared with that at the eastern branch. Moreover, the phase difference at the northern branch is greater than at the two seaward branches, implying that the phase difference is slightly increased after passing through the junction into the northern branch.

Keywords: tidal discharge; phase difference; tidal channel junction; flow division; fluvial acoustic tomography; wavelet analysis

---

1. Introduction

One of the most important features in a tidal channel network is the bifurcations, which are mostly located in the middle section of a delta. In the tidal channel network, there are many branches/ junctions that are commonly bifurcated asymmetrically [1]. Because of the different geometrical shape of branches, the behavioral pattern and magnitude of discharge in the first branch are different from the second branch. In a tidal channel junction where there are three channels connected, the tides that propagate upstream in the channels affect each other, and the tidal energy can propagate in two directions. As a result, the magnitude of discharge and its phases in three tidal channels are not similar [2,3].

An investigation of hydrodynamics at the tidal channel junction related to the interaction between the tidal wave, upstream river, and the geometrical shape of branches has been carried out by a few researchers. For example, Buschman et al. [2], using numerical model analysis, pointed out that the inequality of subtidal flow division is affected by the geometrical shape of the channel such as depth, length, bed roughness, and river discharge. They also emphasized that the flow division of discharge at a junction cannot merely be estimated from the ratio of the wetted cross-sectional areas of the two branches, because the distribution of flow is also affected by spring and neap tide [2]. Sassi et al. [3], using numerical modeling, highlighted the effect of tide on river flow division, where the inequality increases with the bifurcation order. They also found that during the neap tides, the flow may enter the other branch, leading to an unequal discharge distribution. Zhang et al. [4] found that in general, the tides can modify the river discharge distribution over distributaries in the Yangtze estuary. In their numerical result, they also underline that the fortnightly tidal amplitude also contributes to the inequality of subtidal flow division. Moreover, his findings showed that the effect of tidal range on the inequality of flow division is significant. It is important to note that the previous researchers mentioned above focus on subtidal flow division.

However, previous studies mentioned above do not consider the phase difference between the tidal discharge and water level at a tidal channel junction. Indeed, investigating the behavior of phase difference in estuaries still receives little attention, specifically at a junction of estuaries. Horrevoets et al. [5] investigated the phase difference along the single channel of the estuary and suggested that river discharge can control the phase difference. They pointed out that the phase difference in the downstream area is mostly constant, whereas in the upstream part of estuary, the phase difference is not constant. Additionally, the phase difference can become negative in the upstream area. Savenije et al. [6] stated that the phase difference is a basic and significant parameter to describe the tidal wave propagation in an estuary. It is strongly affected by estuary shape and is also a function of the ratio between bank convergence and tidal wave length.

Nevertheless, based on the previous works mentioned above, the temporal variation of phase difference at a tidal channel junction is still unidentified. The change of phase difference in an estuarine system can identify the hydrodynamic processes such as amplification and damping of tidal wave [5,6]. Besides, when the phase difference is near quadrature (close to ~90°), the duration asymmetry of water level can induce asymmetries in tidal current magnitude in estuary channel [7]. More importantly, the phase difference can even influence subtidal transport of flow, sediment, and saltwater in the estuarine system [8,9].

Herein, the chief goal of this work is to shed light on the hydrodynamic aspects of a tidal channel junction monitored by means of an advanced hydro-acoustic system with high-frequency resolution, focusing on the following: (i) the temporal variation of flow division between two seaward branches that are connected to the tidal channel junction, and (ii) identifying the phase difference in tidal discharge in the channels that are connected to the junction.

To further explore the phase difference at a junction, two unidentified issues have been taken into account: (i) the temporal variation of the phase difference in each branch connected to the junction, and the dominant factors that influences the phase difference between the two seaward branches; (ii) the influence of spring-neap tide on the phase difference.

The structure of the conducted study is as follows. Section 2 introduces the field site, the overview of the fluvial acoustic tomography system (FATS), acquisition of FATS data, acoustic Doppler current profiler (ADCP) measurements, bathymetry survey, index-velocity method, and the wavelet analysis. Section 3 presents the results comprising the bathymetry, index velocity relation, tidal discharge result at the junction, discharge validation, streamflow division analysis between the eastern and western branches, and phase difference at the tidal channel junction. Section 4 presents the discussion. The last section outlines the main conclusions.

2. Materials and Methods

2.1. Field Site

The Ota River estuary is a small-scale estuary with a multi-channel network that connects to the Hiroshima Bay, located in Hiroshima City, Japan. Historically, the channel network of Ota River was formed on the Ota River delta [10]. The Ota River has important environmental qualities, and the downstream area is also known as a habitat of many creatures, particularly oysters.

As shown in Figure 1a, the Ota river has an apex junction with an asymmetric branch pattern and a north–south oriented branch. It consists of the Ota River floodway (the western-most branch), Tenma River, Kyu Ota River, and the Motoyasu and Enko Rivers (the eastern-most branch) as the five main branches.

The channel network of the Ota River estuary is characterized as a mixed-semidiurnal and mesotidal system with a tidal range varying from 2 to 4 m. The catchment area of the Ota River is approximately 1710 km^2. The freshwater discharge that is monitored at the Yaguchi gauging station mostly varies between 20 to 50 m^{3}s^{-1} during the low-flow condition, except during periods of heavy rainfall.

There are two sluice gates at the Ota River floodway positioned near the first bifurcation channel (see Figure 1a). The sluice gates of Gion consist of a movable weir with three gates that only open 10% during normal operation; the Oshiba gates consist of a movable weir with three gates that are completely open throughout the year.

The studied junction is in the middle part of the Ota River network channel, approximately 2.5 km downstream from the main bifurcation channel (apex junction) and approximately 5.8 km from the river mouth, as shown in Figure 1b. The tidal channel junction consists of three branches: the northern branch (upstream Kyu Ota River), the eastern branch (downstream Kyu Ota River), and the western branch (Tenma River). As an estuary system with multi-channel network, the saltwater intrusion in the Ota River can reach 11.5 km upstream from the river mouth. The bed materials of the channel network mostly consist of sand containing less silt and clay.

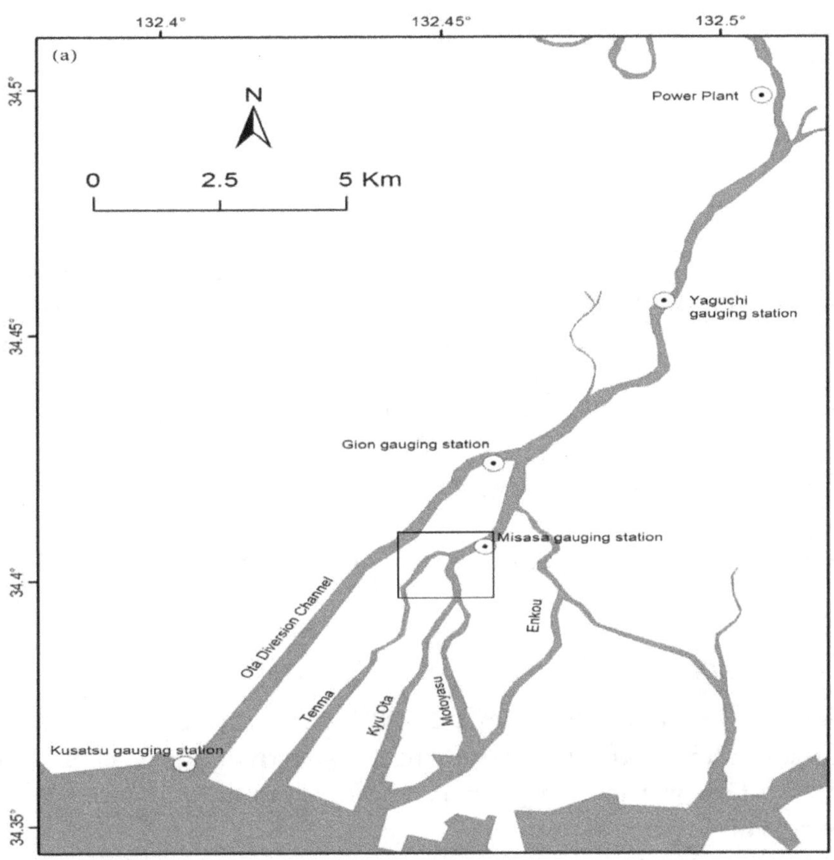

Figure 1. *Cont.*

Figure 1. (**a**) Ota river channel network map. The black box denotes the study area. (**b**) Observation site and deployment of the fluvial acoustic tomography system (FATS) at the eastern, western, and northern branches. The yellow lines represent the transmission lines between the two transducers of FATS. S1–S2, T1–T2, and M1–M4 denote the locations of the transducers for FATS measurements. The red lines denote the acoustic Doppler current profiler (ADCP) transects for the comparison with FATS.

2.2. *Measurement Method*

2.2.1. The Overview of the Fluvial Acoustic Tomography System

The FATS developed at Hiroshima University was used to continuously measure the time series of tidal discharge at the tidal channel junction. The advantages of FATS are that it enables the measurement of discharge continuously in the long-term period that is suitable to be used even in such a tidal river with salt water intrusion, extremely shallow water depth area, and a larger width-to-depth ratio. The sound rays of FATS are capable of covering the whole cross-section of the stream and capturing the depth- and range-averaged velocity. As a result, the FATS can accurately estimate the cross-sectional averaged velocity [11]. To obtain the discharge (Q), the cross-sectional averaged velocity is multiplied by cross-sectional area (A_0), which varies in time and corresponds to water level fluctuation. For detailed information about the FATS measurement method and its reliability, readers can refer to previous studies [11–14].

2.2.2. Acquisition of FATS Data

Continuous measurements of discharge by the FATS were carried out from 8 to 20 June 2017 and 9 to 29 June 2017 for the western and eastern branches, respectively (see Table 1). Two pairs of FATS were deployed at the western and eastern branches located near a tidal channel junction, as shown in Figure 1b. The acoustic pulses (central frequency: 30 and 53 kHz) were transmitted concurrently from both transducers every 30 s. The transducers were installed at the height of 0.2 m above the channel bed using stands.

Table 1. Summary of the fluvial acoustic tomography system (FATS) experiment period.

Date of FATS Experiment	Deployment Site	Method of Discharge Estimation
9 to 29 June 2017	Eastern branch	Index-velocity method as described by Kawanisi et al. [11]
8 to 20 June 2017	Western branch	Index-velocity method as described by Kawanisi et al. [11]
6 to 29 November 2017	Northern branch	Two-crossing transmission lines as described by Kawanisi et al. [13] and Bahreinimotlagh et al. [14]

On the other hand, the monitoring of tidal discharge at the northern branch was conducted from 6 to 29 November 2017 to obtain additional data related to the phase difference investigation at the junction. In this case, we use two pairs of FATS with a two-crossing transmission lines configuration to estimate tidal discharge, as proposed by Kawanisi et al. [13] and Bahreinimotlagh et al. [14]. This method is applied to show the significant variation that occurs during field investigations.

The coordinates of the transducers and the distance between transducers for the western, eastern, and northern branches are presented in Table 2.

Table 2. Coordinates of transducers.

Code	River branch	Transducers	Latitude (°)	Longitude (°)	The Distance between Transducers (m)
Eastern branch	Kyu Ota river	S_1	34°24′00.60″	132°27′08.64″	246.363
		S_2	34°23′52.68″	132°27′07.32″	
Western branch	Tenma river	T_1	34°24′03.80″	132°26′43.63″	158.836
		T_2	34°24′01.09″	132°26′38.34″	
Northern branch	Kyu Ota river	M_1	34°24′20.36″	132°27′18.00″	289.744
		M_2	34°24′20.09″	132°27′06.66″	
		M_3	34°24′23.49″	132°27′14.52″	224.639
		M_4	34°24′16.68″	132°27′11.38″	

2.2.3. Index-Velocity Method

Application of the index velocity method for calculating continuous records of discharge has become increasingly common, particularly since the development of acoustic-based instruments such as acoustic velocity meters (AVMs) and horizontal acoustic Doppler current profilers (H-ADCPs) and acoustic tomography systems [11,15–17]. In this study, the index velocity method (IVM) is used to estimate the discharge based on the regression equation obtained from the relationship between the velocity parameter of ADCP measurement and velocity along the transmission line of FATS measurement, as described in the previous work of Kawanisi et al. [11]. Subsequently, the discharge can be performed as a product of the regression equation and cross-sectional area, which is a function of water depth.

2.2.4. ADCP Measurements and Cross-Sectional Area Determination

In this study, Teledyne RDI StreamPro ADCP was used to provide and establish reference discharge data for validating FATS measurements between two seaward branches. Each ADCP campaign was conducted approximately every 5–8 min to collect the data at the bridge near the transducers (see Figure 1b). However, for the validation of FATS measurement in the northern branch, the ADCP campaign was carried out along the transmission line between two transducers. StreamPro ADCP was set to operate in water mode 12 with a bin size of 12 cm, the number of bin sizes set to 30, with 6 pings per ensemble. In this experiment, ADCP discharges can be obtained using WinRiver II software, in which each of the bad bin ensemble parts did not exceed 2%, while bad ensembles did not exceed 3% in each transect.

To obtain the cross-sectional area (*Ao*) as a function of water level, water level data from three branches were recorded every 10 min using water level loggers (Hobo U20-001-01-Ti, Onset Co.) that were attached to the transducer stands. To obtain water level estimates, the measured pressure values should be normalized with respect to atmospheric pressure, as recorded by the barometer deployed on the riverbank. The accuracy of the water level sensors and the barometer was ±5 and ±3 mm H_2O, respectively. The post-processing of the water pressure and barometric data as parameter inputs to the software program Hoboware Pro [18].

2.2.5. Bathymetry Survey

The bathymetric survey of the three branches was conducted using an autonomous boat equipped with global positioning system (GPS) device (resolution 1/10000 s) and a single beam echo sounder at a frequency of 200 kHz (resolution: 0.01 m). Transect lines were identified according to the site accessibility conditions that were assessed in the field inspection. Thus, in the case of our monitoring program, we performed bathymetry transects for 53 sections, as depicted in Figure 2a. The directions of these lines are perpendicular to the stream direction, with the distance of each line being approximately 50 m (i.e., the distance between the left and right banks ≈50 m). Each transected line represents two points with the north–east coordinates based on the Japanese datum standard system. Output data of bathymetry transects obtained by the boat were stored as x, y, and z positions, where x and y represent the horizontal positions obtained from GPS recording, and z was the water depth measured by the echo sounder. The survey started at the upstream junction and finished at the downstream junction of both branches.

2.2.6. Wavelet Analysis

Wavelet transformation is an advanced analysis method in signal processing particularly used in investigating hydrodynamic processes in an estuary such as a tidal wave and its interaction with the river discharge [3,8,19]. In this study, three functions of wavelet were used to analyze the interaction between tidal discharge and water level, that is, continuous wavelet transformation (CWT), cross-wavelet transformation (XWT), and wavelet coherence (WTC).

CWT is used to detect variations in time series and their simultaneous representation in the time–frequency space, and also to determine the dominant period. Moreover, CWT analysis can display temporal and spatial evolutions of tidal frequency spectra [19]. XWT was utilized to analyze the phase difference between two-time series. The phase difference is basically the phase angle represented by the arrow, that is, the arrow pointing right means both signals travel in the same direction (in-phase), the arrow pointing left means both signals travel in the opposite direction (anti-phase), and the arrow pointing down means the first time series of signal leads the second one by 90°. WTC is used to find significant coherence between the two-time series and to show the confidence level. Details of wavelet are referred to in Grinsted et al. [20].

3. Results

3.1. Bathymetry at the Tidal Channel Junction

Figure 2a shows the bathymetry in three branches at a tidal channel junction where the bed level varies between −1.5 to −4.5 T.P. m, but mostly dominated by −3 T.P. m and −2 T.P. m for the eastern and western branches, respectively. From Figure 2a,b, it is revealed that the eastern branch is deeper and wider than the western branch. Moreover, sedimentation was found in the western branch around 800 m from transect 1. Similarly, the bathymetry in the northern branch shows that the width of the northern branch is wider compared with those of the eastern and western branches. Moreover, the bed level of the northern branch is relatively constant from −2.5 T.P. m to −3.0 T.P. m.

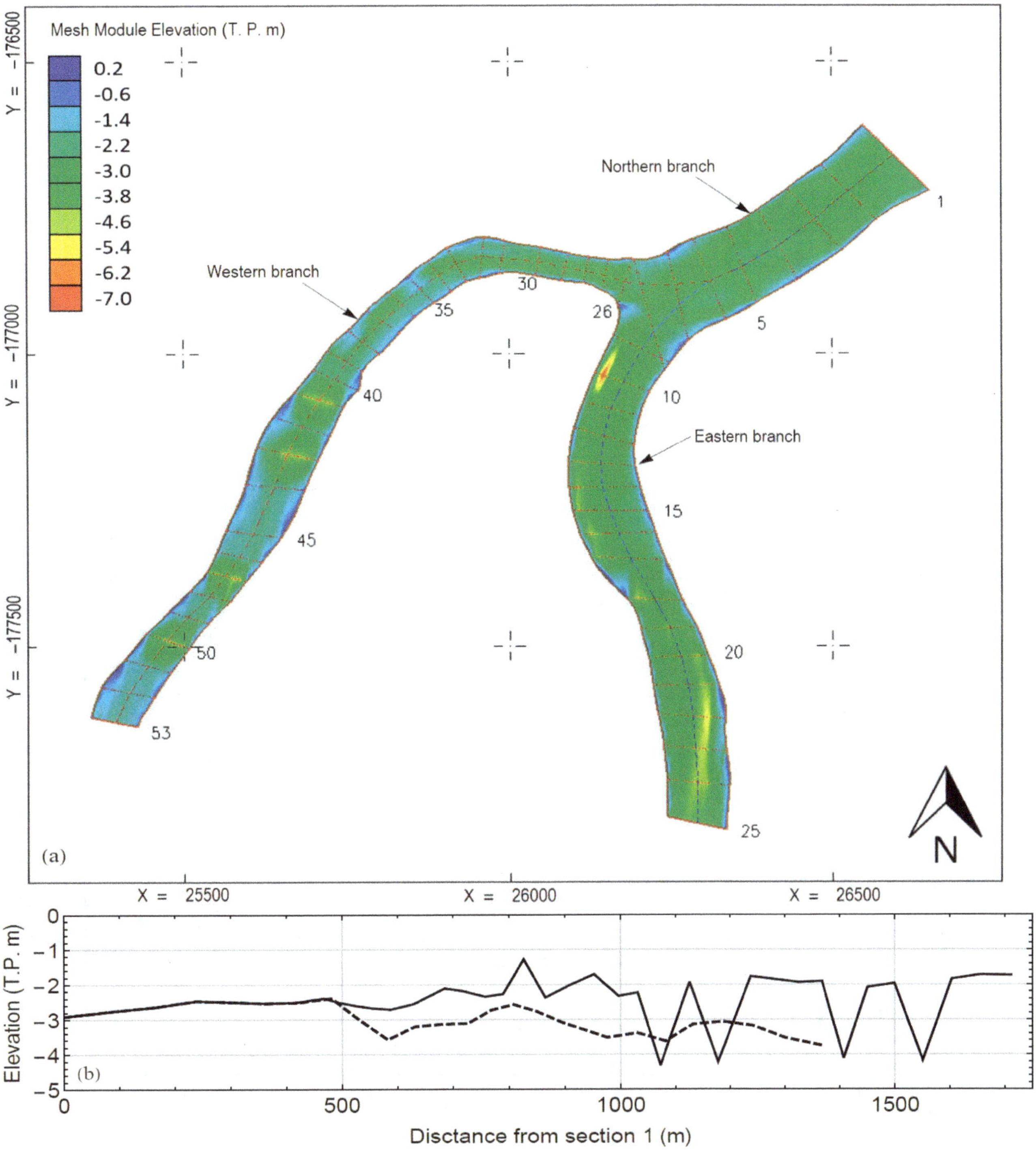

Figure 2. (**a**) Bathymetry map around the bifurcation. The elevation is measured as a vertical distance (T.P. m). The listed numbers in the range of 1–53 denote the transects of the cross-sectional line measured by the Coden boat RC–S3. (**b**) The longitudinal distributions of mean bed level. The black line and the black dashed lines represent the mean bed levels from the northern to the western and eastern branches, respectively.

3.2. Establishing an Index Velocity Rating and Validation of Discharge Measurement Obtained by FATS

We calculated tidal discharge by regressing the FATS index velocity with the velocity of ADCP, yielding discharge after multiplying it with the oblique cross-sectional area (A_O) along the transmission line, which varies in time as a function of water depth. Figure 3a,b show the linear regression relationships between the velocity of ADCP (Q_{ADCP}/A_O) and the FATS velocity along the transmission line (u_m) for the eastern and western branches, respectively. Both regressions show a high correlation with the R^2 ~0.99, though the data used for validation, particularly at the eastern branch, are limited. However, it is necessary to emphasize that having an additional number of ADCP transects in the eastern branch location is very difficult because of the high number of activities that take place along that branch (e.g., water taxi, fishing), which constrained our validation endeavors.

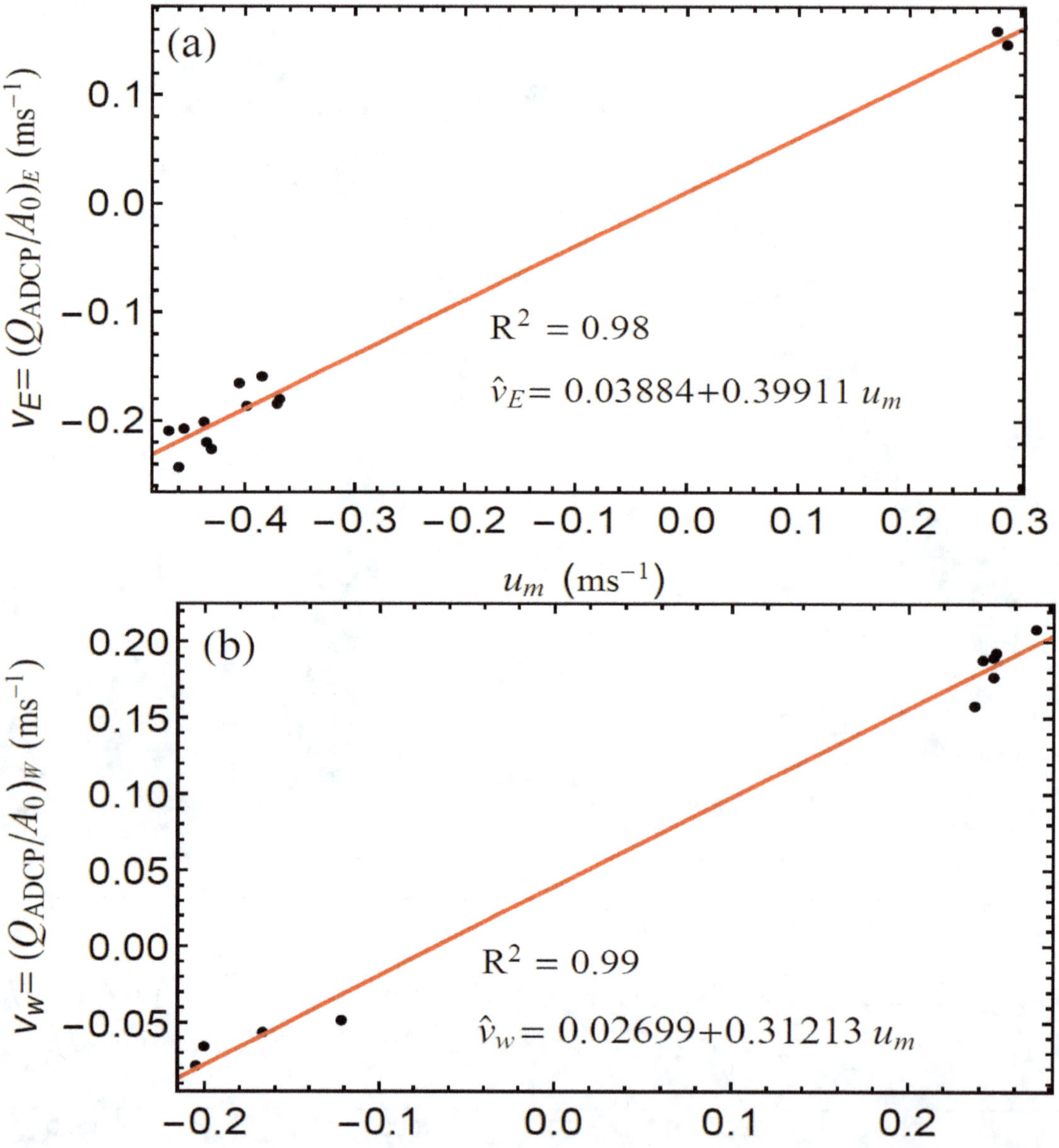

Figure 3. Index velocity relations for the (**a**) eastern and (**b**) western branches.

The index-velocity equations used for the calculation of the cross-sectional average velocity at eastern branch ($\hat{v}_E$) and the western branch ($\hat{v}_W$) are as follows:

$$\hat{v}_E = 0.03884 + 0.39911\,u_m, \tag{1}$$

$$\hat{v}_W = 0.02699 + 0.31213\,u_m. \tag{2}$$

Thus, the discharge of FATS for the western and eastern branches can be respectively computed as follows:

$$Q_E = \hat{v}_E \times A_0, \tag{3}$$

$$Q_W = \hat{v}_W \times A_0. \tag{4}$$

The FATS estimates were validated by the moving-boat ADCP measurements, as shown in Figure 4a–c for the eastern, western, and northern branches, respectively. The time series of tidal discharge for the eastern and western branches were obtained from Equations (3) and (4), respectively, whereas the discharge for northern branch was obtained from the two-cross paths of the FATS method adopted from Kawanisi et al. [13] and Bahreinimotlagh et al. [14].

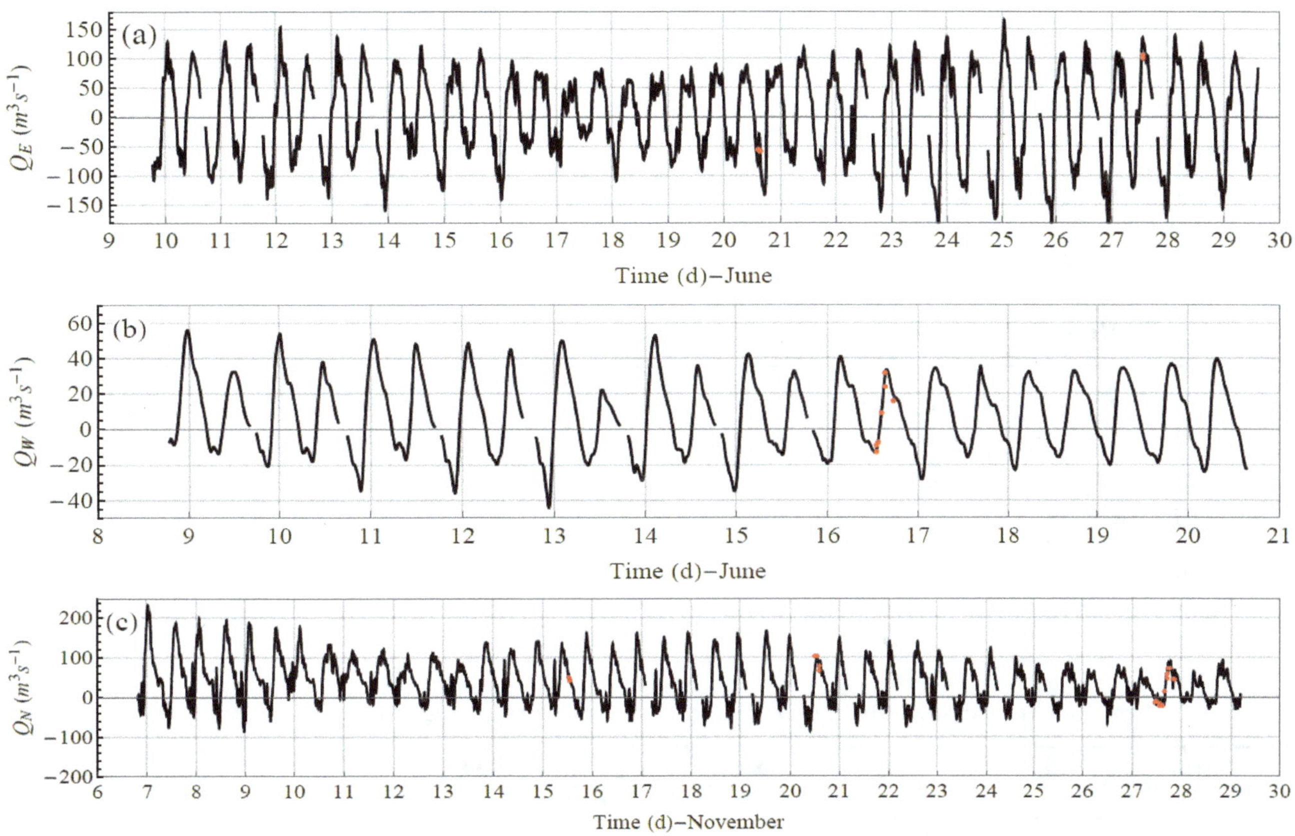

Figure 4. Temporal variations of (**a**) tidal discharge at the eastern branch, (**b**) tidal discharge at the western branch, and (**c**) tidal discharge at the northern branch. The red dots denote the discharge obtained using moving-boat ADCP measurements. Discontinuations of the discharge time series correspond to the missing period, as the result of the transducers were not covered by water during low tide.

It is important to note that particularly in the eastern branch, the validation using ADCP measurement is limited as a result of ship traffic. The relative difference of discharges estimated from ADCP measurements and FATS are mostly ranged from ~0.4% to ~10%, as shown in Table 3, Table 4, and Table 5. The results are consistent with those of previous works, where the relative differences between FATS and ADCP estimates range from 1% to 10% [12–14]. Moreover, we can evaluate how closely the FATS estimates match the ADCP results using the equation of root mean square error (RMSE):

$$\text{RMSE} = \sqrt{\frac{\sum_{i=1}^{n} [Q_{\text{FATS}}(t_i) - Q_{\text{ADCP}}(t_i)]^2}{n}}, \tag{5}$$

where $Q_{ADCP}(t_i)$ and $Q_{FATS}(t_i)$ correspond to the observed and calculated discharge at time t_i, respectively; and n is the number of data points. Thus, based on the data obtained from Table 3 to Table 5, the root means square error (RMSE) at the eastern branch is 5.0 m^3s^{-1} or only ~1.67% of the discharge range, whereas the RMSE of the western branch is 1.68 m^3s^{-1} or ~1.7% of the discharge range. Similarly, the root means square error (RMSE) at the northern branch is 5.3 m^3s^{-1} or ~1.65% of the discharge range. Thus, FATS measurement results are reliable estimates of the discharge in a tidal estuary.

Table 3. Comparison between FATS and acoustic Doppler current profiler (ADCP) measurements at the eastern branch.

Date	Local Time	Q_{FATS} (m^3s^{-1})	Q_{ADCP} (m^3s^{-1})	$\Delta Q = Q_{FATS} - Q_{ADCP}$ (m^3s^{-1})	$\Delta Q/Q_{ADCP}$ (%)
June 20, 2017	14:44:05	−57.56	−55.67	−1.89	3.40
June 20, 2017	14:55:13	−58.53	−55.67	−2.86	5.14
June 20, 2017	15:07:31	−59.09	−58.81	−0.28	0.48
June 20, 2017	15:16:31	−60.31	−58.41	−1.90	3.25
June 20, 2017	15:27:11	−65.81	−58.18	−7.63	13.11
June 27, 2017	13:29:25	108.38	99.55	8.83	8.87
June 27, 2017	13:41:38	109.60	104.65	4.95	4.73

Table 4. Comparison between FATS and ADCP measurements at the western branch.

Date	Local Time	Q_{FATS} (m^3s^{-1})	Q_{ADCP} (m^3s^{-1})	$\Delta Q = Q_{FATS} - Q_{ADCP}$ (m^3s^{-1})	$\Delta Q/Q_{ADCP}$ (%)
June 16, 2017	12:45:11	−11.75	−12.2	0.45	-3.69
June 16, 2017	12:59:19	−10.26	−9.13	−1.13	12.38
June 16, 2017	13:09:38	−8.05	−7.37	−0.68	9.23
June 16, 2017	13:14:28	9.80	8.91	0.89	9.99
June 16, 2017	14:14:28	25.19	24.26	0.93	3.83
June 16, 2017	14:53:59	31.85	31.47	0.38	1.21
June 16, 2017	15:13:13	31.17	27.17	4.00	14.72

Table 5. Comparison between FATS and ADCP measurements at the northern branch.

Date	Local Time	Q_{FATS} (m^3s^{-1})	Q_{ADCP} (m^3s^{-1})	$\Delta Q = Q_{FATS} - Q_{ADCP}$ (m^3s^{-1})	$\Delta Q/Q_{ADCP}$ (%)
November 15, 2017	12:46:48	52.98	49.68	3.3	6.64
November 15, 2017	13:22:43	43.67	42.95	0.72	1.68
November 20, 2017	12:15:03	88.61	103.34	−14.73	−14.25
November 20, 2017	13:14:28	95.68	104.48	−8.80	−8.42
November 20, 2017	14:06:36	78.95	78.64	0.31	0.39
November 27, 2017	11:37:19	−10.07	−10.75	0.68	−6.33
November 27, 2017	13:13:52	−17.80	−19.41	1.61	−8.29
November 27, 2017	14:34:03	−18.16	−20.15	1.99	−9.88
November 27, 2017	16:09:35	14.32	15.24	−0.92	−6.04
November 27, 2017	16:57:07	57.83	57.73	0.10	0.17
November 27, 2017	17:35:10	73.36	73.61	3.75	5.09
November 27, 2017	18:20:39	77.10	71.55	5.55	7.76
November 27, 2017	19:49:15	41.15	45.65	−4.50	−9.86
November 27, 2017	20:24:41	48.85	44.98	3.87	8.60

3.3. Time Series of Tidal Discharge in the Western, Eastern, and Northern Branches

Figure 5a shows the temporal variations of the water levels and discharges in the eastern and western branch. The time series of tidal discharge for the eastern and western branches were obtained from Equations (3) and (4), respectively. The tidal discharge shows the semidiurnal characteristic, that is, the high tide and low tide occur two times a day. The discharges in the branches show distinct characteristics, where the eastern and western branches range from −157 to 155 m^3s^{-1} and from −42 to 58 m^3s^{-1}, respectively. From the comparison of the tidal discharges at the eastern and western branches, as shown in Figure 5a, it is revealed the tidal discharge of the eastern branch is approximately two to six times greater than that of the western branch during flood and ebb tides. This is caused by the difference in the geometry of both branches, that is, the water depth and the channel width of the eastern branch are much larger than that of the western branch, as can be seen in Figure 2. Moreover, the pattern of tidal discharge at the western branch is also much more asymmetric compared with that of the eastern branch, where the discharge during the ebb tide was higher than during the flood tide.

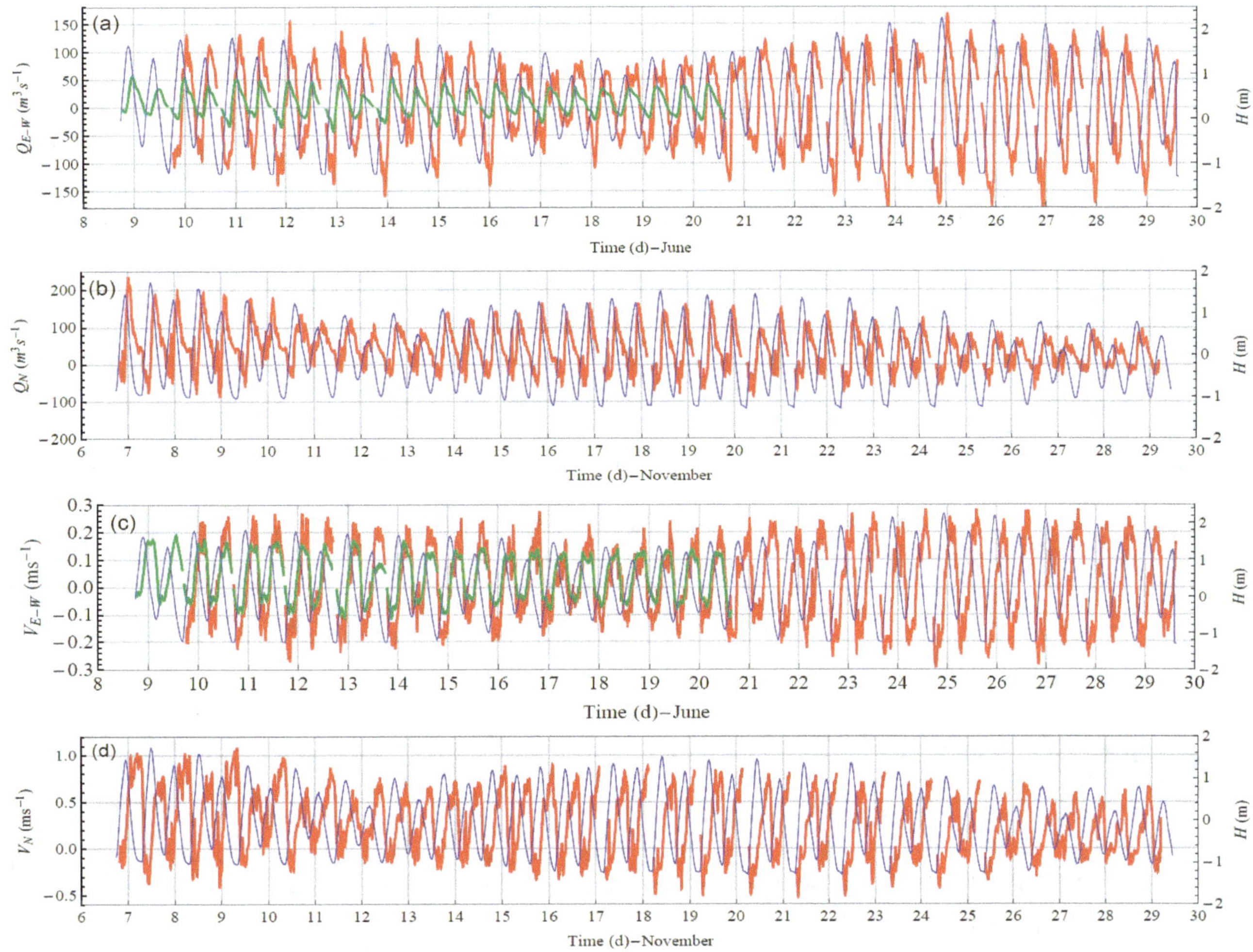

Figure 5. Temporal variations of (**a**) water level at the eastern branch (blue line), tidal discharge at the eastern branch (red line), and tidal discharge at the western branch (green line). (**b**) Water level at the northern branch (blue line) and tidal discharge at the northern branch (red line). (**c**) Water level at the eastern branch (blue line), tidal velocity at the eastern branch (red line), and tidal velocity at the western branch (green line). (**d**) Water level at the northern branch (blue line) and tidal velocity at the northern branch (red line). Positive discharge and velocity coincide with the seaward flow.

Similarly, Figure 5b shows the temporal variation of water level and tidal discharge at the northern branch located at the upstream junction, where the tidal discharge ranged from −90 to 230 m^3s^{-1}. In addition, Figure 5c shows the temporal variation of the water level and tidal velocity for the eastern and western branch, whereas Figure 5d describes the tidal velocity and water level at the northern branch. The temporal variation of tidal velocity at the eastern, western, and northern branch ranges from −0.28 to 0.3 ms^{-1}, −0.12 to 0.19 ms^{-1}, and −0.5 to 1 ms^{-1}, respectively.

3.4. Subtidal Discharge Division between the Western and Eastern Branches

The purpose of quantifying subtidal discharge division is to explore the variation of flow division and the inequality of streamflow during the spring and neap tide. In this study, we use the subtidal discharge obtained by applying a Battle–Lemarie filter of wavelet function. The division of subtidal discharge between the eastern and western branches at a tidal channel junction can be quantified using the discharge asymmetry index (ψ), as proposed by Buschman et al. [2].

$$\psi = \frac{\langle Q_E \rangle - \langle Q_W \rangle}{\langle Q_E \rangle + \langle Q_W \rangle} \quad (6)$$

The tidal discharge asymmetry index (ψ) is zero for an equal discharge division, $\psi = 1$ when all river water flows through the eastern branch, and $\psi = -1$ when the western branch carries all the discharge. Therefore, a positive value is obtained when the discharge at the eastern branch is greater than that at the western branch, and vice versa.

Only data from 10 and 21 June, 2017 were used to calculate the discharge asymmetry index (ψ) using Equation (6), because of the difference in lengths of the data obtained from the FATS measurements between the eastern and western branches. Figure 6 shows the discharge asymmetry index and subtidal discharge variation in the eastern and western branch from 10 to 21 June, suggesting that the temporal variations in subtidal discharge of the eastern branch are slightly larger than those of the western branch, except within 18 and 19 June, the variation was almost the same during the neap tide. Figure 6, which captures a spring-neap fluctuation, shows the temporal variation in the discharge asymmetry index. The discharge asymmetry index fluctuates in the range of −0.02 to 0.26 from 10 to 21 June, indicating the inequality of flow division between the eastern and western branch, where the eastern branch can distribute larger discharge than that in the western branch during spring tide; the flow division between the western and eastern branch is nearly equal during the neap tide (18 to 19 June).

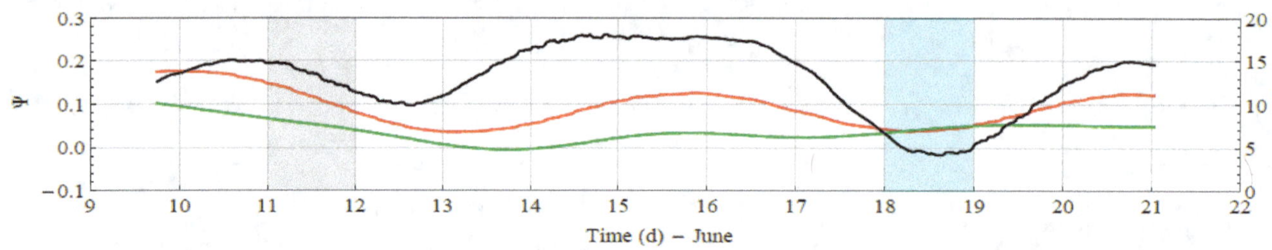

Figure 6. Discharge asymmetry index (black line). Temporal variations of subtidal discharges in the eastern (red line) and western branch (green line). The grey and blue highlighted boxes represent the spring and neap tide, respectively.

3.5. The Wavelet Analysis for the Interaction between the Water Level and Tidal Discharge in the Three Branches

In this section, three functions of wavelet analysis are used to investigate the time-series signal between tidal discharge and water level in three branches connected to a channel junction. In this study, XWT is used to determine the interaction between tidal discharge and water level, that is, propagating, standing wave, and mixed wave. Tidal discharge (Q) varies along the channel in estuaries and co-varies with the tidal wave. There are two methods to determine the type of wave in estuarine channels, that is, phase lag and phase difference [21,22]. The phase lag is the time lag between high water slack (HWS, i.e., when the discharge is zero) and high water (HW, i.e., when the water level is maximum) [8,21]. In contrast, the phase difference is the time lag between the peak tidal discharge and the HW [21,22]. Therefore, there is a relationship between the phase difference and the phase lag, that is, $\varepsilon + \varphi = \frac{\pi}{2}$, where ε is the phase lag and φ is the phase difference [8,23]. In this study, we use phase difference to describe the relationship between tidal discharge and water level [22,24]. If $\varphi = 90°$ is a standing wave, $\varphi = 0°$ is propagating, and $0° < \varphi < 90°$ is a mixed wave.

Figure 7 shows three figures consisting of CWT, XWT, and WTC analyses between the tidal discharge and water level at the eastern branches. From the figures, the dark red color background denotes a strong power spectrum, suggesting a dominant tidal signal. The vertical bar shows the energy of the tidal domain. From the CWT and WTC analyses, the semidiurnal signal is more dominant than the diurnal signal. Moreover, the semidiurnal domain is always present during the spring and neap tide, whereas the diurnal domain is only present during spring tide, but fades during the neap tide. Thus, the tide of the Ota River is characterized by the mixed-semidiurnal tide. Besides the semidiurnal and diurnal signal, we can also identify the quarter-diurnal in the eastern branch, which is discontinuous and weak. In contrast, the fortnightly signal cannot be detected. From XWT analysis, the phase difference between the tidal discharge and the water level at the eastern branch can be denoted by the black arrows pointing straight down at ~78°.

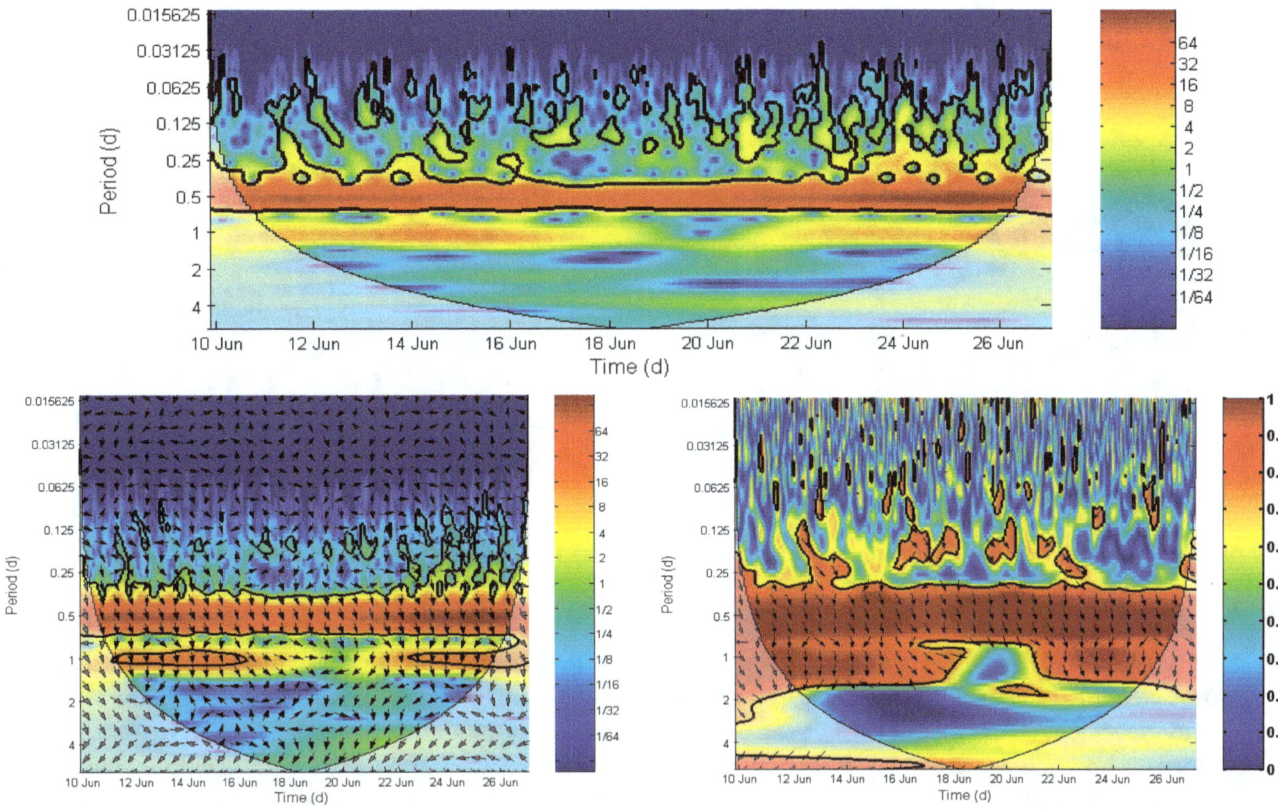

Figure 7. Wavelet analyses in eastern branch: (top) continuous wavelet transformation; (bottom left) cross-wavelet transform; (bottom right) wavelet coherence. The black contours represent the 0.95 confidence level against red noise, and the cone of influence (COI) where edge effects might distort the picture is shown as a lighter shade. The relative phase relationship is shown as arrows (with in-phase pointing right, anti-phase pointing left, and the first time series leading the second one by 90° pointing down). The wavelet power is $\log_2(A^2/v)$, where A is the wavelet amplitude and v is the variance of the original tidal signals. The y-axis in the CWT and XWT is on a $\log_2$ scale.

Similarly, Figure 8 shows three plots consisting of CWT, XWT and WTC analyses between the tidal discharge and water level at the western branches. The semidiurnal domain is always present, whereas the diurnal domain is only present during spring tide but diminishes during the neap tide. We can recognize the quarter-diurnal in the western branch which is discontinuous and weak, but more pronounced compared to the eastern branch. The fortnightly signal is also cannot be identified in the western branch.

more pronounced compared to the eastern branch. The fortnightly signal is also cannot be identified in the western branch.

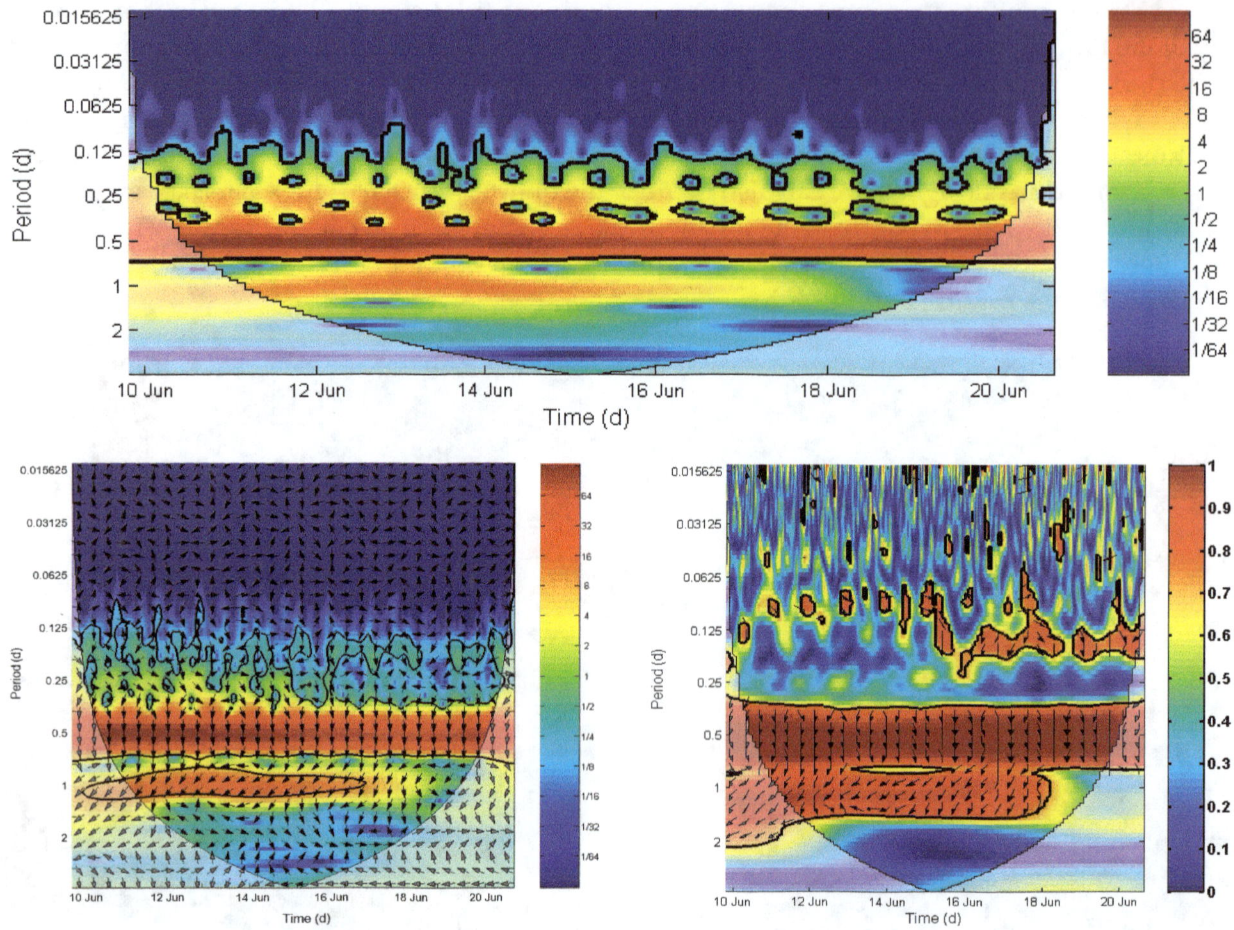

Figure 8. Wavelet analyses in western branch: (top) continuous wavelet transform. (bottom left) cross-wavelet transform. (bottom right) wavelet coherence. The black contours represent the 0.95 confidence level against red noise and the cone of influence (COI) where edge effects might distort the picture is shown as a lighter shade. The relative phase relationship is shown as arrows (with in-phase pointing right, anti-phase pointing left, and the first time series leads the second one by 90° pointing down). The wavelet power is $\log_2(A^2/v)$, where A is the wavelet amplitude, v is the variance of the original tidal signals. The y-axis in the continuous wavelet transformation (CWT) and cross-wavelet transformation (XWT) is on a $\log_2$ scale.

From XWT analysis, the phase difference between the tidal discharge and the water level at the western branch shows that the black arrows are pointing down at ~88.5°.

The phase difference in tidal discharge between the western and eastern branches was further examined to ensure that the tidal discharge phase of the eastern branch is greater than that of the western branch. Figure 9 confirms that the tidal discharge phase at the western branch is larger than the discharge phase at the eastern branch, with a phase difference of ~11°. Moreover, it also seems that the geometry of the branch has a greater effect on the behavior of the phase difference at the western branch compared with the eastern branch.

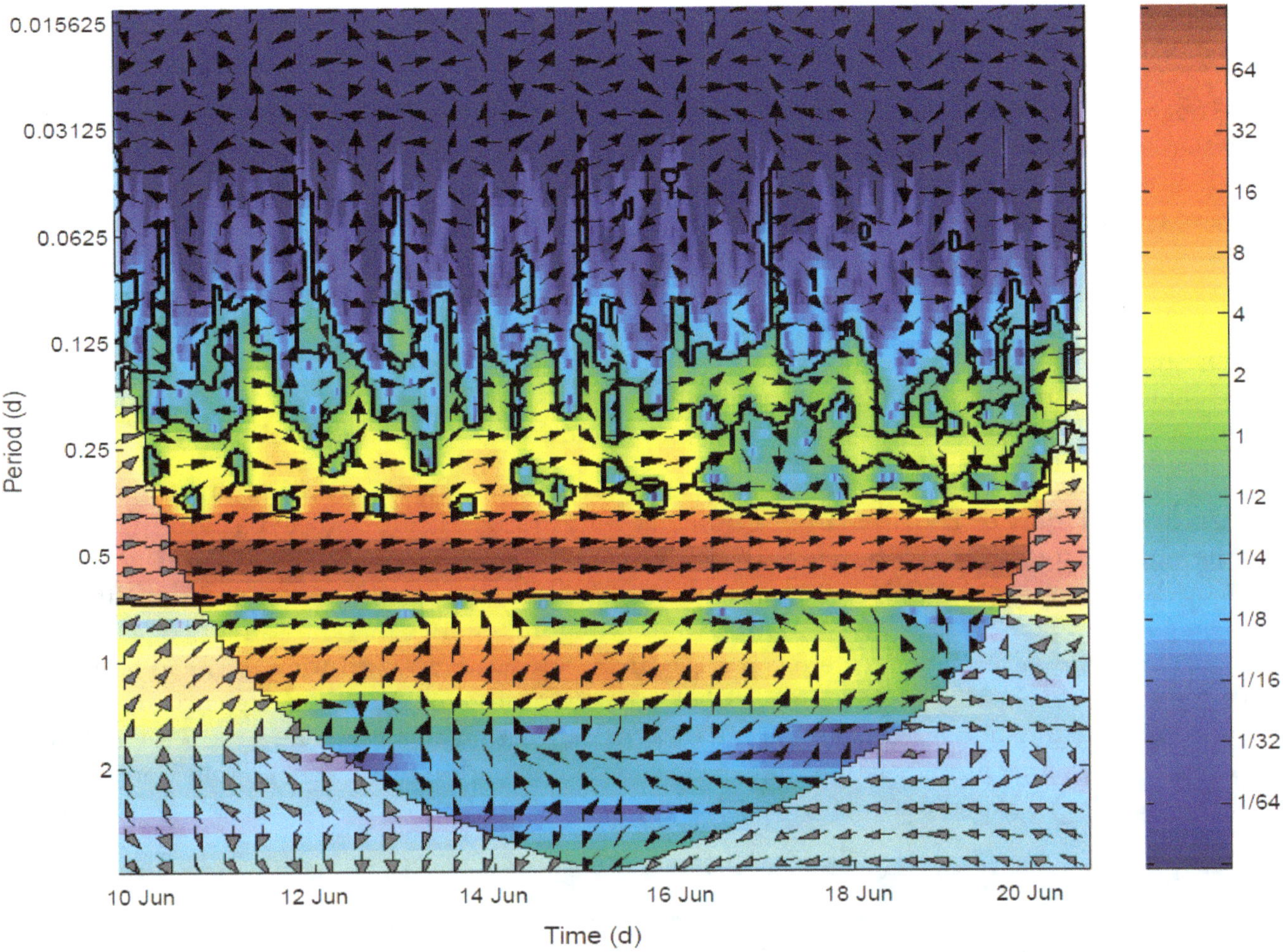

Figure 9. The phase difference in the tidal discharge phase between the eastern and western branches. The black contours represent the 95% confidence level for red noise, and the cone of influence (COI) where edge effects might distort the picture is shown as a lighter shade. The relative phase relationship is shown as arrows (with in-phase pointing right, anti-phase pointing left, and the first time series leading the second one by 90° pointing down). The wavelet power is $\log_2(A^2/v)$, where A is the wavelet amplitude and v is the variance of the original tidal signals. The y-axis is on a $\log_2$ scale.

As mentioned in Section 2.2.2, to gain a comprehensive understanding of phase difference behavior around the junction, an additional observation campaign at the northern branch using FATS is needed to collect the tidal discharge continuously. With this, the phase difference before and after the junction can be investigated completely (see Figure 1b).

The phase difference during semidiurnal is ~90°. This result confirms that the phase difference at the northern branch shows slightly increased behavior compared with the seaward branches (eastern and western branches). This result indicates that the behavior of phase difference is slightly increased after passing through the junction into the northern branch (a landward branch of the junction).

4. Discussion

4.1. Interpretation of Flow Division between Two Seaward Branches

From the bathymetry map presented in Figure 2, the depth and width of the eastern branch are greater than those of the western branch. However, the inequality of flow division at the tidal channel junction cannot be calculated using only a ratio of the wetted cross-sectional areas between the two branches [2], because the hydrodynamic process at the tidal channel junction is also strongly influenced by the spring and neap tide, as well as by the asymmetrical geometry shape between the two seaward branches.

The response of tidal dynamic to the local geometry between two seaward branches shows distinct characteristics, where the asymmetric discharge at the western branch is more pronounced compared with that at the eastern branch. The asymmetric tidal discharge observed at the western branch indicates the ebb discharge is larger than the flood discharge. On the contrary, the characteristic of tidal discharge at the eastern branch is nearly equal during the ebb and flood tide. As we can see in Figure 2, the western branch is shallower and narrower than the eastern branch. Therefore, the propagating tidal wave in shallow and narrow branches, and modulated with the river discharge, can generate the nonlinear effect as a result of interaction with the local topography [9]. This nonlinear effect leads to the generation and development of shallow-water constituents, which cause tidal distortion and asymmetry [25]. This asymmetric tidal discharge can induce a seaward sediment movement [26].

Regarding the flow division characteristic at the tidal channel junction of Ota River, our findings are depicted in Figure 6. The discharge asymmetry index ranges from −0.02 to 0.26 during the 10 to 21 June, suggesting that the eastern branch has the capability to deliver greater amounts of subtidal discharge, approximately 55–63% compared with the western branch. However, the flow division is nearly equal during the neap tide (i.e., 18 to 19 June). Figure 6 shows that the equality of asymmetry index between the eastern and western channels can be observed clearly during the neap tide period. Nevertheless, the inequality of flow division is obviously prominent during the spring tide duration. The fluctuation in the subtidal discharge at the eastern branch is greater than that at the western branch; the subtidal discharge at the eastern branch decreases during the neap tide, so that both subtidal discharges are nearly equal during the neap tide.

4.2. Interpretation of the Interaction between the Tidal Discharge and Water Level in the Three Branches

CWT analysis identifies the semidiurnal and diurnal signal in three branches connected to the junction. The quarter-diurnal signal can be detected, but is not dominant, because the signal energy is weak and discontinuous. This happened because the shallow water effect in the branch does not influence tidal distortion of D2, and thereby cannot produce enough energy to transform from D2 to D4 [19,25]. Moreover, the tidal channel junction is located in the middle of estuary, which is still near to the river mouth, implying that the tide needs to travel more in the landward direction to generate D4.

In contrast, the fortnightly tidal signal does not appear in the CWT analysis in the relationship between the tidal discharge and water level—possibly because of two reasons, that is, (i) the time-series data are too short for capturing fortnightly signal; or (ii) freshwater discharge is very low. For comparison with the work of previous researchers, for example, the work of Leonardi et al. [8], the fortnightly signal does not exist in their wavelet analysis over period of 1.5 months, with the freshwater discharge ranging from 110 to 8235 m^3s^{-1}. On the contrary, Sassi et al. [3] could recognize the fortnightly tide using longer time-series data over the period of six months, with the freshwater discharge ranging from 3000 to 5000 m^3s^{-1}. However, in our case, the time-series data are less than one month, with the freshwater discharge in the Ota estuary during normal condition ranging from 20 to 40 m^3s^{-1}. Thus, in the case of the Ota River, the absence of a fortnightly signal in the interaction between the water level and tidal discharge is the result of the short time-series data [19].

Two main questions arise about the characteristic of phase difference at the junction: (i) the first regarding the temporal variation of the phase difference between the two seaward branches (the eastern and western branches). As a follow-up question, what is the dominant factor that influences the phase difference between the two seaward branches? and (ii) is the phase difference influenced by the spring and neap tide?

To answer the first question, we must first determine the phase difference between the two branches. The phase difference between the tidal discharge and water level at the eastern and western branch for semidiurnal are ~78° and ~88.5°, indicating a mixed wave and mimic standing wave, respectively. It is important to note that the phase difference range of 78° to 88.5° implies that the tidal discharge leads the water level. The discharge phase difference between the eastern and western branches shows that the discharge phase at the western branch leads by ~15 min compared with the discharge phase at the eastern branch. In addition, the phase difference between the tidal velocity and water level for the eastern and western branches is ~71° and ~80°, respectively (figures not shown). Thus, the phase difference between the velocity and water level is consistent with the phase difference between the discharge and water level, where the phase difference in the western branch is larger than that in the eastern branch. We want to investigate further the reason that the phase difference at the western branch is higher than at the eastern branch. According to Savenije [21], the phase difference is strongly affected by the convergence shape effect, rather than river discharge and bottom friction. Accordingly, we quantify the convergence levels between the eastern and western branches using the exponential function form with three important parameters; namely, the mean water depth, river width, and cross-sectional area, as exemplified by Savenije et al. and Cai et al. [21,27].

Figure 10 shows the geometry of both the eastern and western branches. The geometry analysis presented in Figure 10 (right) obviously indicates that the western branch is more convergent compared with the eastern branch, as shown in Figure 10 (left), resulting in a larger phase at the western branch. From the results of phase difference characteristic and the geometry analysis between two seaward branches, it can be inferred that the amplitude of tidal wave at the western branch near the junction is more magnified compared with at the eastern branch [6,7].

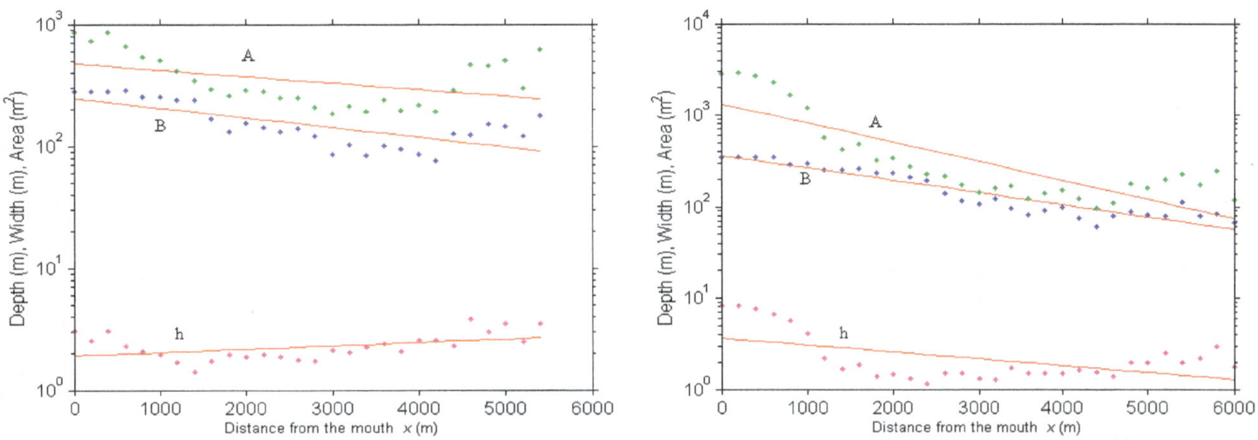

Figure 10. Geometry of the eastern branch (left) and western branch (right); showing the longitudinal variation of the cross-sectional area (A), the width (B), and the depth (h).

Figure 11 shows that the phase difference between the tidal discharge and water level at the northern branch (a landward branch of the junction) is ~90°, indicating a standing wave. Thus, it is important to note that the phase difference at the northern branch is slightly higher than that of the two seaward branches. This evidence indicates that the phase difference is slightly increased after passing through the junction into the northern branch. However, further investigation is needed to confirm this evidence.

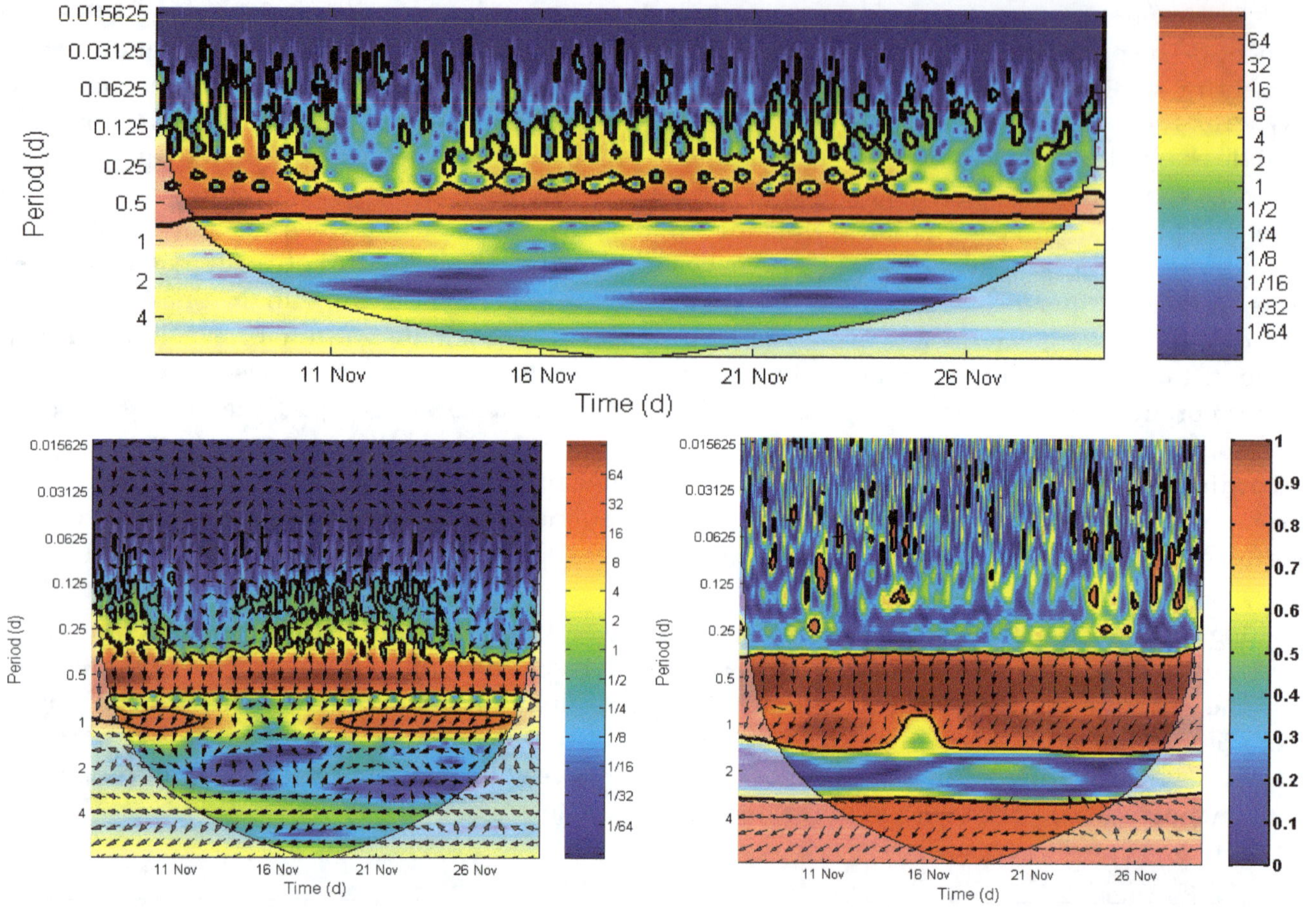

Figure 11. Wavelet analyses in the northern branch: (top) continuous wavelet transform; (bottom left) cross-wavelet transform; (bottom right) wavelet coherence. The black contours represent the 0.95 confidence level against red noise, and the cone of influence (COI) where edge effects might distort the picture is shown as a lighter shade. The relative phase relationship is shown as arrows (with in-phase pointing right, anti-phase pointing left, and the first time series leading the second one by 90° pointing down). The wavelet power is $\log_2(A^2/v)$, where A is the wavelet amplitude and v is the variance of the original tidal signals. The y-axis in the CWT and XWT is on a $\log_2$ scale.

The answer for the second question is that the effect of spring and neap tide on the phase difference in the semidiurnal domain is negligible. As shown in Figure 9, the phase difference between two seaward branches changes slightly during the spring and neap tide. The phase difference between the tidal discharge and water level is relatively constant, where during the spring tide, the phase difference is ~11°, whereas during the neap tide, the phase difference is ~9°. Therefore, the change of the phase difference between spring and neap tide is not significant. This phenomenon is probably because the phase difference is a function of a ratio between bank convergence and tidal wave length [21]. Moreover, the phase difference can be affected by two factors, that is, the river discharge [8,21] and topography of the estuarine channel, also known as estuary shape [21].

The upstream discharge from Yaguchi gauging station during the experiment is relatively constant, ranging from 20 to 40 m^3s^{-1} during June 2017, though the freshwater discharge from Yaguchi gauging station increased slightly from 20 to 90 m^3s^{-1} in November 2017. Thus, during this condition, the tidal processes of the Ota River can be classified as a tidally dominated estuary [8]. Additionally, this finding is still more or less consistent with the work of Leonardi et al. [8], who pointed out that the phase difference always tends to be close to 90° as long as the characteristic of flow is bidirectional.

5. Conclusions

Temporal variations of streamflow were successfully measured by FATS instruments at three branches connected to a tidal channel junction during a low-flow condition. To investigate the flow division of subtidal discharge between the two seaward branches (eastern and western branches), the discharge asymmetry index was used. Moreover, to characterize the tidal discharge behavior and its relationship with the water level and modulated with the freshwater discharge, the phase difference is investigated.

The response of tidal dynamic to the local geometry between two seaward branches shows that the asymmetric discharge at the western branch is more pronounced compared with the eastern branch. The asymmetric tidal discharge at the western branch indicates the ebb discharge is larger than the flood discharge. This happens because the interaction between tidal propagation and the local geometry of western branch that modulated by river discharge could generate the nonlinear effect. This nonlinear effect leads to generation and development of shallow-water constituents, which cause tidal distortion and asymmetry. This asymmetric tidal discharge can induce a seaward sediment movement. In contrast, the characteristic of tidal discharge at the eastern branch is nearly equal during the ebb and flood tide.

Regarding the flow division characteristic at the tidal channel junction of Ota River, the discharge asymmetry index varies from −0.02 to 0.26 during the studied period, suggesting that the eastern branch has the capability to deliver greater amounts of subtidal discharge, approximately 55–63% compared with the western branch, except during the neap tide where the flow division is nearly equal between the eastern and western branch. The temporal change in the flow division is induced by the fluctuation in the subtidal discharge at the eastern branch. The subtidal discharge decreases during the neap tide, so that both subtidal discharges are nearly equal.

The wavelet analyses of the temporal variation of the phase differences between the two seaward branches show slightly different behavior. A mimic standing wave characteristic ($\Delta\varphi = \sim 88.5°$) occurs at the western branch, whereas a mixed wave characteristic ($\Delta\varphi = \sim 78°$) occurs at the eastern branch. It is shown that the phase difference between tidal discharge and the water level is relatively constant during the neap and spring tide. The discharge phase at the western branch leads the discharge phase at the eastern branch by ~15 min. Additionally, the geometry analysis between the two seaward branches reveals that the western branch is more convergent compared with the eastern branch, and thus causes a larger phase difference compared with the eastern branch. As a result, the amplitude of tidal wave at the western branch near the junction is more magnified compared with the eastern branch.

Using the same analysis, we can say the phase difference between the tidal discharge and water level at the northern branch (a landward branch of the junction) indicates a standing wave characteristic ($\Delta\varphi = \sim 90°$), which is slightly higher than the phase difference in the two seaward branches. This evidence implies that the phase difference is slightly increased after passing through the junction into the northern branch. Further study is needed to explore the characteristics of the phase difference in the larger area of the multi-channel estuary.

Although this study deals with the Ota River as a case-specific example, the presented results and discussion are not only aimed at demonstrating the dynamic of flow division and relationship between tidal discharge and water level at a junction, but also take advantage of utilizing the innovative hydro-acoustic system (FATS) for measuring discharge in high temporal resolution and exploring the hydrodynamic process at a tidal channel junction. Moreover, our findings should increase the knowledge for common hydrodynamic processes in multi-channel estuaries that have nearly similar environments over the world.

Author Contributions: Conceived the study, M.M.D.; Data curation, M.M.D. and M.B.S.; Validation, M.M.D.; Supervision, K.K.; Writing—original draft, M.M.D.; Writing—review & editing, M.B.S. and K.K.

Acknowledgments: The authors are grateful to Noriaki Gohda of the Hiroshima University/Aqua Environmental Monitoring Limited Liability Partnership (AEM-LLP) for his technical support. Our thanks are also addressed to the members of the Department of Civil and Environmental Engineering, Hiroshima University, for their assistance with the collection of field data in the estuary.

References

1. Edmonds, D.A.; Slingerland, R.L. Stability of delta distributary networks and their bifurcations. *Water Resour. Res.* **2008**, *44*, 1–13. [CrossRef]
2. Buschman, F.A.; Hoitink, A.J.F.; Van Der Vegt, M.; Hoekstra, P. Subtidal flow division at a shallow tidal junction. *Water Resour. Res.* **2010**, *46*, 1–12. [CrossRef]
3. Sassi, M.G.; Hoitink, A.J.F.; de Brye, B.; Vermeulen, B.; Deleersnijder, E. Tidal impact on the division of river discharge over distributary channels in the Mahakam Delta. *Ocean Dyn.* **2011**, *61*, 2211–2228. [CrossRef]
4. Zhang, W.; Feng, H.; Hoitink, A.J.F.; Zhu, Y.; Gong, F. Estuarine, Coastal and Shelf Science Tidal impacts on the subtidal fl ow division at the main bifurcation in the Yangtze River Delta. *Estuar. Coast. Shelf Sci.* **2017**, *196*, 301–314. [CrossRef]
5. Horrevoets, A.C.; Savenije, H.H.G.; Schuurman, J.N.; Graas, S. The influence of river discharge on tidal damping in alluvial estuaries. *J. Hydrol.* **2004**, *294*, 213–228. [CrossRef]
6. Savenije, H.H.G.; Veling, E.J.M. Relation between tidal damping and wave celerity in estuaries. *J. Geophys. Res.* **2005**, *110*, 1–10. [CrossRef]
7. Savenije, H.H.G.; Toffolon, M.; Haas, J.; Veling, E.J.M. Analytical description of tidal dynamics in convergent estuaries. *J. Geophys. Res.* **2008**, 1–27. [CrossRef]
8. Leonardi, N.; Kolker, A.S.; Fagherazzi, S. Advances in Water Resources Interplay between river discharge and tides in a delta distributary. *Adv. Water Resour.* **2015**, *80*, 69–78. [CrossRef]
9. Nidzieko, N.J. Tidal asymmetry in estuaries with mixed semidiurnal/diurnal tides. *J. Geophys. Res. Ocean.* **2010**, *115*, 1–13. [CrossRef]
10. Gotoh, T.; Fukuoka, S.; Tanaka, R. *Evaluation of Flood Discharge Hydrographs and Bed Variations in a Channel Network on the Ota River Delta*, 357th ed.; Chavoshian, A., Takeuchi, K., Eds.; IAHS: Wallingford, UK, 2013; ISBN 9781907161353.
11. Kawanisi, K.; Al Sawaf, M.B.; Danial, M.M. Automated real-time stream flow acquisition in a mountainous river using acoustic tomography. *J. Hydrol. Eng.* **2018**, *23*, 1–7. [CrossRef]
12. Razaz, M.; Kawanisi, K.; Nistor, I.; Sharifi, S. An acoustic travel time method for continuous velocity monitoring in shallow tidal streams. *Water Resour. Res.* **2013**, *49*, 4885–4899. [CrossRef]
13. Kawanisi, K.; Razaz, M.; Yano, J.; Ishikawa, K. Continuous monitoring of a dam flush in a shallow river using two crossing ultrasonic transmission lines. *Meas. Sci. Technol.* **2013**, *24*, 055303. [CrossRef]
14. Bahreinimotlagh, M.; Kawanisi, K.; Danial, M.M.; Al Sawaf, M.B.; Kagami, J. Application of shallow-water acoustic tomography to measure flow direction and river discharge. *Flow Meas. Instrum.* **2016**, *51*, 30–39. [CrossRef]
15. Sassi, M.G.; Hoitink, A.J.F.; Vermeulen, B.; Hidayat, H. Discharge estimation from H-ADCP measurements in a tidal river subject to sidewall effects and a mobile bed. *Water Resour. Res.* **2011**, *47*, 1–14. [CrossRef]
16. Levesque, V.A.; Oberg, K.A. *Computing Discharge Using the Index Velocity Method*; Techniques and Methods 3-A23; U.S. Geological Survey: Reston, VA, USA, 2012; Volume 148.
17. Hidayat, H.; Hoitink, A.J.F.; Sassi, M.G.; Torfs, P.J.J.F. Prediction of Discharge in a Tidal River Using Artificial Neural Networks. *J. Hydrol. Eng.* **2014**, *19*, 1–8. [CrossRef]
18. Kempema, E.; Stiver, J.; Ettema, R. *Effects of Warm CBM Product Water Discharge on Winter Fluvial and Ice Processes in the Powder River Basin*; 2010; Available online: http://www.uwyo.edu/owp/_files/project27finalreport.pdf (accessed on 23 April 2017).
19. Guo, L.; Van Der Wegen, M.; Jay, D.A.; Matte, P.; Wang, Z.B.; Roelvink, D.; He, Q. River-tide dynamics: Exploration of nonstationary and nonlinear tidal behavior in the Yangtze River estuary. *J. Geophys. Res. C Ocean.* **2015**, *120*, 3499–3521. [CrossRef]
20. Grinsted, A.; Moore, J.C.; Jevrejeva, S. Application of the cross wavelet transform and wavelet coherence to geophysical time series. *Nonlinear Process. Geophys.* **2004**, *11*, 561–566. [CrossRef]
21. Savenije, H.H.G. *Salinity and Tides in Fluvial Estuaries*, 1st ed.; Elsevier Science: Amsterdam, The Netherlands,

2005; ISBN 9780444521071.

22. Friedrichs, C.T.; Aubrey, D.G. Tidal propagation in strongly convergent channels. *J. Geophys. Res.* **1994**, *99*, 3321–3336. [CrossRef]
23. Cai, H.; Savenije, H.H.G.; Toffolon, M. Linking the river to the estuary: Influence of river discharge on tidal damping. *Hydrol. Earth Syst. Sci.* **2014**, *18*, 287–304. [CrossRef]
24. Jay, D.A. Green's law revisited: Tidal long-wave propagation in channels with strong topography. *J. Geophys. Res.* **1991**, *96*, 20585. [CrossRef]
25. Lu, S.; Tong, C.; Lee, D.; Zheng, J.; Shen, J.; Zhang, W.; Yan, Y. Propagation of tidal waves up in Yangtze Estuary during the dry season. *J. Geophys. Res. Ocean.* **2015**, *120*, 6445–6473. [CrossRef]
26. Guo, L.; Van der Wegen, M.; Roelvink, J.A.; He, Q. The role of river flow and tidal asymmetry on 1-D estuarine morphodynamics. *J. Geophys. Res. Earth Surf.* **2014**, *119*, 2315–2334. [CrossRef]
27. Cai, H.; Toffolon, M.; Savenije, H.H.G.; Yang, Q.; Garel, E. Frictional interactions between tidal constituents in tide-dominated estuaries. *Ocean Sci.* **2018**, *14*, 769–782. [CrossRef]

7

Experimental Optimization of Gate-Opening Modes to Minimize Near-Field Vibrations in Hydropower Stations

Yong Peng [1], Jianmin Zhang [1,*], Weilin Xu [1] and Matteo Rubinato [2]

[1] State Key Laboratory of Hydraulics and Mountain River Engineering, Sichuan University, Chengdu 610065, China; pengyongscu@foxmail.com (Y.P.); xuwl@scu.edu.cn (W.X.)

[2] Civil and Structural Engineering Department, University of Sheffield, Sheffield S1 3JD, UK; m.rubinato@sheffield.ac.uk

* Correspondence: zhangjianmin@scu.edu.cn.

Abstract: Multi-Horizontal-Submerged Jets are successfully applied to dissipate energy within a large-scale hydropower station. However, notable near-field vibrations are generated when releasing high discharges through the gates, which is generally typical in a flooding case scenario. Under these conditions, the magnitude of the vibrations varies when applying different gate-opening modes. To investigate and find optimized gate-opening modes to reduce the near-field vibration, multiple combinations were tested by varying gate-opening modes and hydraulic conditions. For each of the tests conducted, fluctuating pressures acting on side-walls and bottoms of a stilling basin were measured. The collected datasets were used to determine the maximum and minimum fluctuating pressure values associated with the correspondent gate-opening mode and a detailed comparison between each of the gate-opening modes was completed. The paper presents the quantitative analysis of the discharge ratio's effect on fluctuating pressures. It also investigates the influence of different gate-opening modes by including side to middle spillways and upper to lower spillways configurations. The flow pattern evolutions triggered by each different gate-opening mode are discussed and optimal configurations that minimize near-field vibrations at high discharges are recommended to support both the design of new systems and assessment of the performance of existing ones.

Keywords: multi-horizontal-submerged jets; energy dissipation; near-field vibration; fluctuating pressure; gate-opening modes

---

1. Introduction

Since 2000, over 300 hydropower projects have been completed or are under construction in China [1]. Flood discharge and energy dissipation are crucial to maintaining the safety of these hydropower installations. Furthermore, lessons must be drawn from the past dam disasters, e.g., Vajont Dam [2,3] and Banqiao Dam [4]. Decades ago, hydropower stations were not constructed in proximity to residential areas, and as a result, research at that time mainly focused on the stability and the functionality of the hydraulic structures. However, in recent years, due to increasing urbanization [5], some large hydropower stations had to be located in the proximity of residential areas. Moreover, due to the climatic scenario [6], it is likely that in the future there will be numerous and more frequent intense rainfall events, and consequently large volumes of water may need to be released from dam reservoirs. Therefore, it is essential to provide solutions to a need for increasing the stability of these hydraulic structures and surrounding areas, minimizing environmental impacts due to extreme flow discharges [7].

The purpose of this work is to optimize a recently developed novel technique for energy dissipation, Multi-Horizontal-Submerged Jets (simplified as MHSJ). MHSJ have been studied and improved by many researchers in recent years [8–17]. It has been demonstrated that MSHJ possesses three main advantages when compared with the hydraulic jump, which is the traditional energy dissipater: (1) lower atomization; (2) a higher rate of energy dissipation; (3) more flexibility of gate-opening modes [12–14]. Thus, it seems feasible to implement MHSJ to tackle the problem of energy dissipation within large hydropower stations that are typically characterized by higher water heads and larger flowrates. The adoption of MHSJ was tested by Deng et al. [12] and successfully applied as energy dissipation in large-scale hydropower stations [14].

However, despite positive progress on solving energy dissipation and erosion problems, there is still a lack of studies on the considerable flow-induced vibrations that occur in this kind of structure and propagate towards the near-field when releasing the greater flowrates. These near-field vibrations can be considered a serious threat for the stability of the structure, and hence this phenomenon needs to be further investigated. Yin and Zhang [18] and Yin et al. [19] studied the relationship between the discharge flow and the induced vibrations and simulated the results with a finite element method. Yin et al. [19] found that at higher flowrates during flooding situations, the flow is complex and unsteady, and this is the main cause of near-field vibrations. Therefore, if the vibrations are caused by releasing higher flowrates, optimizing gate-opening modes could reduce this effect and near-field vibrations could decrease correspondingly. To date, this issue has not been investigated in detail and to address this gap, this paper presents an experimental study to explore the relationship between complex flow scenarios and gate-opening modes in order to recommend an optimal solution to minimize the formation of vibrations.

2. Methods

The experimental facility, shown in Figure 1, was constructed based on a large-scale hydropower station (Xiangjiaba hydropower station. It sits on the Jinsha River, a tributary of the Yangtze River in Yunnan Province and Sichuan Province, Southwest China). The experimental facility follows Froude similarity, i.e., the ratio between inertial and gravitational forces remains constant, with the scale of 1:80. The experimental model is composed of the upper and lower reaches of the river, a dam, mid-level channels, spillways and two stilling basins with identical layout (the difference between two stilling basins is the their location on the dam, being parallels). The single stilling basin includes 5 mid-level channels and 6 spillways. Both of mid-level channels and spillways were arranged alternately and division piers were located between them.

In the model, the outlets of spillways are horizontal with symmetrical contraction of 1 m and width of 8 m. Also, the mid-level channels' outlets are horizontal, without any contraction, and their width is 6 m. The division piers are 3 m wide. The offset of mid-level channels, spillways, and division piers from the bottom of the stilling basin are 8 m, 16 m, 26 m, respectively. The test section (L = 15 m, W = 2 m and H = 1 m) is made of plexiglass. The modelled stilling basin is 3 m long and a gate is adopted to control tail-water level downstream of the model. To measure the fluctuating pressure within the system, 135 measurement taps for pressure transducers are located in the right-side wall and bottom of the stilling basin, as well as in the right-side wall and bottom of spillway (1# spillway in Figure 1) and mid-level channel (1# mid-level channel in Figure 1). For each test, the fluctuating pressure was measured synchronously by sensors for pressure transducers at each measurement point. The accuracy of the pressure sensors utilized in this study is 0.1%, with a 0–50 kPa working range and a frequency response of 100 Hz.

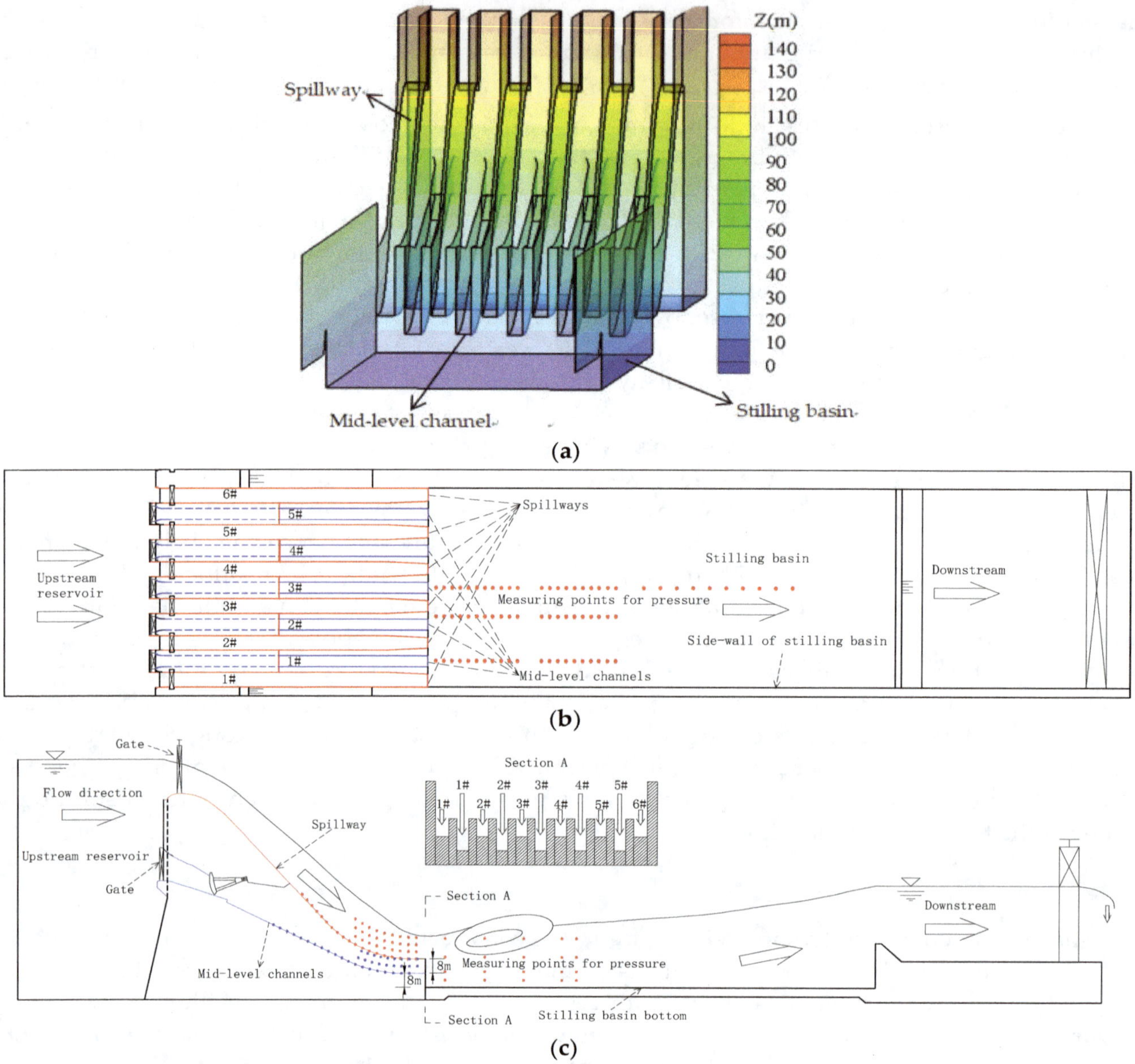

Figure 1. Sketch of experimental arrangement (single stilling basin). (**a**) 3D view of gate session; (**b**) plan view; (**c**) side view.

Multiple experimental flooding scenarios were completed within the experimental facility, varying flowrate and changing gate-opening modes. The area under investigation includes the: (i) stilling basin bottom (BB); (ii) stilling basin side-wall (BS); and (iii) piers, spillways and mid-level channels (PSO). Discharges considered varied gradually between 800 and 24,500 m^3/s. The downstream water elevation was measured within the model for each test.

3. Results and Discussion

3.1. Relation between Fluctuating Pressure and Flood Discharge

After calculating the root-mean-square (RMS) of the fluctuating pressure measured for each hydraulic condition tested, the arithmetic mean of RMS values obtained for every part of the BB, BS and PSO were calculated individually and these final values are used in the following sections.

Figure 2 shows (a) the maximum and minimum RMS of fluctuating pressure on BB against the flood discharge scenario and (b) the gate-opening modes considered. In Figure 2a, the best fitting line and its correspondent equation are displayed. The configuration of gate-opening modes has a significant impact on the magnitude of the fluctuating load acting on the BB during the release of flows in the range of 2000 to 12,000 m^3/s. This is a clear indicator that the fluctuating load can be reduced effectively by optimizing gate-opening modes. As a consequence, the near-field vibrations induced by releasing higher flow associated with flooding scenario can also be reduced. However, this cannot be confirmed when the flood discharge increases to 15,000 m^3/s into a single stilling basin (as shown in Figure 2a) due to the limited adjustment of the gate-opening modes available within this experimental facility. Additionally, Figure 2b highlights that the use of side spillways seems favorable to reduce fluctuating pressure on the BB.

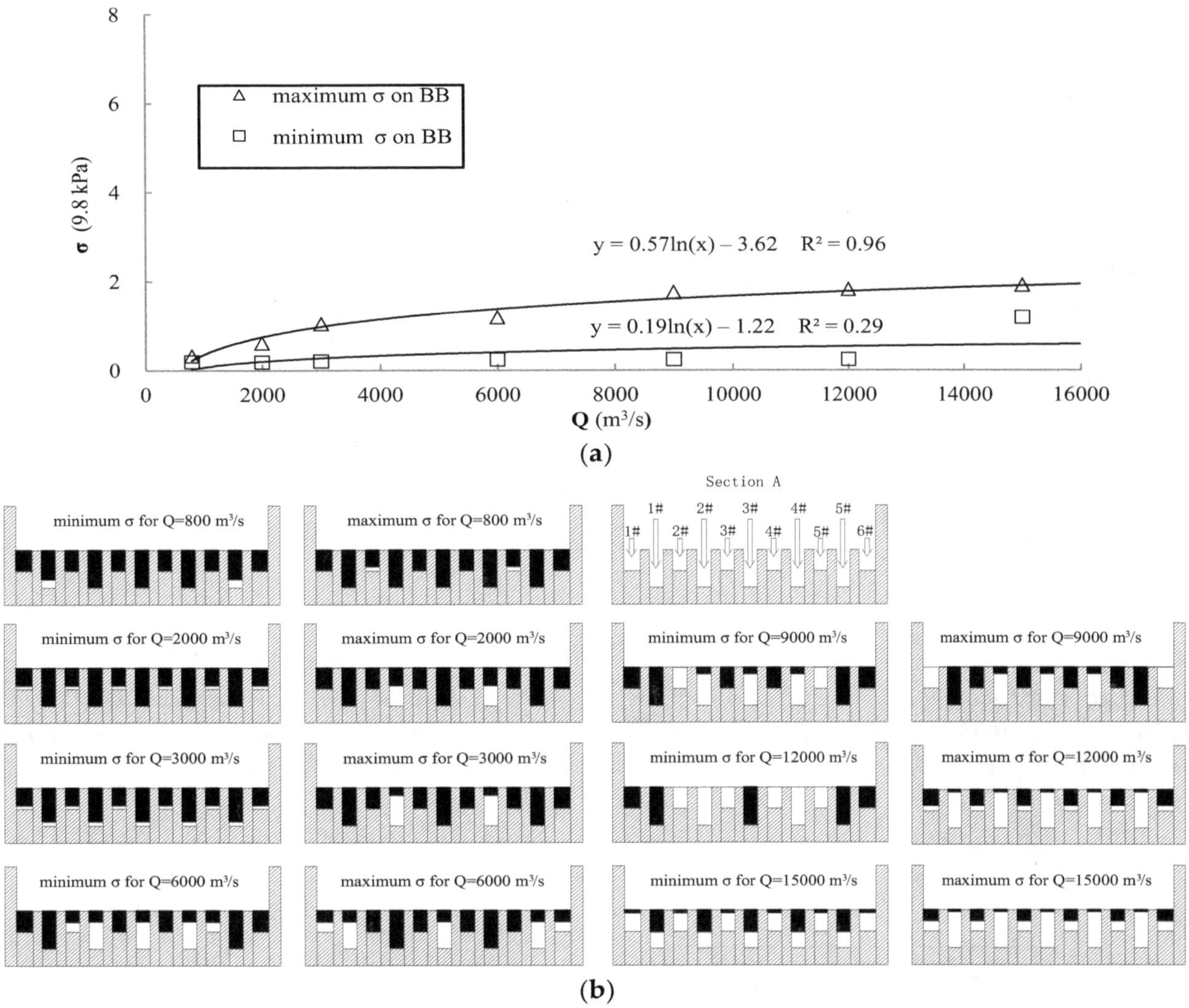

Figure 2. (**a**) Maximum and minimum RMS of fluctuating pressure on BB and (**b**) corresponding gate opening modes with flood discharge. (Black indicates closed and white indicates open).

Figure 3 displays the maximum and minimum RMS of fluctuating pressure on the BS during flood discharge scenarios, similar to Figure 2. Outcomes displayed in Figure 3 are clear indicators that it is beneficial to reduce pressure values in the BS by reducing the discharge associated with the side spillways as they are directly connected with BS, and hence cause a higher impact. For example, the maximum is more than 7 times higher than the minimum on BS, with Q of 12,000 m^3/s.

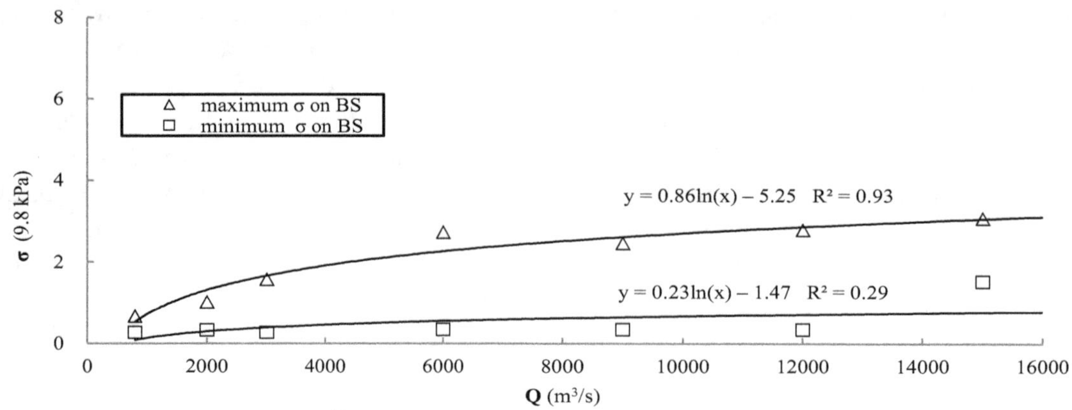

(a)

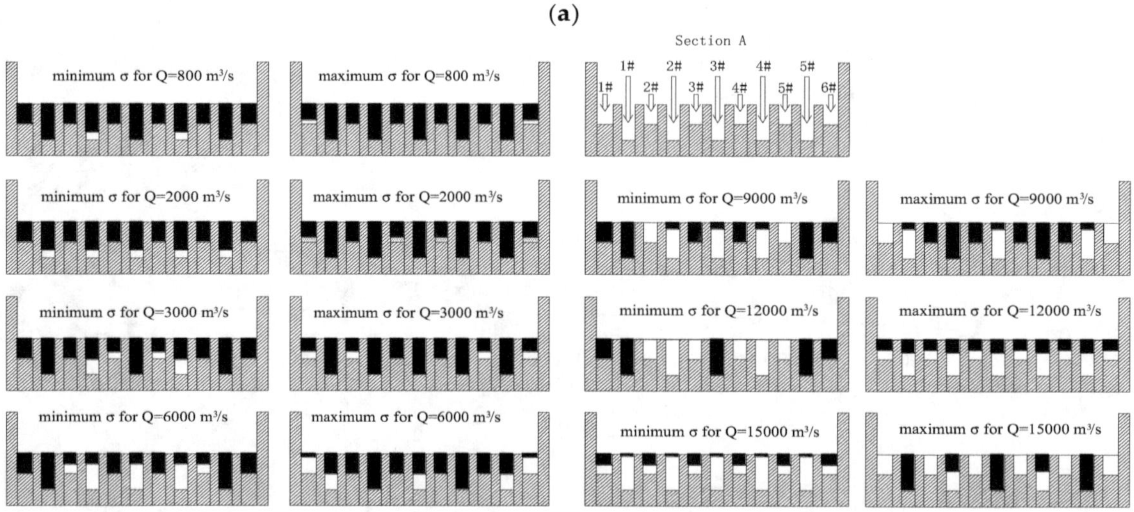

(b)

Figure 3. (**a**) Maximum and minimum RMS of fluctuating pressure on BS and (**b**) corresponding gate-opening modes with flood discharge. (Black indicates closed and white indicates open).

Figure 4 indicates maximum and minimum RMS of fluctuating pressure on the PSO and corresponding gate opening modes. Although fluctuating pressure on the PSO is larger than those on the BB and BS, the main sources of near-field vibration are the BB and BS. The effect of PSO is relatively small according to Yin and Zhang [18] and Yin et al. [19].

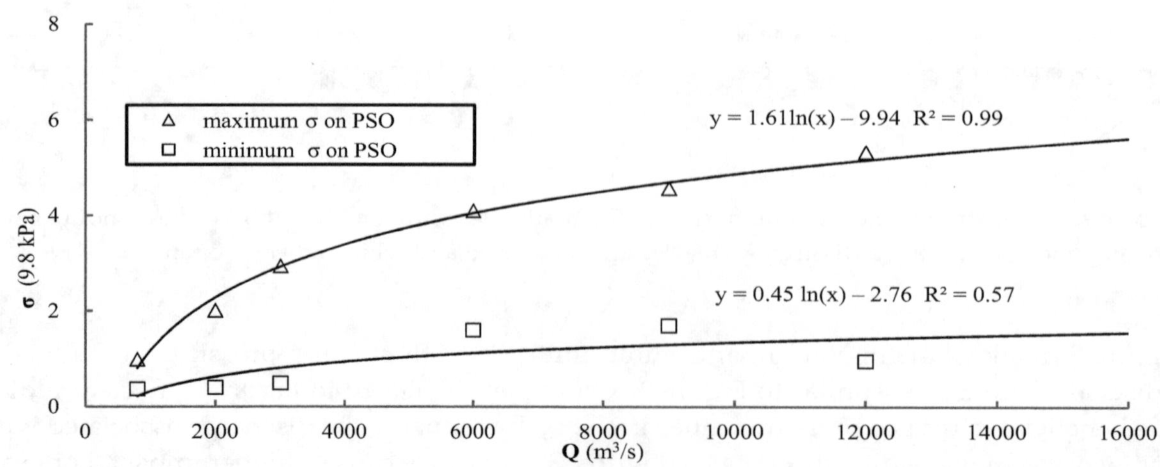

(a)

Figure 4. *Cont.*

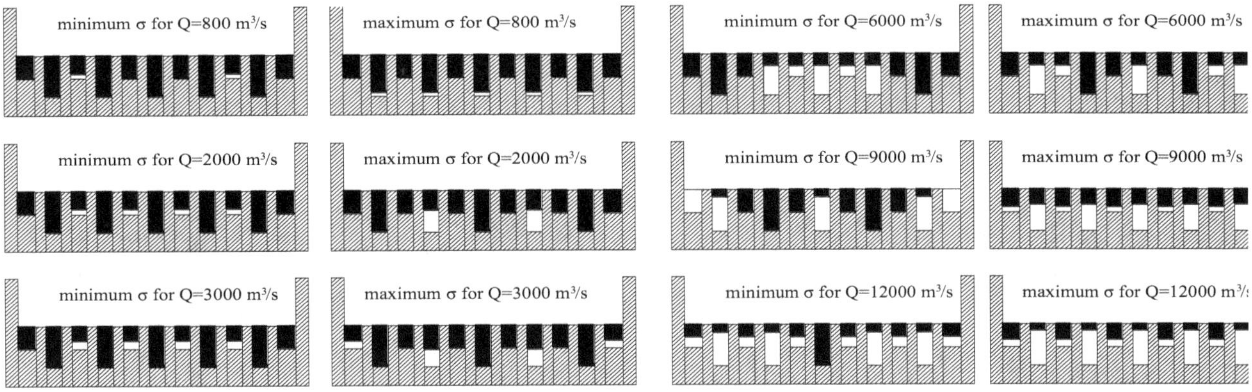

Figure 4. (**a**) Maximum and minimum RMS of fluctuating pressure on PSO and (**b**) corresponding gate-opening modes with flood discharge. (Black indicates closed and white indicates open).

3.2. *Effect of Flow Discharge Ratio and Configuration Adopted to Release the Flow on Fluctuating Pressure*

To quantitatively study the effect of different flowrates, more experiments were carried out. Specifically, three kinds of gate-opening modes were tested at higher flowrates typical of flooding scenarios: (i) mid-level channels separately, (ii) spillways separately, and (iii) combining both previous cases (i) and (ii). Based on measured fluctuating pressure and flow pattern evolution of these three different configurations, recommended gate-opening modes are proposed for various flood discharge rates.

The following notations are used to simplify the understanding of outcomes described in the upcoming sections: Q is total flood discharge for single stilling basin (m^3/s); Q_u and Q_d are the total discharge of spillways and mid-level channels, respectively (m^3/s); Q_{ds} and Q_{dm} are the total discharge of two side mid-level channels (1# and 5# as shown in Figure 1) and three middle ones (2–4#), respectively (m^3/s); Q_{us} and Q_{um} are total flood discharge of two side spillways (1# and 6#) (m^3/s) and four middle ones (2–5#) (m^3/s).

Figure 5 illustrates the effect of the ratio Q_{ds}/Q_{dm} on fluctuating pressure on the BB and BS. It indicates that fluctuating pressure tends to increase on the BS and to reduce on the BB as Q_{ds} increases slightly if the mid-level channels are used separately. This result suggests that mid-level channels should be opened uniformly to avoid this disruption.

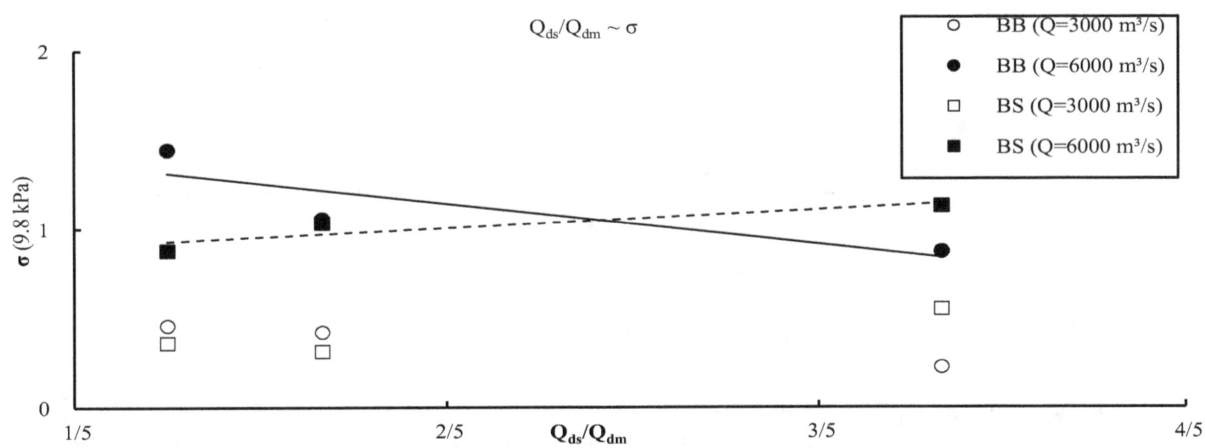

Figure 5. Effect of the ratio Q_{ds}/Q_{dm} on RMS of fluctuating pressure on BB and BS.

Figure 6 shows the effect of the ratio Q_{us}/Q_{um} on fluctuating pressure on BB and BS. By slightly increasing Q_{us}, if spillways are used separately, Figure 6 shows that the corresponding fluctuating pressure generated increases on the BS and reduces on the BB. This effect is consistent if applied to the mid-level channels configuration. Therefore, the same as the mid-level channels, spillways should be opened uniformly to reduce the magnitude of pressure on the BB.

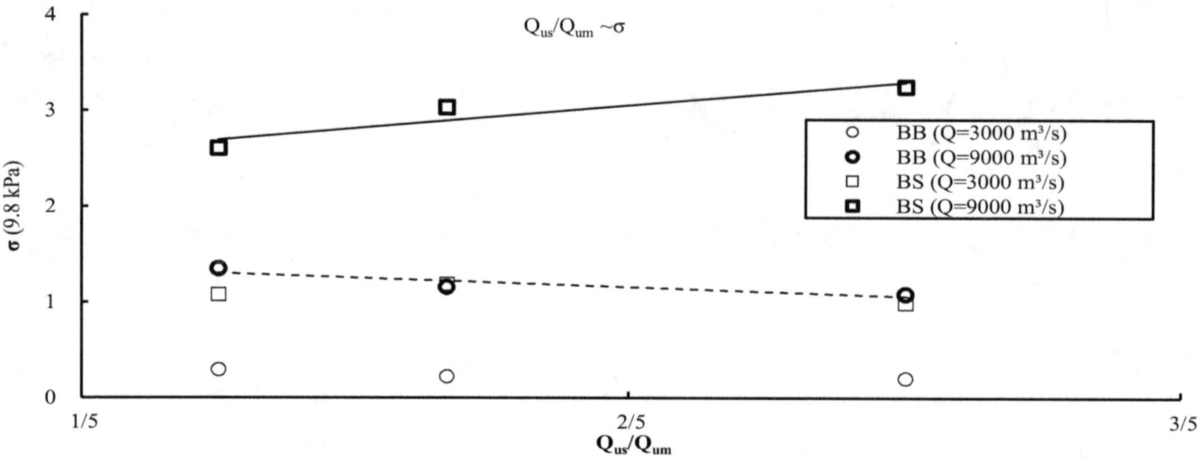

Figure 6. Effect of the ratio Q_{us}/Q_{um} on RMS of fluctuating pressure on BB and BS.

Figure 7 displays the effect of the ratio Q_u/Q_d on fluctuating pressure on the PSO, BB, and BS. When combination of the spillways and mid-level channels is adopted, fluctuating pressure increases on the BS and reduces on the BB and PSO as Q_u increases maintaining constant Q. Additionally, by applying this configuration, fluctuating load varies more rapidly on the PSO, and this effect is clearer as Q increases. When the ratio Q_u/Q_d is between 2 and 3, values of fluctuating pressure in all of three areas of study are relatively small.

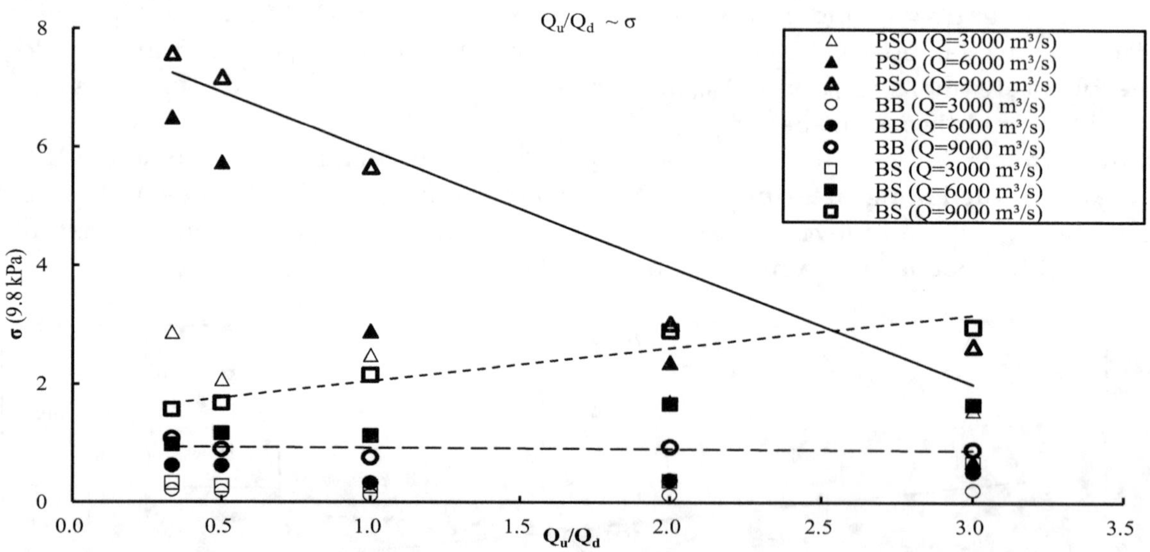

Figure 7. Effect of the ratio Q_u/Q_d on RMS of fluctuating pressure on BB, BS and PSO.

3.3. Effect of Flow Discharge Ratio and Configuration Adopted to Release the Flow on Flow Patterns in the Stilling Basin

In this section, the flow patterns in the stilling basin generated by the different hydraulic conditions and gate-opening configurations tested are discussed. As observed by previous studies [10], there is a correlation between the flow fluctuations and the flow patterns in the stilling basin. This explains why there is a significant difference, considering the same discharge flowrate, between fluctuating pressure on the PSO, on BB, and on BS. Figure 8 displays some examples of flow patterns observed in the stilling basin when the mid-level channels were used separately for diverse discharge flowrates. This is an indicator that various jets tend to quickly merge together and are likely to swing in the stilling basin if two or three of the middle mid-level channels are opened symmetrically. The formation of these flow patterns generates an increase in fluctuating pressure on the side-walls of the stilling basin. This phenomenon is mainly due to the outlet' width of each mid-level channel that is much smaller than the width of the stilling basin. When replicating this hydraulic condition of using mid-level channels separately, a backflow characterized by an "S" shape emerges in the stilling basin. The flow stability is relative good in the case of combination of 1#, 3#, 5# orifices (Figure 8c), or 1#, 2#, 4#, 5# orifices (Figure 8d), but the flow is not stable when all 1–5# orifices (Figure 8f) are fully open. This result suggests that mid-level channels can only be used individually for small flowrates during flooding events, and mid-level channels should only be opened symmetrically.

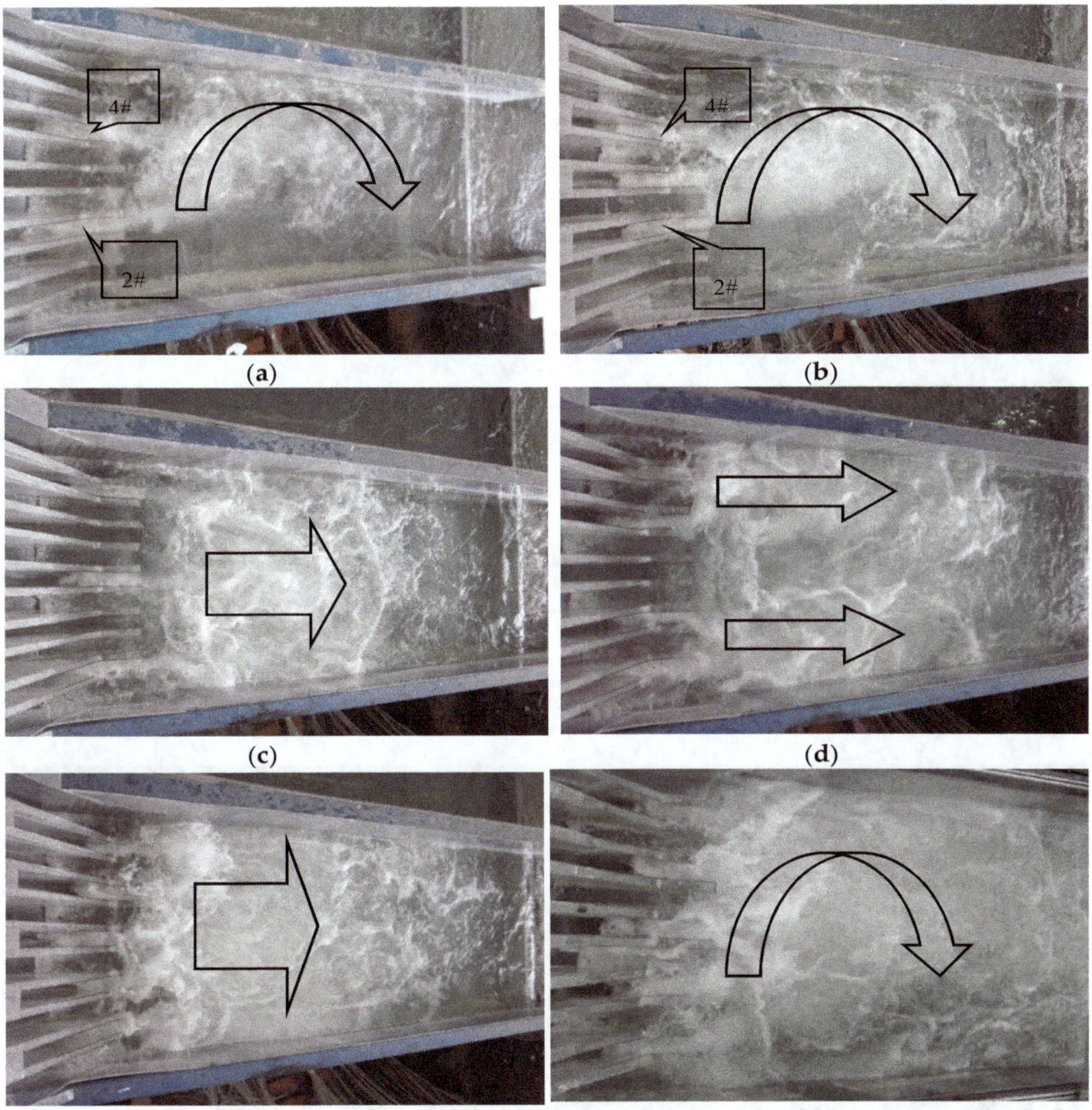

Figure 8. Flow pattern evolution in the stilling basin with opening of mid-level channels separately for different flow rates. (**a**) Opening partly of 2# and 4# $Q = 2000\ m^3/s$; (**b**) Opening partly of 2#, 3# and 4# $Q = 4000\ m^3/s$; (**c**) Opening fully of 1#, 3# and 5# $Q = 5300\ m^3/s$; (**d**) Opening fully of 1#, 2#, 4# and 5# $Q = 7000\ m^3/s$; (**e**) Opening partly of 1–5#, $Q = 7000\ m^3/s$; (**f**) Opening fully of 1–5#, $Q = 8874\ m^3/s$.

Figure 9 displays flow patterns that tend to develop when spillways are opened separately. It shows that the flow pattern is stable when the spillways are uniformly opened and separately regardless of the gate-opening height. Moreover, flow patterns are still stable if the middle spillways are opened symmetrically by closing side spillways. Finally, Figure 10 shows flow patterns generated when combining discharging orifices and spillways. In this case, flows discharged and flow in the stilling basin fully mix, and the flow patterns produced confirm the stability of the flow in the stilling basin. This stability continues until the outlet section of the stilling basin.

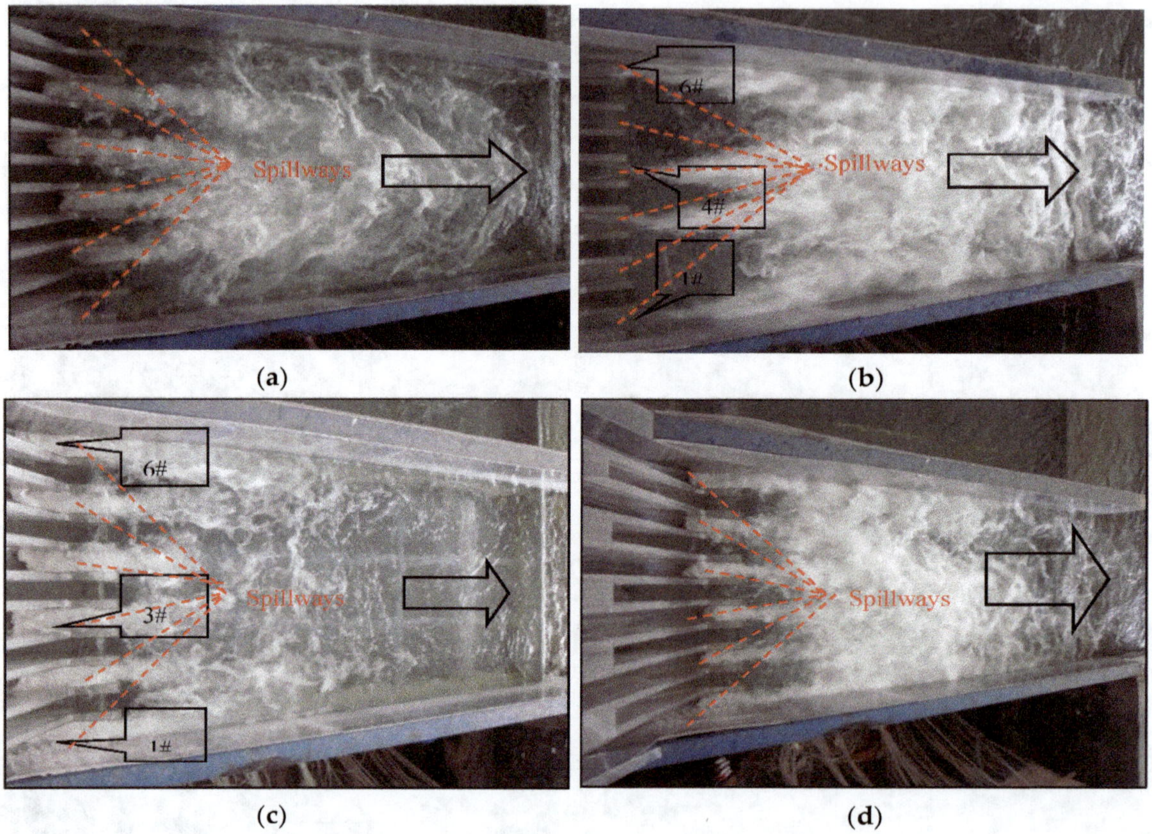

Figure 9. Flow pattern evolution in the stilling basin with opening of spillways separately for different flowrates. (**a**) Opening fully of 2–5#, $Q = 4000$ m^3/s; (**b**) Opening fully of 1#, 3#, 4#, 6#, $Q = 8634$ m^3/s; (**c**) Opening partly of 1–6#, $Q = 2000$ m^3/s; (**d**) Opening fully of 1–6#, $Q = 5868$ m^3/s.

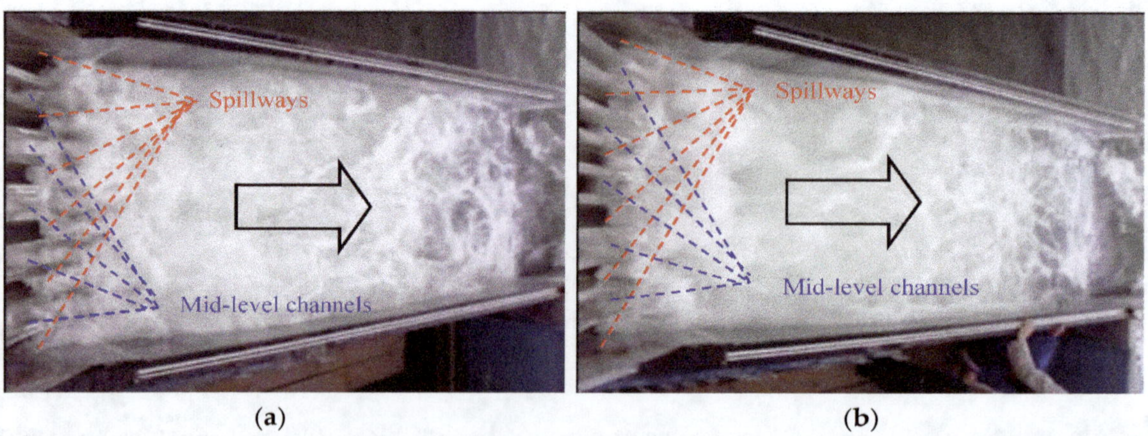

Figure 10. Flow pattern evolution in the stilling basin for combining mid-level channels with spillways for different flowrates. (**a**) Opening fully of 1–6# spillways and partly of 1–5# mid-level channels $Q = 17{,}800$ m^3/s, (**b**) Opening fully of 1–6# spillways and 1–5# mid-level channels $Q = 24{,}200$ m^3/s.

To summarize, flow patterns are stable in the stilling basin when combinations of mid-level channels and spillways are applied. The lack of clear and repeatable flow patterns in the stilling basin generated under different scenarios when mid-level channels are opened separately suggest that such a scenario should be avoided. Despite this, there is a consistent good agreement between the flow patterns and the measured fluctuating pressure values acting on the overflow walls.

4. Conclusions

According to the experimental results, the relationship between gate-opening modes and fluctuating pressure on the overflowing walls was explored and optimal gate-opening modes were recommended. Based on the above study, the following conclusions can be stated:

(1) Gate-opening modes are very flexible for MHSJ. Moreover, proper and strategical adjustment of gate-opening modes can be very effective in reducing fluctuating loads on overflow walls (e.g., fluctuating pressure can be reduced by nearly 50% for the same flood discharge by optimizing the opening mode of the gates).

(2) To improve the overall performance and stability of the hydraulic structure, both spillways and mid-level channels should be opened uniformly, but side spillways should be opened slightly less to reduce fluctuating pressure on the side-wall of the stilling basin. The discharge ratio Q_u/Q_d (the ratio of spillway discharge to mid-level channel discharge) should be kept between 2–3 when side spillways and mid-level channels are used together.

(3) The flow patterns in the stilling basin show remarkable differences depending on gate-opening modes. No repeatable shape was observed when trying different configurations. It is recommended to open the mid-level channels partially and uniformly when discharging orifices are used separately. The flow patterns are always stable when spillways are used separately regardless of full or partial opening. The flow patterns are relatively stable in the stilling basin when a combination of spillways and mid-level channels are used to release a high flowrate typical of flooding scenarios.

(4) It is proposed to use spillways and mid-level channels separately for middle orifices and small flowrates released. The combination of mid-level channels and spillways can be used for large discharge.

Author Contributions: J.Z. and W.X. conceived and designed the experiments; Y.P. performed the experiments; J.Z. and Y.P. analyzed the data; Y.P. and M.R. wrote the paper.

Abbreviations

BB	stilling basin bottom
BS	stilling basin side-wall
H	height dimension (m)
L	length dimension (m)
MHSJ	Multi-Horizontal-Submerged Jets
PSO	piers, spillways and mid-level channels
RMS	root-mean-square of fluctuating pressure
W	width dimension (m)
Q	discharge (m^3/s)
Q_d	total discharge of mid-level channels (m^3/s)
Q_{ds}	total discharge of two side mid-level channels(1# and 5#) (m^3/s)
Q_{dm}	total discharge of three middle mid-level channels (2–4#) (m^3/s)
Q_u	total discharge of spillways (m^3/s)
Q_{um}	total flood discharge of four middle spillways (2–5#) (m^3/s)
Q_{us}	total flood discharge of two side spillways (1# and 6#) (m^3/s)
σ	RMS of fluctuating pressure (9.81 kPa)

References

1. Wu, J.H.; Ai, W.Z.; Zhou, Q. Head loss coefficient of orifice plate energy dissipater. *J. Hydraul. Res.* **2010**, *48*, 526–530.
2. Manenti, S.; Pierobon, E.; Gallati, M.; Sibilla, S.; D'Alpaos, L.; Macchi, E.; Todeschini, S. Vajont Disaster: Smoothed Particle Hydrodynamics Modeling of the Postevent 2D Experiments. *J. Hydraul. Eng.* **2015**, *142*, 05015007. [CrossRef]
3. Vacondio, R.; Mignosa, P.; Pagani, S. 3D SPH numerical simulation of the wave generated by the Vajont rockslide. *Adv. Water Resour.* **2013**, *59*, 146–156. [CrossRef]
4. Xu, Y.; Zhang, L.; Jia, J. Lessons from Catastrophic Dam Failures in August 1975 in Zhumadian. *China Geocongr.* **2008**, *178*, 162–169.
5. Tan, Y.T.; Xu, H.; Zhang, X.L. Sustainable urbanization in China: A comprehensive literature review. *Cities* **2016**, *55*, 82–93. [CrossRef]
6. Intergovernmental Panel on Climate Change (IPCC). *Summary for Policymakers, Climate Change 2014: Impacts, Adaptation and Vulnerability—Contributions of the Working Group II to the Fifth Assessment Report*; IPCC: Geneva, Switzerland, 2014; pp. 1–32.
7. Zhang, J.M.; Peng, Y.; Xu, W.L. *Study on the Cause of Vibration Induced by Flood Discharge and Its Countermeasures in Xiangjiaba Hydropower Station*; Technical Report; Sichuan University: Chengdu, China, 2013.
8. Zhang, J.M.; Li, Y.L.; Yang, Y.Q.; Xu, W.L.; Zeng, X.H.; Cheng, H. Investigation on energy dissipation of multiple submerged horizontal jets with pressure flow. *Water Resour. Hydropower Eng.* **2004**, *35*, 30–33. (In Chinese)
9. Zhang, J.M.; Chen, J.G.; Xu, W.L.; Peng, Y. Characteristics of the vortex structure in multi-horizontal submerged jets stilling basin. *Proc. Inst. Civ. Eng.-Water Manag.* **2013**, *167*, 322–333. [CrossRef]
10. Li, Y.L.; Hua, G.C.; Zhang, J.M.; Yang, Y.Q.; Hu, Y.H. Analysis on energy dissipation of the single-lever with multi-strand and multi-lever with multi-strand horizontal submerged jets. *Chin. J. Hydrodyn.* **2006**, *1*, 26–31. (In Chinese)
11. Li, Y.L.; Hua, G.C.; Zhang, J.M.; Yang, Y.Q. Factors affecting the hydraulic characteristics of horizontal submerged jets. *Adv. Water Sci.* **2006**, *17*, 761–766. (In Chinese)
12. Deng, J.; Xu, W.L.; Zhang, J.M.; Qu, J.; Yang, Y. A new type of plunge pool—Multi-horizontal submerged jets. *Sci. China (E Technol. Sci.)* **2008**, *51*, 2128–2141. [CrossRef]
13. Deng, J.; Xu, W.L.; Zhang, J.M.; Qu, J.X.; Yang, Y.Q. A new type of stilling pool—Multi-strand horizontal submerged jets. *Sci. China (E Technol. Sci.)* **2009**, *39*, 29–38. (In Chinese)
14. Huang, Q.J.; Feng, S.R.; Li, Y.N.; Wu, J.H. Experimental study on energy dissipation characteristics of multi-horizontal submerged jets. *Chin. J. Hydrodyn.* **2008**, *23*, 694–701. (In Chinese)
15. Chen, J.G.; Zhang, J.M.; Xu, W.L.; Li, S.; He, X.L. Particle image velocimetry measurements of vortex structures in stilling basin of multi-horizontal submerged jets. *J. Hydrodyn. Ser. B* **2013**, *25*, 556–563. [CrossRef]
16. Chen, J.G.; Zhang, J.M.; Xu, W.L.; Wang, Y.R. Scale effects of air-water flows in stilling basin of multi-horizontal submerged jets. *J. Hydrodyn. Ser. B* **2010**, *22*, 788–795. [CrossRef]
17. Chen, J.G.; Zhang, J.M.; Xu, W.L.; Wang, Y.R. Numerical simulation investigation on the energy dissipation characteristics in stilling basin of multi-horizontal submerged jets. *J. Hydrodyn.* **2010**, *22*, 732–741. [CrossRef]
18. Yin, R.G.; Zhang, J.H. Vibration source and shock absorption scheme research of near-field vibration caused by flood discharge and energy dissipation of a stilling pool. *Geo Shanghai Int. Congr.* **2014**, *237*, 107–116.
19. Yin, R.G.; Zhang, J.H.; Liu, X.K. 3D dynamic FEM analysis of near-field vibration caused by flood discharge and energy dissipation of a stilling pool. *J. Water Resour. Arch. Eng.* **2014**, *12*, 72–76. (In Chinese)

8

Experimental Investigation of Coherent Vortex Structures in a Backward-Facing Step Flow

Fangfang Wang [1,*], Ang Gao [1], Shiqiang Wu [1,*], Senlin Zhu [1], Jiangyu Dai [1] and Qian Liao [2]

1 State Key Laboratory of Hydrology-Water Resources and Hydraulic Engineering, Nanjing Hydraulic Research Institute, Nanjing 210029, China; gaoang@whu.edu.cn (A.G.); slzhu@nhri.cn (S.Z.); jydai@nhri.cn (J.D.)

2 Department of Civil Engineering and Mechanics, University of Wisconsin-Milwaukee, Milwaukee, WI 53201, USA; liao@uwm.edu

* Correspondence: ffwang@nhri.cn (F.W.); sqwu@nhri.cn (S.W.)

Abstract: Coherent vortex structures (CVS) are discovered for more than half a century, and they are believed to play a significant role in turbulence especially for separated flows. An experimental study is conducted for a pressured backward-facing step flow with Reynolds number (Re_h) being 4400 and 9000. A synchronized particle image velocimetry (PIV) system is developed for measurement of a wider range of velocity fields with high resolution. The CVS are proved to exist in the separation-reattachment process. For their temporal evolution, a life cycle is proposed that vortices form in the free shear layer, develop with pairings and divisions and finally shed at the reattachment zone, and sometimes new vortical structures are restructured with recovery of flow pattern. The CVS favor the free shear layer with frequent pairings and divisions particularly at the developing stage around x/h = 2~5 (x: distance from the step in flow direction, h: step height), which may contribute to the high turbulent intensity and shear stress there. A critical distance is believed to exist among CVS, which affects their amalgamation (pairing) and division events. Statistics show that the CVS are well organized in spatial distribution and show specific local features with the flow structures distinguished. The streamwise and vertical diameters (D_x and D_y) and width to height ratio (D_x/D_y) all obey to the lognormal distribution. With increase of Re_h from 4400 to 9000, D_x decreases and D_y increases, but the mean diameter ($D=0.5 \times (D_x + D_y)$) keeps around (0.28~0.29) h. As the increase of Re_h, the vortical shape change toward a uniform condition, which may be contributed by enhancement of the shear intensity.

Keywords: coherent vortex structure; backward-facing step; synchronized PIV; separation and reattachment; free shear layer; vortical evolution

1. Introduction

Flow separation is common in various applications in our daily life (airfoils at large attack angle, spoiler flows, flow behind a vehicle, flow inside a combustor/condenser, flows in an enlarged pipe or a channel with steps, and flows around a boat or a building, etc.). Backward-facing step (BFS) is one basic model and fundamental in geometry and engineering design [1–5]. Flow separation behind the step changes transport of momentum, heat, and mass, and increases fluid mixing [6], and it also possibly leads to additional flow resistance and noise, and even for stalling of an aircraft under critical conditions. In a combustor or chamber, effective mixing is preferred and can be designed with detailed BFS geometric considerations [5,7–9]. In addition, the enlarged boundary in a pipe or channel sometimes generates unfavorable boundary erosion and structural vibration, and sometimes energy dissipation caused by structural changes in hydraulic engineering arouses people's curiosity. In field of scouring downstream of a structure, motions and fluctuations of the vortices will cause sediment transport and river-bed scour [10]. However, because of the irregularity and complexity of turbulence, the flow characteristics, various mechanisms, and controlling studies on this flow are still worth further efforts with developments of measurement techniques [1–5,11]. In particular, coherent vortex structures (CVS) are discovered to play an important role and may be a promising way to investigate this flow [1,3–6,11–16]. However, quantitative studies of the CVS (e.g., the conception, identification, structure, scale, and evolution, etc.,) are deficient, and it is still difficult to fully understand the hydrodynamic mechanisms and unsteady interactions in many applications (e.g., pipe cavitation, earth dam break, riverbed and bank erosion, sand dune evolution, and energy dissipation).

Figure 1 shows the typical flow structures in a two-dimensional BFS. Flow separates at the step edge (step height, h) and reattaches the bottom wall at some distance downstream, forming a recirculation region behind the step. This process consists of most physical features of a separation-reattachment flow, such as the free shear layer, the reattachment zone, and the redeveloping boundary layer [5,11,13,16–19]. The recirculation region is extensively studied in the previous researches. Length of this region (recirculation length, X_r) keeps a relative stable value of (7 ± 1) h for a developed turbulent flow, while it increases for a laminar flow and decreases for a transitional flow with the increase of Reynolds number [20]. The time-averaged flow structure is believed to be stable in a developed turbulent flow such as the free shear region (FSR), the corner region (CR), the redeveloping region (RR), and the reattachment zone [11]. These regions show various flow characteristics. The FSR is generated from the step top corner with strong shear intensity and finally limited by the bottom wall. The CR is located at the step corner with a secondary backflow. The RR shows a new development for the turbulent boundary layer after reattachment. The reattachment zone is defined as the locus of points, where the limiting streamline of the time-averaged flow rejoins the surface [21]. For the temporal flow, perturbation of the step generates successive CVS which promotes the fluid mixing and intensifies the unsteady interaction. The CVS shows four typical developing stages with various characteristic frequencies in the flow direction: forming, developing, shedding, and redeveloping [11].

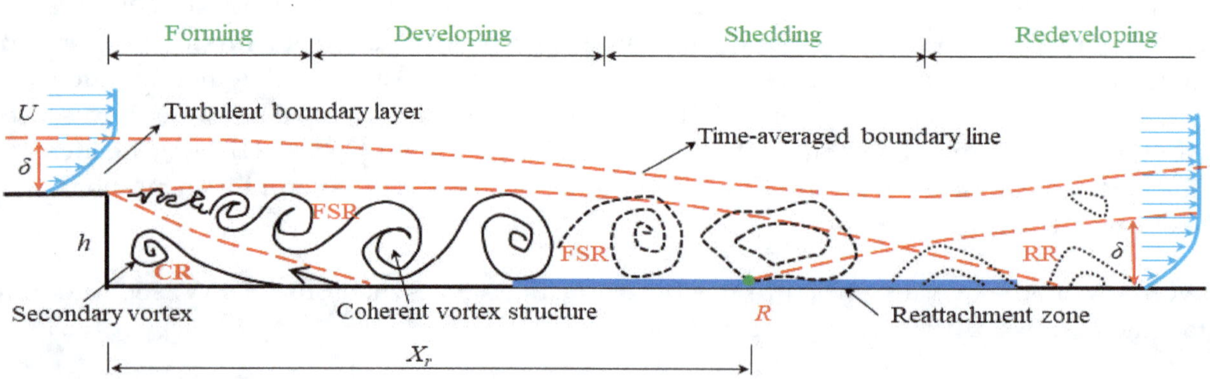

Figure 1. Simplified representation of a two-dimensional backward-facing step (BFS) flow: (FSR: free shear region; CR: corner region; RR: redeveloping region; R: mean reattachment point).

The complexities of BFS flows are summarized by Roos and Kegelman (1986) [12]: (1) development of CVS in the free shear layer; (2) substantial unsteadiness of the entire separated flow, especially near the reattachment zone and particularly at low frequencies; (3) slow relaxation to equilibrium of the reattached shear layer, and the third one is also related to the former two. Recent decades, researchers are working on the CVS and unsteady motions in the complex flows.

Flow visualization is first used to investigate this flow. Tani et al. (1961) [22] showed aluminum-power pictures of an instantaneous BFS flow and found that short exposure pictures reveal some distinct vortex structures (CVS) in the recirculation (Figure 2a), while pictures of long exposure only indicate something like that shown in the mean dividing streamlines. The CVS present formation, merging, and increasing irregularity with the increase of shear layer turbulence and persistence of the structures downstream of reattachment (Figure 2b–d) [6,14]. The CVS are believed to be rolled up in the strong shear layer, which grows with a vortex pairing mechanism until it approaches the reattachment zone [6,12,14,15,22]. In the reattachment zone, the CVS are torn roughly into two in the neighborhood of the reattachment point (*R*), where part of the flow is deflected upstream into the recirculation region to supply the entrainment and another part continues downstream [16]. McGuinness (1978) [23] believes that some CVS are swept upstream with the recirculating flow, while others proceed downstream. The fact that CVS move alternately up and downstream is also supported by the measurements of Chandrsuda (1975) [24] and Kim et al. (1978) [25]. In first part of the FSR, CVS and their evolution are usually explained by the Kelvin-Helmholtz instability that proposed in a plane mixing layer. However, BFS flow shows some differences [1,12]: (1) The free shear layer has a finite length due to the reattachment, which limits the development of the CVS; (2) The free shear layer supports an adverse pressure gradient and is curved; (3) Strong interaction between the unsteady shear layer and the reattachment wall generates substantial acoustic and convective disturbances, that feed back through the recirculation to influence the shear layer at separation. The peak value of turbulence intensity in a BFS flow is 10~20% higher than that of a plane mixing layer [1]. Additionally, local forcing with certain frequency is proved to contribute to amalgamation of the CVS and fluid mixing (Figure 2b,c), which also shortens X_r [6].

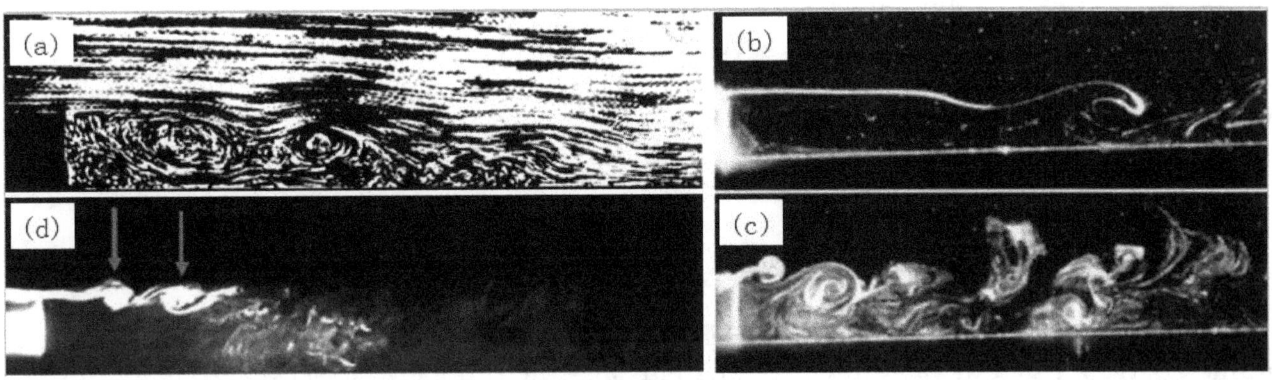

Figure 2. Visualizations of BFS flow: (**a**) Tani et al. (1961) [22]; (**b**) Chun et al. (1998), without excitation [6]; (**c**) Chun et al. (1998), with excitation [6]; (**d**) Kostas et al. (2002) [14].

Various quantitative measurements are applied to obtain velocity or pressure information such as hot-wire, x-wires, hydrogen bubble, laser doppler velocimetry, and particle image velocimetry (PIV) [11,16,20,26–30]. Many of the measurements can only measure and analyze point-by-point characteristics or time-averaged flow fields information such as X_r, turbulent shear stress, turbulent intensity, backflow, and characteristic frequency, etc. Recent developments of experimental measurements and numerical simulations allowed in-depth investigation of the flow dynamics. For example, PIV makes it possible to obtain some velocity fields and flow structures with temporal vortical evolutions. Scarano et al. (1999) [29] investigate the trajectories and sizes of CVS with $Re_h = 5000$, utilizing a pattern recognition analysis. They concluded that the length scale of the CVS ranges from 0.12 h to 0.44 h, and those with smaller diameters occur more frequently. They suggest that the clockwise CVS arise from a Kelvin–Helmholtz instability in the FSR, while the counter-clockwise CVS in the CR are due to a three-dimensional breakdown of the clockwise ones. Staggered vortex trains and pairings within the FSR are identified in the instantaneous PIV measurements by Kostas et al. (2002) [14]. They propose that the vortical interactions are most likely to contribute to the high Reynolds stresses and turbulent kinetic energy production. Furthermore, the CVS seem to be largely responsible for the persistence of the streamwise turbulent intensity (u'^2) and the shear stress ($u'v'$) in the flow downstream of reattachment, while the vertical turbulent intensity (v'^2) is predominantly by the fine scale structures. Hudy et al. (2007) [15] further propose a "wake model" that the CVS grow in place before reaching a height equivalent to h, and at this position the CVS shed and accelerate downstream to the ultimate convection speed. This opinion compares the evolution of this flow structure to that in the wake of bluff bodies, which is a little different with the traditional "shear-layer model."

Some other researches deal with the unsteadiness of separation flow with some low-frequency motions [6,11,13,31–35]. Four motions are summarized in separated flows including the BFS configuration that respectively correspond to flapping (FP), oscillation of X_r (OX), Kelvin-Helmholtz instability (KH), and pairings of the Kelvin-Helmholtz vortices (PKH). These motions are associated with the main instability mechanisms occurring in the shear layer and the recirculation region. The FP is the universal frequency, implying an absolute instability. While the OX appears obviously around the reattachment zone. The KH and PKH particularly concentrate at the shear layer [11]. Hu et al. (2016) [35] shows the temporal CVS in their numerical simulation and proposes a shear layer model and a shedding model. The former model is coincided with the Kelvin-Helmholtz instability mechanism with Strouhal number ($S_t = fh/U_0$, U_0 is the maximum velocity and f is the characterized frequency) of 0.2, and the later one is characterized by shedding frequency of $S_t = 0.074$. An obviously declining trend for the characteristic frequency along the streamwise behind the step is showed by Lian et al. (1993) [27] and Wang et al. (2019) [11]. They suggest this declining trend to be caused by the evolution of CVS.

Overall, some new advances have focused on the unsteady dynamics and CVS in BFS flow. However, it is still difficult to illustrate their essential developments and evolutions. Quantitative studies of the basic definition and features of the CVS are still deficient. In this study, we present the CVS particularly in the temporal streamlines and attempt to quantitatively investigate their characteristics and evolutional laws.

2. Methods

2.1. Experimental Model

A water tunnel flow system with a BFS model was built (Figure 3) and some configuration parameters were listed in Table 1. A rectangular pressured tunnel was set horizontally (top wall slope, $i = 0$). It enlarges halfway and forms a vertical step at the bottom surface with h = 50 mm. The expansion ratio, $E_r = H_d/H_u$, equals to 2, where H_d and H_u are respectively water depth upstream and downstream the step. The tunnel width was suggested to be no less than 10 h to keep the flow bidimensionality [16], therefore $A_r = L_z/h = 10$ was set. The channel lengths of upstream and downstream the step were L_{xu} = 44 h and L_{xd} = 50 h, respectively. Honeycomb tubes were set near the inlet of the channel to obtain fully developed turbulent flow before the step.

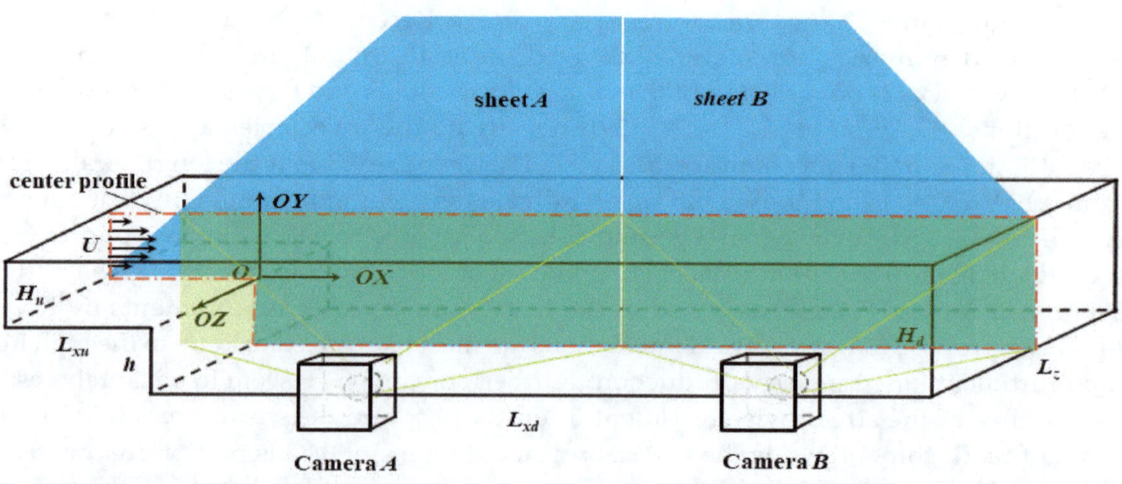

Figure 3. Schematic of the test section arrangements with a synchronous particle image velocimetry (PIV) system.

Table 1. Configuration parameters of the BFS model.

Parameters	Values	Units
h	50	mm
L_{xu}	2220	mm
L_{xd}	2500	mm
H_u	50	mm
H_d	100	mm
L_z	500	mm
i	0	°
E_r	2:1	-
A_r	10:1	-

The flow tunnel was made of transparent PMMA (polymethyl-methacrylate) to provide a clear view for the PIV system. PIV was employed to measure the velocity fields at the central profile (Figure 3).

2.2. Measurement System

According to the previous literatures, X_r behind step was about (4.2~8.6) h when flow was fully turbulent, and it was even longer for a lower Reynold number [1,11,20,30]. To study the evolution process of the CVS, a global view of velocity fields with no less than 10 h (about 60 cm) behind the step was expected. A large-scale field of view was difficult to realize with high vector density and high precision. To obtain the global velocity field, some previous studies took pictures of different subareas and matched the time-averaged velocity fields spatially [14,29]. However, this was still unavailable for the global view of the instantaneous velocity fields with CVS' evolution process. This study developed a synchronous PIV system, by which a doubled and effective view of the instantaneous velocity field was achieved (Figure 3, Sheets A and B).

The synchronous PIV system mainly consisted of laser module, image acquisition module, and control module. The laser module was a laser whose ray scanned to a light sheet optics with thickness of about 1 mm. The laser power energy was 3 W and the maximum scanning frequency reached to 40 Hz. The image acquisition module was made of a charge coupled device (CCD) camera with a resolution about 2560 pix × 2048 pix. Hollow sphere glass particles were seeded into the water with an average diameter of 15 μm. Two light sheets (A and B) and cameras (A and B) were synchronously controlled by a control module. This PIV system successfully made a global view of an instantaneous velocity field of approximate 12 $h \times 5$ h with a resolution of 7.35 pix/mm. Interrogation area of 20 pix × 15 pix (2.722 mm × 2.042 mm) was used when computing velocity vectors. The measurement accuracy was within 1% of the maximum velocity.

2.3. Flow Conditions

The experiment was carried out with a constant water head for each case. The Reynolds number, $Re_h = U_b h/v$, was defined by the step height and the averaged velocity (U_b) located at 1 h upstream the step, and v was the kinematic viscous coefficient of water at 20 °C.

Figure 4 shows the relationship between X_r and Re_h. Three typical flow regions were identified: the laminar flow (region I, $Re_h < 740$), transition flow (region II, $740 < Re_h < 4070$), and turbulent flow (region III, $Re_h > 4070$). A relative stable recirculation region with a constant X_r value (4.2~8.6 h) occurs in the turbulent flow (III) [1,20,36–38]. The flow structures in this flow region (III) were proposed to be stable, while the X_r values in another two flow patterns (I and II) were both changeable. Since most flow are with high Reynolds number in real applications, the turbulent flow region (III) were particularly favored here. For comparison, two cases with $Re_h = 4400$ and $Re_h = 9000$ were analyzed (Table 2). In addition, the good agreement of the data points in Figure 4 also verify the reliability of the measuring instrument.

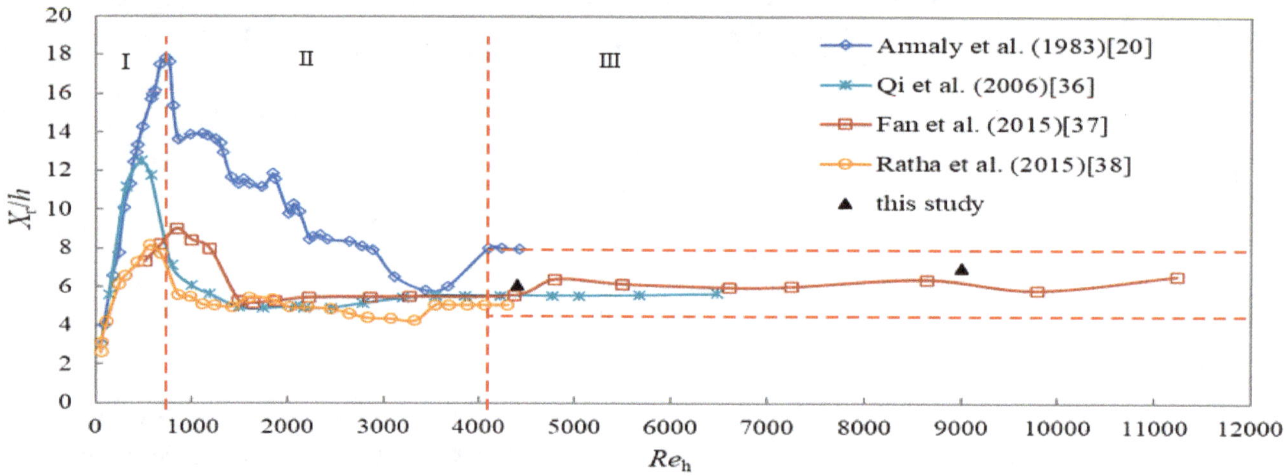

Figure 4. The reattachment length (X_r) vs. Reynolds number (Re_h).

Table 2. Flow conditions in the experiment.

h (cm)	U_b (m/s)	E_r	A_r	Re_h	X_r/h
5	0.088	2:1	10	4400	6.1
5	0.182	2:1	10	9000	7.0

3. Results and Discussions

3.1. Vortex Visualization and Definition

The vortex structures are visualized by the ghosting pictures of the inspired particles, and the pictures are also compared with the instantaneous streamlines (Figure 5). Comparison shows that the two figures present similar flow phenomena. The main flow diffuses and forms a recirculation region behind the step. The instantaneous flow is turbulent with the main flow interacting with the recirculation region.

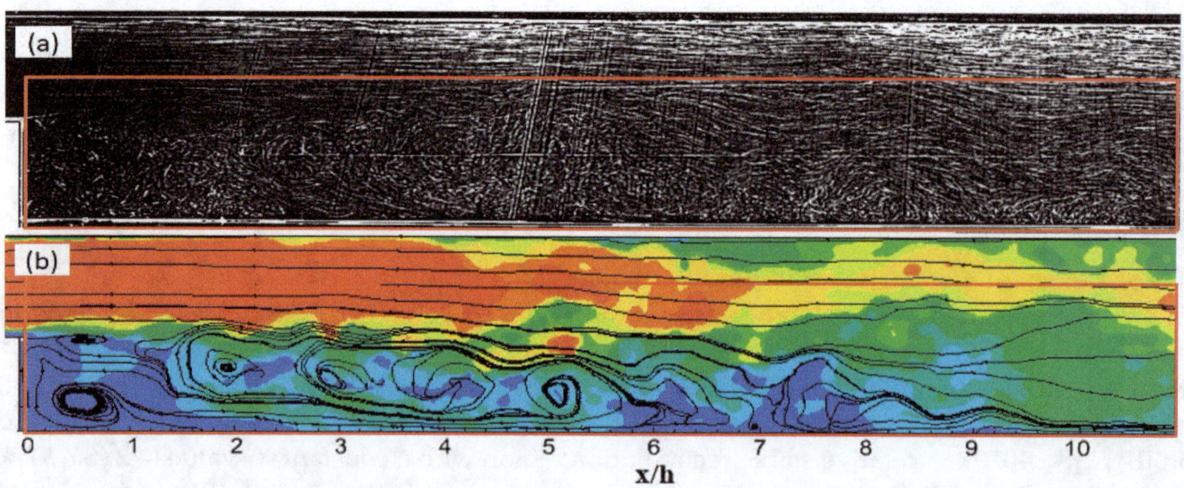

Figure 5. Coherent vortex structures (CVS) in the instantaneous flow: (**a**) visualization, (**b**) streamlines with velocity contours.

Particularly, the streamlines show some spanwise and large-scale vortex structures behind the step. They are well organized in time and space. They successively roll up in the FSR, move downstream and approach to the reattachment zone, with maximum length scale of the same order with the step height. These vortex structures present mostly an elliptical shape from a side view particularly in the early developing stage, and then break down when approaching the reattachment zone. These flow phenomena are consistent with the visualization of Tani et al. (1961) in Figure 2a [22].

One may ask if these elliptical streamlines are vortices or not. Lugt (1979) [39] defined a vortex as "multitude of material particles rotating around a common center." Robinson (1991) [40] further concluded that a vortex exists when instantaneous streamlines mapped onto a plane normal to the vortex core exhibit a roughly circular spiral pattern, when viewed from a reference frame moving with the center of the vortex core. These two definitions both propose two major features ("a common center" and "rotating"), and the later one further points to use streamlines to show a vortex. Although the concept of CVS has been proposed for centuries, its physical definition is still ambiguous. The vortex structures in the flow visualizations or by streamlines have the basic vortical features, particularly the streamlines show some elliptical patterns which rotates around a common center. Thus, they are consistent with the vortical definitions. Here, we define these vortex structures shown by streamlines to be the coherent vortex structures.

To quantitatively describe these CVS, some basic parameters such as the locations and patterns are defined (Figure 6). The common center of streamlines is regarded as the vortex core ($Z(x, y)$), representing location of a vortex. The typical shape is regarded close to an ellipse with streamwise diameter (D_x), and vertical diameter (D_y). Thus, D_x/D_y represents width to height ratio and $D = 0.5 \times (D_x + D_y)$ indicates the mean scale. Two CVS are separated by an equilibrium point, called the saddle point.

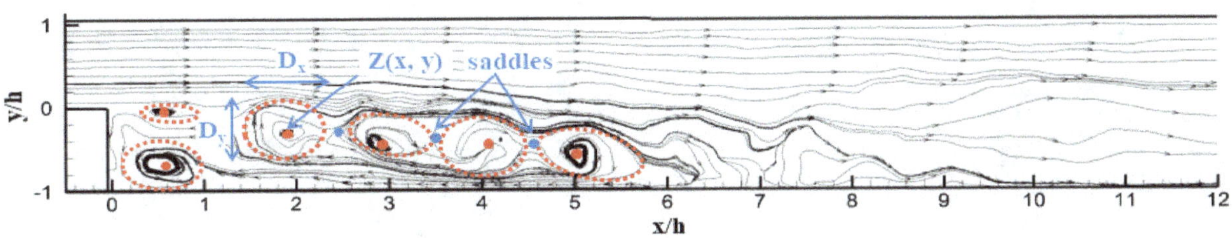

Figure 6. Parameter definition for the CVS (D_x—streamwise diameter, D_y—vertical diameter, $Z(x, y)$—vortex center, blue point—saddle point).

3.2. Vortex Motions with Developing Stages

Figure 7 shows the spatial and temporal evolution process of the CVS in streamline pictures. Totally 12 instantaneous streamline pictures with an interval time of 0.5 s are compared ($T = tU_b/h$ = 0.00~9.68). These streamlines are generated with the corresponding velocity matrix by TECPLOT 2013.

In the first picture (T = 0.00), eight representative CVS (#1~#8) are indexed, which show various patterns and characteristics with flow structures (flow regions and developing stages see Figure 1). An analysis for these CVS' features based on the flow structures divided is conducted here.

(1) Vortex #1 is located in the forming stage (the first stage, x = (0~2) h in FSR) just behind the step. Strong free shear layer is developing from the step corner with increasing thickness. Successive vortex structures generate and grow with similar size (diameter < 0.5 h) when travelling downstream (green lines). Finally, they are mixed by larger and fully developed ones at the end of this stage (x = (1~2) h). Their average speed toward downstream is about (0.3~0.6) U_b in this case.

(2) Vortex #2~#5 occur in the developing stage (the second stage, x = (2~5) h in FSR). The CVS are fully developed and energetic with larger length scales that close to h. Abundant amalgamation (pairing) and division events happen in this stage (red arrow lines). Because of the high mixing rate and flow unsteadiness, the turbulent intensity and shear stress reach a peak there [1,11,14]. This response may be a proof that the CVS is an important contributor for the Reynolds stress. Compared with the former stage, their travelling speeds reduce to approximately (0.05~0.3) U_b before attaching to the reattachment zone (red lines).

(3) Vortex #6~#7 appear in the shedding stage (the third stage, $x = (5\sim8)\,h$ in the reattachment zone). Flow is strongly disturbed by the bottom wall in this stage. The streamlines are twisted sometimes when approaching to the wall. This may be induced by the enhancement of the three-dimensional characteristics. The CVS slow down and mostly attach on the bottom wall (blue lines), and sometimes they vanish there. Beyond R, some streamlines begin to roll up and generate some new vortex structures in a relatively smaller length scale and weaker rotating intensity. These vortex structures usually keep for a short time.

(4) Vortex #8 is usually found in the corner region ($x = (0\sim2)\,h$ in CR). It is induced by the recirculation, where the backflow separates from bottom wall and induces a counter-clockwise vortex [14,20,27,35]. The temporal evolution shows that the separation streamline is usually generated at the separated point of the backflow leaving the wall. The diameter of this vortex is about 0.5 h [11]. This vortex has a much weak rotating intensity. And it travels upstream at the speed of about 0.1 U_b in this case (yellow lines, $Re_h = 4400$).

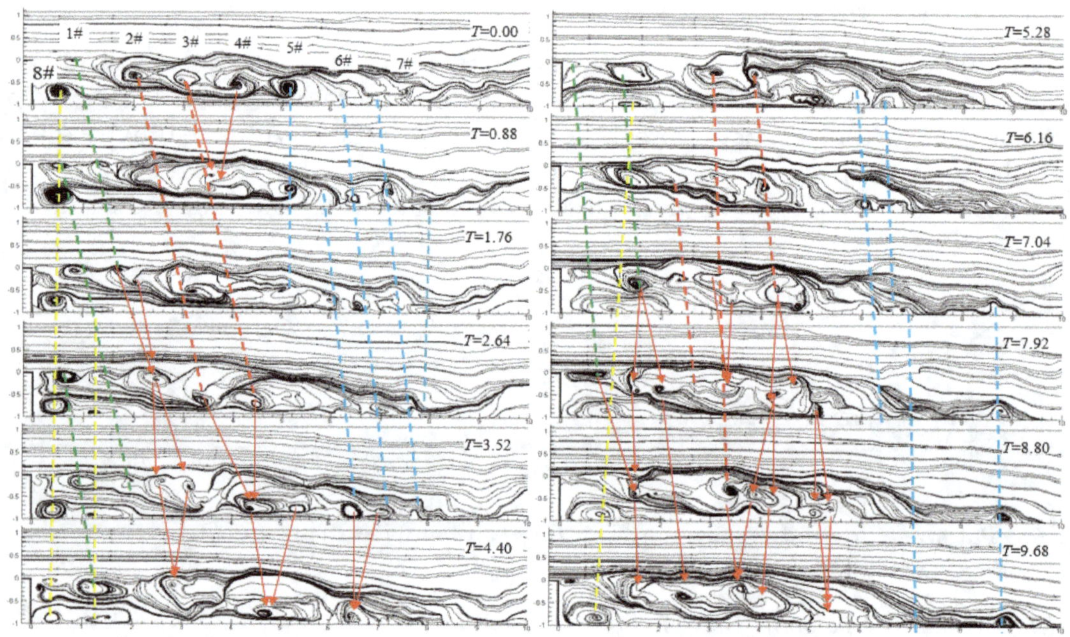

Figure 7. Motions and evolutions of the CVS with $Re_h = 4400$ (Dash lines with green, red, blue and yellow represent vortex trajectory in different developing stages; Arrow lines represent the amalgamation and division behaviors; $T = 0.00\sim9.68$).

There are totally dozens of CVS in each streamline picture. The evolution shows a life process of forming, developing, shedding, and redeveloping when travelling downstream. Their characteristics such as size, rotation direction and intensity, and evolution behaviors vary with the flow regions and developing stages. In particular, amalgamation (pairing) and division behaviors often occur in the second developing stage.

3.3. Vortex Trajectory and Evolution

The life process of the CVS is also reflected in the temporal evolution of an individual vortex structure. The blue line tracks the trajectory of a typical vortex center and other color lines show some break up and amalgamation (Figure 8). In this figure, the blue diamond point means the vortex center and the arrows show its moving direction. It starts nearby the step corner, and then travels toward the reattachment zone. It initially wanders in the strong shear layer, forming a relatively narrow path in the first stage (forming stage, 0~2 h). When arriving to the end of this stage, the center "jump" at about $x = 1.5\,h$, which can be explained by the amalgamation (merged by the vortex in later stage) shown in Figure 7. Then the vortex lingers and fluctuates particularly around $x = 2.2\ h$ and $x = 3.4\ h$ in the developing stage, where frequent amalgamation and division events occur there. The branches lines with other colors indicate vortex structures breaking up from the original one. When travelling to the end of this stage ($x = 4\sim5\ h$), the original vortex could not be found in the streamline pictures.

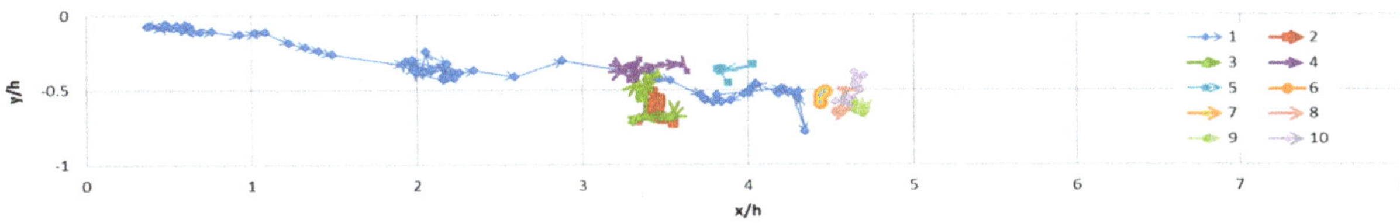

Figure 8. Trajectory and evolution of vortex centers (Re_h = 4400).

To further investigate the evolution events, a typical pairing process of two CVS is analyzed (Figure 9). To our knowledge, there is no experimental analysis for this process. A critical distance, λ_c, between two vortex patches was numerically studied by the contour dynamics method, and a distance parameter, λ, was defined when considering two scheduled circular vortex patches with the same diameters and shown by certain vorticity contour [41,42]:

$$\lambda = L/r\ (L > 2\ r) \tag{1}$$

where L is the distance between the two vortex patches, and r is the radius of each vortex patch.

When $\lambda > \lambda_c$, these two vortex patches rotate and change in shape, and their shape is close to a tear drop with the decrease of λ. When $\lambda < \lambda_c$ (λ_c = 3.5), these two vortex patches start to interact with each other. With the decrease of λ, they will get tangled up and finally they merge together.

In this experimental study, a similar parameter λ' is defined between two vortex structures with equivalent radiuses of r_1 and r_2, respectively (Equation (2)). A critical value is also expected when the CVS start to pair. It should be pointed out that these CVS are defined by streamlines instead of vorticity contours, and they are mostly elliptical but not circular in shape.

$$\lambda' = 2\ L'/(r_1 + r_2)\ (2\ L' > r_1 + r_2) \tag{2}$$

where L' is the distance between two vortex centers, and r_1 and r_2 are of the equivalent radiuses of two CVS, respectively.

As is shown in Figure 9, two CVS keep independent initially when they stay at a sufficiently far distance. As they are getting closer, they change in shape and start to interact with each other and finally come to an amalgamation when their distance is less than the critical value ($\lambda' < \lambda_c$). This indicates that critical distance still exists in current CVS. From the point view of the direction of the streamlines, these two CVS have an opposite streamline direction, one converges to the center (right) and another originates from the center (left). The former one is usually stable, while the other one is unstable. They finally merge into a stable one (streamlines converge to the vortex center).

3.4. Vortex Center and Diameter Distributions

Temporal evolution of the specific vortex structure is particularly disclosed above, their features of spatial distribution and length scales are to discuss here. Totally more than 2000 CVS were extracted from 201 instantaneous streamline pictures (picture no. 100–300), and the distributions of vortex centers ($Z(x, y)$) and diameters parameters (D_x, D_y, D, D_x/D_y) were statistically analyzed (Figures 10 and 11).

Figure 10 shows the distribution of the vortex centers with Re_h = 4400 and 9000 separately. The distribution seem to be regular, that fit well with the divided flow regions and developing stages (Figure 1). The vortex centers mainly distribute in the FSR, CR, and RR. In other words, these CVS are directly and closely related to the flow separation and reattachment characteristics. And they present specific features in different regions, going through the process of forming, developing, shedding, and redeveloping when travelling downstream. This statistical distribution also supports well the above discussion on vortex center trajectory and vortex evolution. With increase of Re_h from 4400 to 9000, the distribution of vortex centers is almost unchanged in space. It provides a good explanation for the various features with the specific flow structures.

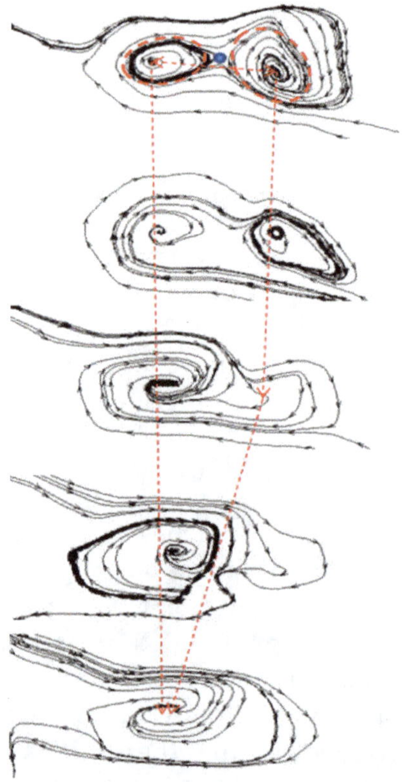

Figure 9. A vortex pairing process.

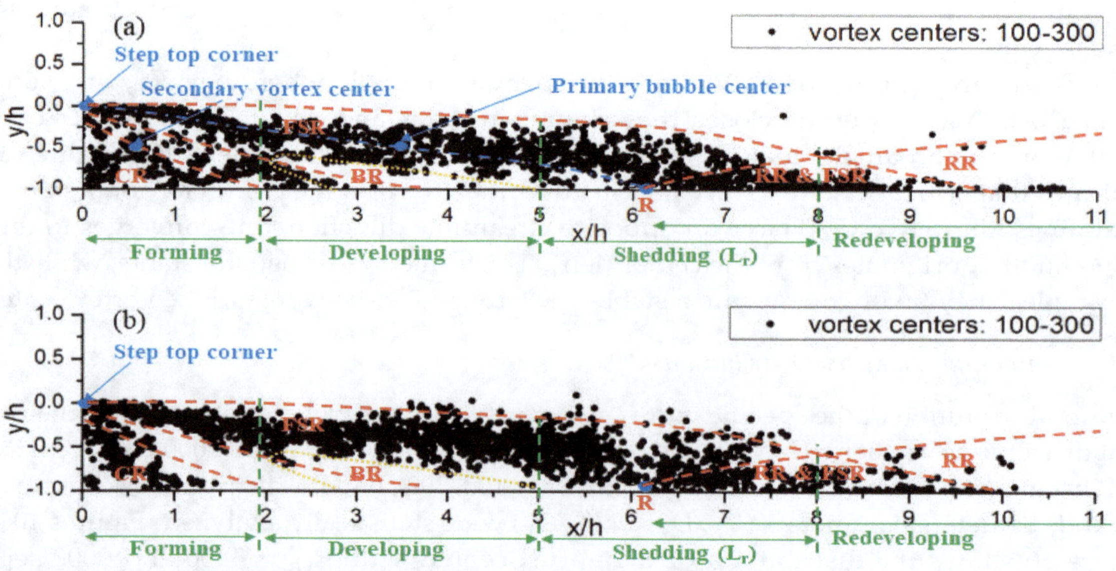

Figure 10. Distribution of the CVS with different Re_h: (**a**) Re_h = 4400 [11] and (**b**) Re_h = 9000; (FSR: free shear layer region, CR: corner region; BR: backflow region; RR: redeveloping region; R: reattachment point; L_r: length of reattachment zone).

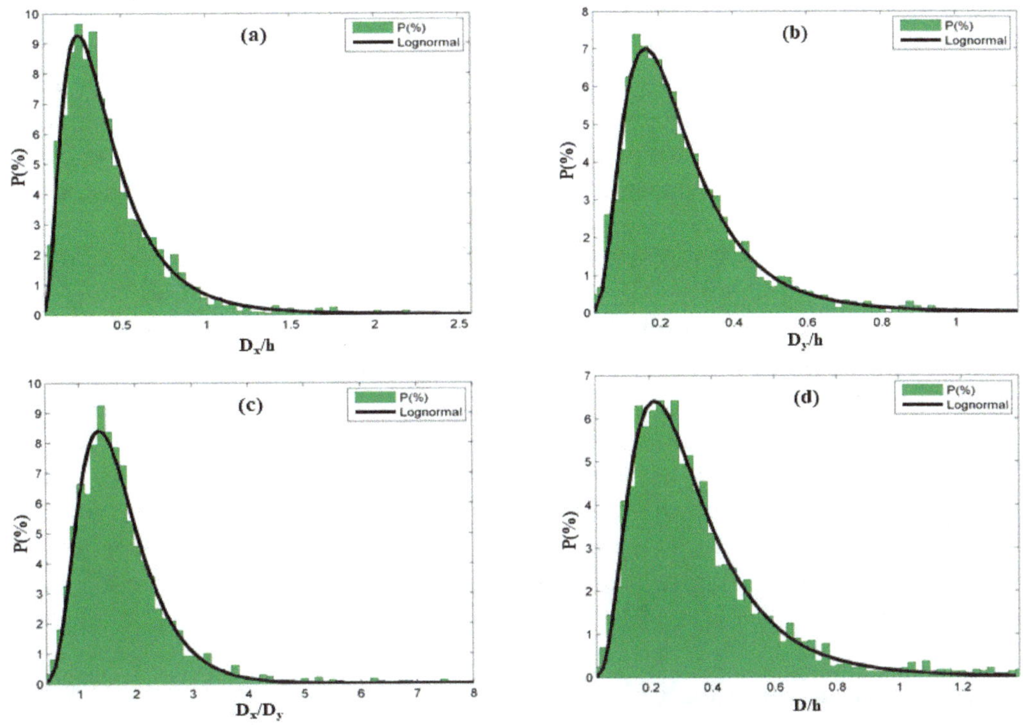

Figure 11. Distribution of length scale parameters ($Re_h = 4400$): (**a**) D_x; (**b**) D_y; (**c**) D_x/D_y; (**d**) D.

Besides the vortex center, the distribution of parameters D_x, D_y, D_x/D_y, and D are also well organized. Over 2000 vortex structures are analyzed. Figure 11 shows that four parameters (D_x, D_y, D_x/D_y, and D) all obey to the lognormal distribution. In other words, their logarithms follow a normal distribution. We know that lognormal distributions are positively skewed with long right tails because of low mean values and high variances in the random variables. The mean value ($\exp(\mu)$) and standard deviation (σ) were used to evaluate the length scale for the CVS, and also compare the influences of Reynolds number ($Re_h = 4400$ and 9000) and geometric model (BFS and turbulent boundary layer, TBL). The mean value ($\exp(\mu)$) and standard deviation (σ) were obtained by data fitting and are compared in Table 3.

Table 3. Comparison of the length scale parameters (BFS and TBL).

Literature	Configuration	Parameter	μ	Exp (μ)	σ
This study	BFS $Re_h = 4400$	D_x/h	−1.03	0.36	0.61
		D_y/h	−1.49	0.22	0.56
		D_x/D_y	0.41	1.51	0.38
		D/h	−1.22	0.29	0.56
	BFS $Re_h = 9000$	D_x/h	−1.13	0.32	0.70
		D_y/h	−1.43	0.24	0.68
		D_x/D_y	0.33	1.40	0.37
		D/h	−1.26	0.28	0.68
Wang (2009) [43]	TBL $Re_b = 4677$	D_x/h_w	−1.37	0.25	0.69
		D_y/h_w	−2.15	0.12	0.73
		D_x/D_y	0.78	2.18	0.64

With the increase of Re_h from 4400 to 9000, these four variables change with different trends. The streamwise diameter (D_x/h) decreases slightly from 0.36 to 0.32, while the vertical diameter (D_y/h) increases from 0.22 to 0.24. As a result, the width to height ratio (D_x/D_y) reduces from 1.51 to 1.40. This means that the long axis of the ellipse becomes shorter and the short axis grows, making the shape more circular with increasing of Re_h. However, the mean diameter (D/h) keeps a relative stable value around 0.28~0.29 for current experimental conditions. The standard deviation σ of variables (D_x/h, D_y/h, and D/h) all increase with increase of Re_h, while that of D_x/D_y has almost no response to Re_h. With the increase of Re_h, previous study showed that the shear intensity behind the step increases, and flow turbulence or unsteadiness is enhanced [20]. This will contribute to the vortex mixing and accelerate vortex evolution. We can understand that the streamwise diameter (D_x) is generally greater than the vertical diameter (D_y) as the streamwise shear direction and vertical wall constraints.

An experimental study of a turbulent boundary layer (TBL) on the distribution for vortex length scales was reported by Wang (2009) [43]. Different to the BFS configuration, there is a free water surface and no disturbance of step in the TBL model, yet the Reynold number, $Re_b = U_b R/\nu = 4677$ (R is hydraulic radius), is the same order with this study. In comparison, the CVS are more circular because of the vortical pattern $D_x/D_y = 1.40$–1.51 in the BFS flow is smaller than $D_x/D_y = 2.18$ in TBL flow. This may be explained by the different features of their shear layers, where the CVS are initially formed. The shear intensity is particularly strong near the bottom wall with a uniform mean velocity direction in TBL. While shear layer in BFS initially generates and develops from the step with a recirculation below. The extra recirculation and less impact of the wall in BFS flow both contribute to the more circular vortical shape. As for the standard deviation (σ), the BFS flow shows a smaller value than that of TBL flow. This means a more concentrated vortex scale distribution in BFS flow.

4. Conclusions and Recommendations

Velocity measurements and CVS investigation for the experiments at a BFS water flow with $Re_h = 4400$ and 9000 were reported. From a viewpoint of CVS defined in the streamline pictures, the vortical spatio-temporal coherences were verified and quantitatively studied. The following conclusions may be drawn:

(1) A synchronized PIV system is necessary for the measurement of a wider range of velocity fields with sufficiently high resolution to study the global evolution process of CVS.

(2) CVS are proved to exist in the separation and reattachment process of BFS flow. They are well organized in the recirculation regions with forming, developing, shedding, and redeveloping in the shear layer. Flow structures with various local characteristics are disclosed and explained by the motions and evolutions of CVS.

(3) The spatial distribution of CVS is almost unaffected by the Reynolds number when flow is fully turbulent. The CVS are particularly active at the developing stage with frequent pairings and divisions, which contribute to the high turbulent intensity and shear stress. And the shedding of CVS can explain the rapid decrease of the Reynolds stress at the reattachment zone.

(4) Statistics show the streamwise (D_x/h), vertical (D_y/h), and mean (D/h) diameters, and width to height ratio (D_x/D_y) of the CVS all obey to the lognormal distribution. The mean diameter (D/h) is about 0.28~0.29 for a BFS flow with $Re_h = 4400$ and 9000. Compared with the flow in a turbulent boundary layer, they have the same order of vortex length scale, while the shape of CVS in BFS is more circular. A critical distance is supposed to exist between two CVS.

Concept of CVS has been put forward for more than half a century and it is proved to play a significant role particularly in the separated flows. With the development of measurement techniques and numerical simulations, more quantitative information should be used to verify the previous hypothesis and suggestions. The quantitative vortex length scales measured in the experiments may help to choose the grid scale in certain simulations considering the large-scale vortex structures (such as the Large Eddy Simulation). In terms of the local energy dissipation at the enlarged step, the significant impact of the CVS cannot be ignored.

Author Contributions: F.W. and S.W. conceived and designed the experiment; F.W. and A.G. performed the experiment and analyzed the data and wrote the manuscript. S.Z., A.G., J.D., and Q.L. contributed to scientific advising and provided a substantial input to improve the paper.

Acknowledgments: The authors would also like to thank Corporation Hawksoft for their assistances in the laboratory measurements.

References

1. Eaton, J.K.; Johnston, J.P. A review of research on subsonic turbulent flow reattachment. *AIAA J.* **1981**, *19*, 1093–1100. [CrossRef]
2. Vogel, J.C.; Eaton, J.K.; Adams, E.W. Combined heat transfer and fluid dynamic measurements behind a backward-facing step. *Am. Soc. Mech. Eng.* **1985**, *107*, 922–929.
3. Jacob, M.C.; Guerrand, S.; Juvé, D.; Guerrand, S. Experimental study of sound generated by backward-facing steps under wall jet. *AIAA J.* **2001**, *39*, 1254–1260. [CrossRef]
4. Hu, R.Y.; Wang, L.; Fu, S. Review of backward-facing step flow and separation reduction. *Sci. Sin. Phys. Mech. Astro.* **2015**, *45*, 124704. (In Chinese)
5. Chen, L.; Asai, K.; Nonomura, T.; Xi, G.; Liu, T. A review of backward-facing step (BFS) flow mechanisms, heat transfer and control. *Therm. Sci. Eng. Prog.* **2018**, *6*, 194–216. [CrossRef]
6. Chun, K.B.; Sung, H.J. Visualization of a locally-forced separated flow over a backward-facing step. *Exp. Fluids* **1998**, *25*, 133–142. [CrossRef]
7. Chen, L.; Zhang, X.R.; Okajima, J.; Maruyama, S. Thermal relaxation and critical instability of near-critical fluid microchannel flow. *Phys. Rev. E* **2013**, *87*, 043016. [CrossRef]
8. Baigmohammadi, M.; Tabejamaat, S.; Farsiani, Y. Experimental study of the effects of geometrical parameters, Reynolds number, and equivalence ratio on ethaneoxygen premixed flame dynamics in non-adiabatic cylindrical meso-scale reactors with the backward facing step. *Chem. Eng. Sci.* **2015**, *132*, 215–233. [CrossRef]
9. Shahi, M.; Kok, J.B.W.; Pozarlik, A. On characteristics of a non-reacting and a reacting turbulent flow over a backward facing step (BFS). *Int. Commun. Heat Mass Transf.* **2015**, *61*, 16–25. [CrossRef]
10. Dodaro, G.; Tafarojnoruz, A.; Sciortino, G.; Adduce, C. Modified Einstein sediment transport method to simulate the local scour evolution downstream of a rigid bed. *J. Hydraul. Eng.* **2016**, *142*, 04016041. [CrossRef]
11. Wang, F.F.; Wu, S.Q.; Huang, B. Flow structure and unsteady fluctuation for separation over a two-dimensional backward-facing step. *J. Hydrodyn.* **2019**. [CrossRef]
12. Roos, F.W.; Kegelman, J.T. Control of coherent structures in reattaching laminar and turbulent shear layers. *Aiaa J.* **1986**, *24*, 1956–1963. [CrossRef]
13. Wee, D.; Yi, T.; Annaswamy, A.; Ghoniem, A.F. Self-sustained oscillations and vortex shedding in backward-facing step flows: Simulation and linear instability analysis. *Phys. Fluids* **2004**, *16*, 3361–3373. [CrossRef]
14. Kostas, J.; Soria, J.; Chong, M. Particle image velocimetry measurements of a backward-facing step flow. *Exp. Fluids* **2002**, *33*, 838–853. [CrossRef]
15. Hudy, L.M.; Naguib, A.; Humphreys, W.M. Stochastic estimation of a separated-flow field using wall-pressure-array measurements. *Phys. Fluids* **2007**, *19*, 1093. [CrossRef]
16. Bradshaw, P.; Wong, F.Y.F. The reattachment and relaxation of a turbulent shear layer. *J. Fluid Mech.* **1972**, *52*, 113–135. [CrossRef]
17. Nadge, P.M.; Govardhan, R.N. High Reynolds number flow over a backward-facing step: Structure of the mean separation bubble. *Exp. Fluids* **2014**, *55*, 1657–1678. [CrossRef]
18. Calomino, F.; Tafarojnoruz, A.; De Marchis, M.; Gaudio, R. Experimental and numerical study on the flow field and friction factor in a pressurized corrugated pipe. *J. Hydraul. Eng.* **2015**, *141*, 04015027. [CrossRef]
19. Calomino, F.; Alfonsi, G.; Gaudio, R.; D'Ippolito, A.; Lauria, A.; Tafarojnoruz, A.; Artese, S. Experimental and numerical study of free-surface flows in a corrugated pipe. *Water* **2018**, *10*, 638. [CrossRef]
20. Armaly, B.F.; Durst, F.J.; Pereira, J.C.F.; Schönung, B. Experimental and theoretical investigation of backward-facing step flow. *J. Fluid Mech.* **1983**, *127*, 473–496. [CrossRef]
21. Simpson, R.L. Review—A review of some phenomena in turbulent flow separation. *J. Fluids Eng.* **1981**, *103*, 520–533. [CrossRef]
22. Tani, I.; Iuchi, M.; Komoda, H. *Experimental Investigation of Flow Separation Associated with a Step or a Groove*; Aeronautical Research Institute, University of Tokyo: Tokyo, Japan, 1961; p. 364.
23. McGuinness, M. Flow with a Separation Bubble—Steady and Unsteady Aspects. Ph.D. Thesis, Cambridge Univ., Cambridge, UK, 1978.

24. Chandrsuda, C. A Reattaching Turbulent Shear Layer in Incompressible Flow. Ph.D. Thesis, Dept. of Aeronautics, Imperial College of Science and Technology, London, UK, 1975.
25. Kim, J.; Kline, S.J.; Johnston, J.P. *Investigation of Separation and Reattachment of a Turbulent Shear Layer: Flow over a Backward-Facing Step*; Thermosciences Div., Dept. of Mechanical Engineering, Stanford Univ., Rept. MD-37: Stanford, CA, USA, 1978.
26. Etheridge, D.W.; Kemp, P.H. Measurements of turbulent flow downstream of a rearward-facing step. *J. Fluid Mech. Digit. Arch.* **1978**, *86*, 22. [CrossRef]
27. Lian, Q.X. An experimental investigation of the coherent structure of the flow behind a backward facing step. *Acta Mech. Sin.* **1993**, *9*, 129–133. (In Chinese)
28. Stevenson, W.H.; Thompson, H.D.; Craig, R.R. Laser velocimeter measurements in highly turbulent recirculating flows. *J. Fluids Eng.* **1984**, *106*, 173–180. [CrossRef]
29. Scarano, F.; Benocci, C.; Riethmuller, M.L. Pattern recognition analysis of the turbulent flow past a backward facing step. *Phys. Fluids* **1999**, *11*, 3808–3818. [CrossRef]
30. Ma, X.; Geisler, R.; Schroder, A. Experimental investigation of separated shear flow under subharmonic perturbations over a backward-facing step. *Flow Turbul. Combust.* **2017**, *99*, 71–91. [CrossRef]
31. Driver, D.M.; Seegmiller, H.L.; Marvin, J.G. Time-dependent behavior of a reattaching shear layer. *AIAA J.* **1987**, *25*, 914–919. [CrossRef]
32. Le, H.; Moin, P.; Kim, J. Direct numerical simulation of turbulent flow over a backward-facing step. *J. Fluid Mech.* **1997**, *330*, 349–374. [CrossRef]
33. Aider, J.; Danet, A.; Lesieur, M. Large-eddy simulation applied to study the influence of upstream conditions on the time-dependant and averaged characteristics of a backward-facing step flow. *J. Turbul.* **2007**, *8*, 1–30. [CrossRef]
34. Kopera, M.A.; Kerr, R.M.; Blackburn, H.M.; Barkley, D. Direct numerical simulation of turbulent flow over a backward-facing step. *J. Fluid Mech.* **2014**, under publication.
35. Hu, R.Y.; Wang, L.; Fu, S. Investigation of the coherent structures in flow behind a backward-facing step. *Int. J. Numer. Methods Heat Fluid Flow* **2016**, *26*, 1050–1068. [CrossRef]
36. Qi, E.R.; Huang, M.H.; Li, W.; Zhang, X. An experimental study on the 2D time average flow over a backward facing step via PIV. *J. Exp. Mech.* **2006**, *21*, 225–232. (In Chinese)
37. Fan, X.J.; Wu, S.Q.; Zhou, H.; Xiao, X.; Wang, Y. Investigation on the Characteristics of Water Flow over a Backward Facing Step under High Reynolds Number with Particle Image Velocimetry. In Proceedings of the International Conference on Industrial Technology and Management Science, Tianjin, China, 27–28 March 2015.
38. Ratha, D.; Sarkar, A. Analysis of flow over backward facing step with transition. *Front. Struct. Civ. Eng.* **2015**, *9*, 71–81. [CrossRef]
39. Lugt, H.J.; Gollub, J.P. *Vortex Flow in Nature and Technology*; Wiley: Hoboken, NJ, USA, 1972.
40. Robinson, S.K. Coherent motions in the turbulent boundary layer. *Annu. Rev. Fluid Mech.* **1991**, *23*, 601–639. [CrossRef]
41. Deem, G.S.; Zabusky, N.J. Stationary "V-States," interactions, recurrence, and breaking. *Phys. Rev. Lett.* **1978**, *41*, 518–518. [CrossRef]
42. Wu, J.Z.; Ma, H.Y.; Zhou, M.D. *Introduction to Vorticity and Vortex Dynamics*; Advanced Education Press: Beijing, China, 1993; pp. 463–466.
43. Wang, L. Experimental Study on Coherent Structure in Open-Channel Flow. Ph.D. Thesis, Tsinghua University, Beijing, China, 2009. (In Chinese).

9

Development of Bubble Characteristics on Chute Spillway Bottom

Ruidi Bai [1], Chang Liu [1], Bingyang Feng [1], Shanjun Liu [1] and Faxing Zhang [1,*]

State Key Laboratory of Hydraulics and Mountain River Engineering, Sichuan University, Chengdu 610065, China; bairuidiscu@163.com (R.B.); sculiuchangvip@163.com (C.L.); 18625801588@163.com (B.F.); drliushanjun@vip.sina.com (S.L.)

* Correspondence: zhfx@scu.edu.cn.

Abstract: Chute aerators introduce a large air discharge through air supply ducts to prevent cavitation erosion on spillways. There is not much information on the microcosmic air bubble characteristics near the chute bottom. This study was focused on examining the bottom air-water flow properties by performing a series of model tests that eliminated the upper aeration and illustrated the potential for bubble variation processes on the chute bottom. In comparison with the strong air detrainment in the impact zone, the bottom air bubble frequency decreased slightly. Observations showed that range of probability of the bubble chord length tended to decrease sharply in the impact zone and by a lesser extent in the equilibrium zone. A distinct mechanism to control the bubble size distribution, depending on bubble diameter, was proposed. For bubbles larger than about 1–2 mm, the bubble size distribution followed a—5/3 power-law scaling with diameter. Using the relationship between the local dissipation rate and bubble size, the bottom dissipation rate was found to increase along the chute bottom, and the corresponding Hinze scale showed a good agreement with the observations.

Keywords: chute aerator; spillway bottom; air concentration; air bubble frequency; air bubble chord length

---

1. Introduction

Chute aerators, regarded as a cost-effective method, have been widely used owing to their proven success as a countermeasure in the Grand Coulee dam since 1960 [1]. Previous studies have provided a significant amount of information on the macroscopic air-water flow properties [2–13], such as the air concentration, air entrainment and detrainment. Rutschmann and Hager [2], combining both model and prototype data, proposed two approaches to calculate the air entrainment coefficient $\beta = q_a/q$, where q_a = air discharge and q = water discharge. The air entrainment coefficient was also investigated by, among others, Shi [14], Pinto [15,16], and Low [17]. Chanson [3] investigated the characteristics of chute aerator flow on the distributions of various bubble parameters (void fraction, velocity, air discharge, turbulent diffusivity, etc.) for different inflow Froude numbers. He also provided a better understanding of the air entrainment processes and found a strong detrainment process occurred in the impact zone. Kramer et al. [5] divided the chute aerator into five zones: inflow region; air detrainment region; minimum air concentration region; air entrainment region; and uniform mixture flow region, and systematically investigated the air detrainment gradients in the far-field of chute aerators. Pfister and Hager [6–8] measured the air concentration and divided the chute aerator into three zones: jet zone, reattachment and spray zone, and far–field zone. The authors also discussed the air entrainment and proposed two comprehensive methods to calculate the average and bottom air concentration development using deflectors and offsets along the chute. Avoiding the upper aeration effect, Bai et al. [9] divided the chute aerator into three zones and found that an intensive air detrainment occurred in the impact zone and the detrained air escaped into the cavity zone. The previous experimental investigations are summarized in Table 1.

Table 1. Experimental investigations of Chute Spillways.

Reference	V_0 (m/s)	h_0 (m)	F_0	Remarks
Chanson [3]	9.2 6.2–11.3 5.3	0.023 0.035 0.081	19.5 10.5–19.5 6.0	Model (W = 0.25 m)
Prifser [6–8]	4.7–7.5	0.041–0.066	7.4–9.3	Model (W = 0.30 m)
Bai [9,10]	4.0–9.0	0.15	3.3–7.4	Model (W = 0.25 m)
Shi [14]	14.0	0.058	18.6	Model (W = 0.20 m)
Pinto [15,16]	19.5–27.7 19.9–32.7 19.9–36.2	0.38–1.06 0.38–1.43 0.38–1.29	7.1–10.3 8.7–11.1 10.3–11.4	Prorotype (Foz do Areia)
Low [17]	4.2–9.5	0.05	6–13.5	Model (W = 0.25 m)

Many studies concerning the minimum or critical air concentration to avoid cavitation erosion have been conducted [18,19]. Peterka [18] found that when the average air concentration was between 0.01 and 0.06, no cavitation erosion was observed. Up to now, there is no reliable design guideline for the distance required between two aerators to produce sufficient bottom air concentration [5]. Based on prototype observations on Russian dams, Semenkovand Lantyaev [20] provided an average bottom air concentration decay of 0.40% to 0.80% per chute meter.

However, air concentration governs the comprehensive performance of the bubbles, including their chord length and frequency form. In particular, Robinson [21], Xu et al. [22], and Brujan [23] adopted advanced techniques to investigate the mechanism of erosion damage and found that the interaction between the air bubbles and cavitation bubbles was the mechanism that assisted in the prevention of erosion damage. These studies indicate that the bubble size and frequency are significant parameters for the mechanism of erosion damage. However, negligible research has been conducted on the spillway bottom air microcosmic characteristics such as the bottom bubble frequency and chord size. Such investigations are difficult because of the complex nature of the flow and strong impaction of the bottom. In the present study, the development of the bottom bubble characteristics was measured and analyzed.

2. Hydraulics Model

The experiments were conducted in a rectangular chute that was 0.25-m wide with a variable bottom slope $5.71° \leq \alpha \leq 18.2°$ (Figure 1). The measured section was 5-m long and consisted of a smooth convergent nozzle with a width of 0.25 m and height of 0.15 m. The approach flow depth was h_0 = 0.15 m and the emergence angle was $0° \leq \theta_0 \leq 14.1°$. The water discharge was supplied by a large tank (3.1-m wide, 3.2-m long, and 9.2-m high) and was measured using a rectangular sharp-crested weir with an accuracy of approximately 1%. In order to gain sufficient velocity, the nozzle was located in the bottom of the tank. A range of approach flow velocities was set from 4.0 m/s to 9.0 m/s. Observations indicated that the approach of the flow was stable. Hence, the Froude number F_0 $(V_0/(gh_0)^{0.5})$ was within the range of 3.3 to 7.4 (Table 2), where g was the gravity. x was the streamwise coordinate along the chute bottom, h_s was the offset height, L was the cavity length, and L_m and L_D were the specific lengths [9].

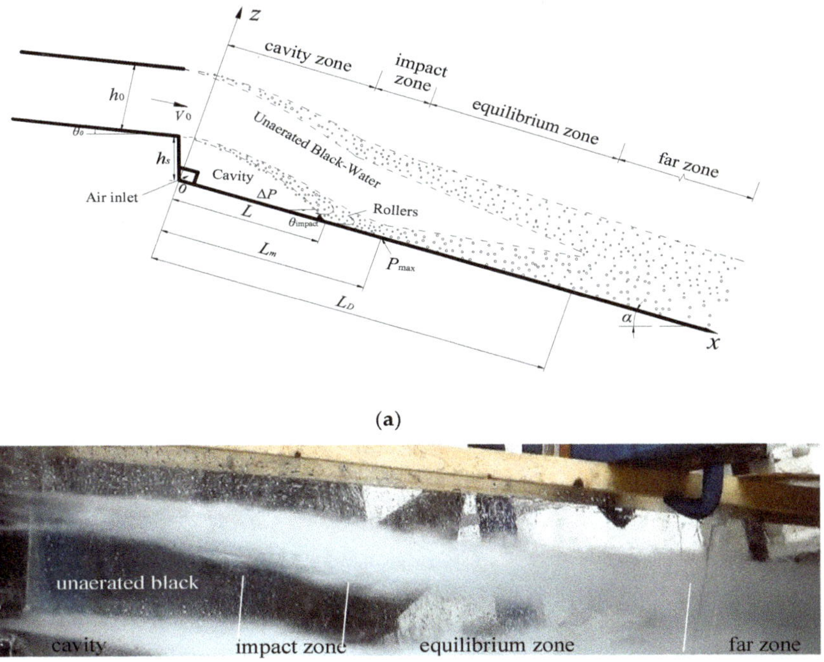

(a)

(b)

Figure 1. Definition sketch with relevant parameters (**a**) and side view of chute aerator flow with high speed camera (**b**).

Table 2. Experimental conditions of chute aerator flow.

Run	V_0 (m/s)	h_s (m)	θ_0 (°)	α (°)	F_0
S1	5.0–7.0	0.03	5.71	12.5	3.8–5.8
S2	4.0–6.0	0.05	12.5	18.2	3.3–4.9
S3	5.0–7.0	0.045	0	5.7	3.8–5.8
S4	4.0–6.0	0.045	5.7	14.1	3.3–4.9
S5	6.0–9.0	0.045	14.1	14.1	4.9–7.4

All measurements were performed on the channel centre line. The air-water flow properties were measured with a phase-detection needle probe. The working principle of a conductivity probe was based on the difference in the voltage indices at the platinum tip between the air and water phases. The response time of this sensor was less than 10 μs. The signals from the conductivity probe were recorded at a scan rate of 100 kHz per for a 40 s scan period. The accuracy of the air concentration was $\Delta C/C = 3\%$, and the bubble frequency was $\Delta f/f = 1\%$, where C was the air concentration and f was the bubble frequency. The length of the impact zone was very short. A few studies in the impact zone have been performed with a space of 0.03 m along the chute bottom.

3. Basic Chute Aerator Flow Properties

Air Concentration and Bubble Frequency Profiles

A typical photo of the offset-aerator flow is shown in Figure 2, where four zones were introduced. The typical distributions of the air concentration and bubble frequency in the different zones are depicted in Figure 2 as a function of the dimensionless distance perpendicular to the bottom, i.e., z/h_0. Owing to a sufficient depth of the approach flow, the unaerated water prevented the upper aeration effect.

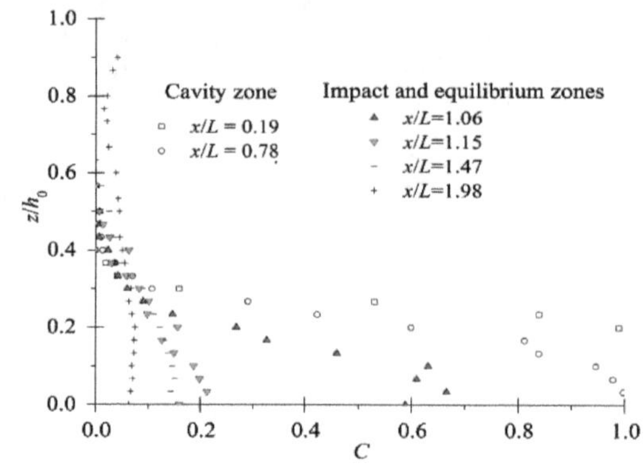

(a)

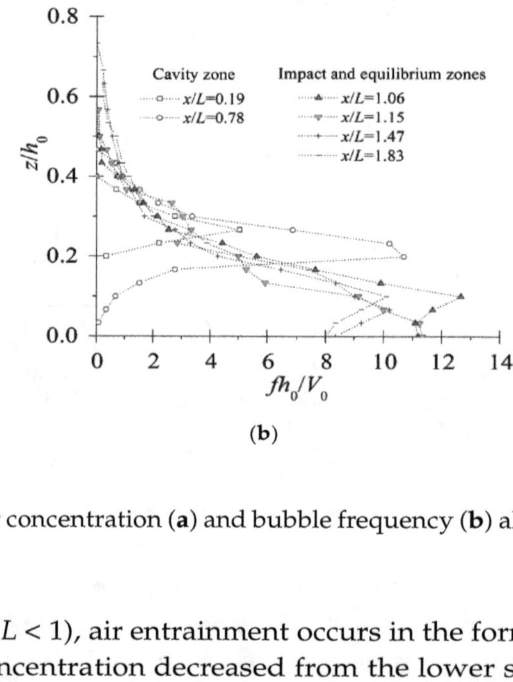

(b)

Figure 2. Distributions of air concentration (**a**) and bubble frequency (**b**) along the chute (S5: $V_0 = 9$ m/s, $F_0 = 7.4$).

In the cavity zone ($0 < x/L < 1$), air entrainment occurs in the form of air bubbles entrapped into the lower flow and the air concentration decreased from the lower surface to the vertical direction. At $x/L = 1$, the lower jet impacted on the chute bottom. Further downstream ($x/L > 1$), the distributions of air concentration were different from the cavity zone. Figure 2a shows that the air concentration distributions downstream of the cavity zone, exhibiting an air concentration peak in the turbulent shear region, whereas the air concentration increases from the chute bottom to its maximum value, and then decreases in the vertical direction.

Note that: (1) for $x > L$, the maximum value air concentration C_m decreased sharply in the impact zone, while in the equilibrium zone, the peak air concentration C_m decreased slightly. (2) Experimental results indicated the existence of a maximum value f_m in bubble frequency (Figure 2b), but its location differed from the locus of the maximum void fraction as noticed by Bai et al. [10]. In the cavity, the location of f_m approached $C = 0.50$ with the development of the low jet aeration. The location of f_m was approximately the same as that of the section maximum air concentration C_m at a section in the impact zone, but lower than that in the equilibrium zone.

For more than 25 experiments conducted in this model, only one set of experiments is shown herein to illustrate the basic chute aerator flow property [9]. However, equations presented hereafter are based on all experiments.

4. Bottom Air Bubble Characteristics

The entrained air from the chute aerator aims to prevent erosion damage. Recent investigations have indicated that the bubble size and frequency are the significant factors preventing erosion damage [22,23]. The air bubbles generated in the cavity zone travelled through the impact zone to continue downstream. For the chute aerator flows, a large amount of air entrained in the cavity zone was only partially transported to the downstream flow. Under the effect of rollers, large bubbles broke into small bubbles before being transported downstream zone simultaneously in the impact zone.

4.1. Bottom Air Concentration and Bubble Frequency

Bottom air concentration C_b was defined experimentally using the values measured closest to the model chute bottom (z = 1–1.5 mm). The experimental data of bottom air concentration C_b is presented in Figure 3. Note the following: (1) the decay rate of C_b was different in the impact ($L < x < L_m$) and equilibrium zones ($L_m \leq x < L_D$); C_b decreased sharply in the impact zone and decreased slightly in the equilibrium zone. In the impact zone, bottom air concentration C_b = 0.64 at x/L = 1.06 and it decreased to 0.18 at x/L = 1.16. In contrast, the bottom air concentration decreased slightly in the equilibrium zone, with concentration C_b = 0.145 at x/L = 1.31 decreasing to 0.064 at x/L = 1.91. (2) With the increase in F_0, the values of C_b increase in the present experiments, in agreement with the results of Pfister and Hager [7]. (3) The comparison of the measured values of C_b with the values from the formula proposed by Pfister and Hager [7] is shown in Figure 3, which displays a similar trend downstream of the chute aerator. However, the present values are generally larger than those of Pfister and Hager [7]. At a sufficient water depth, the detrained air bubbles cannot escape into the atmosphere through the upper surface; it can only escape into the cavity zone, which may cause the level of C_b to be larger than that of previous studies. In addition, for a shorter length of the impact zone, the fewer measurement sections may cause an insufficient sharp decrease. (4) It can be concluded that the air transportation regularity is different in the impact and equilibrium zones.

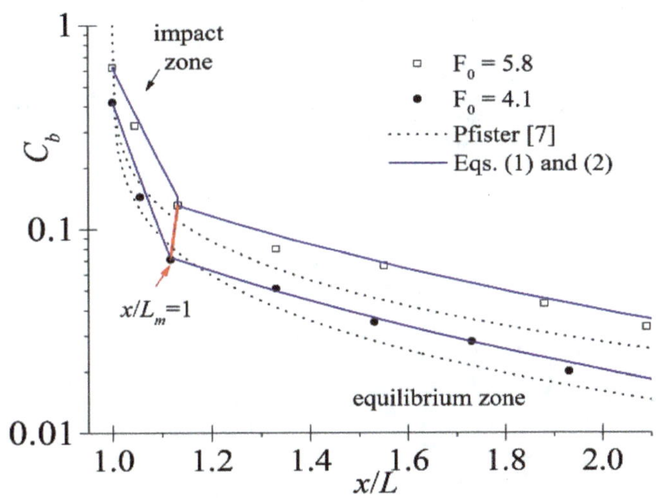

(a)

Figure 3. *Cont.*

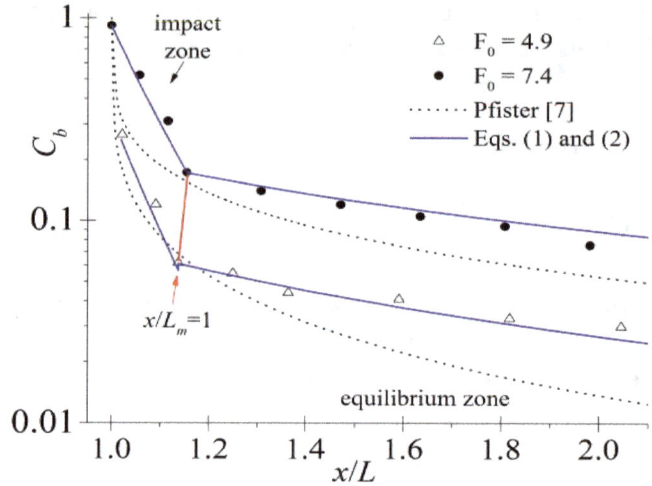

(b)

Figure 3. Bottom air concentration along the chute and compared with Pfister [7] (**a**): S2; (**b**): S5.

From the experimental data analysis, bottom air concentration C_b can be expressed by the following equation (Figure 3):

$$\frac{C_b}{(C_b)_L} = \left(\frac{x}{L}\right)^{0.985F_0-17.5} \text{ for } L < x < L_m,\ R^2 = 0.89 \tag{1}$$

$$\frac{C_b}{(C_b)_{Lm}} = \left(\frac{x}{L_m}\right)^{0.071F_0-1.73} \text{ for } x > L_m,\ R^2 = 0.97 \tag{2}$$

where R is the correlation coefficient, $(C_b)_L$ is the bottom air concentration at $x = L$, $(C_b)_{Lm}$ is the bottom air concentration at $x = L_m$.

The experimental data of bottom bubble frequency $f_b h_0/V_0$ along the chute are displayed in Figure 4. There are a few reports on the chute bottom bubble frequency of the lower jet downstream of the chute aerator. Note the following: (1) $f_b h_0/V_0$ decreased in both the impact and equilibrium zones, whereas decay rate of $f_b h_0/V_0$ differed slightly. For $F_0 = 7.4$, $f_b h_0/V_0 = 11.2$ at $x/L = 1.06$, and it decreased to 10.88 at $x/L = 1.17$ in the impact zone (Figure 4d). In contrast, the bottom bubble frequency decreased slightly more in the equilibrium zone, with $f_b h_0/V_0 = 9.44$ at $x/L = 1.31$ decreasing to 6.22 at $x/L = 1.91$. (2) In contrast, $f_b h_0/V_0$ decreased slightly in the impact zone, and there was a large variation in air bubble concentration C_b. (3) Furthermore, $f_b h_0/V_0$ increased with increasing F_0.

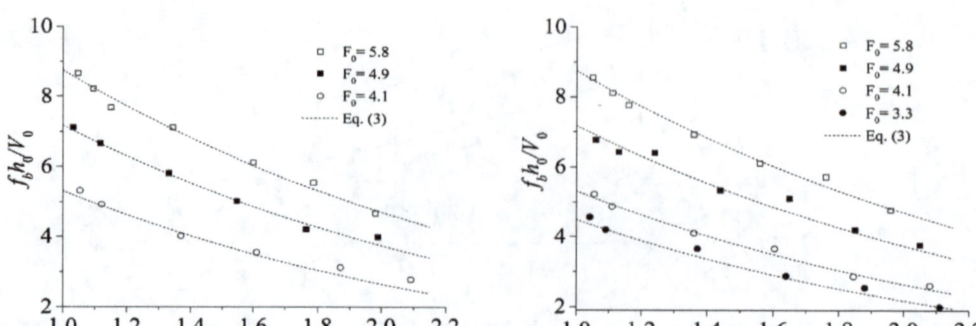

(a) (b)

Figure 4. *Cont.*

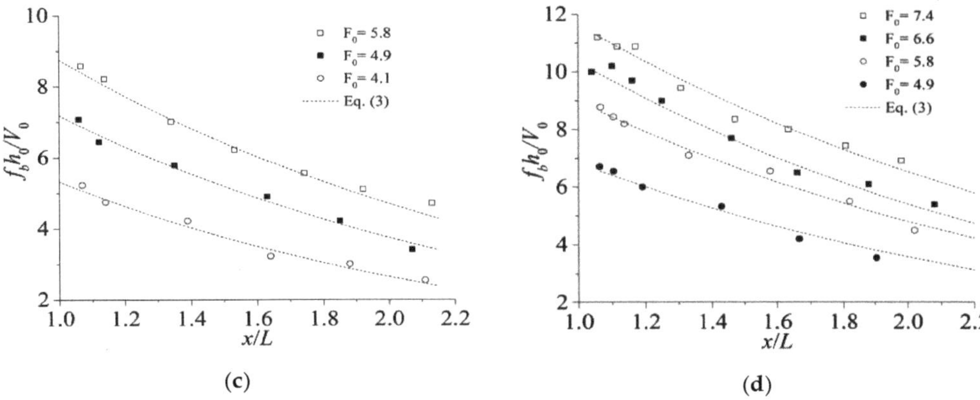

Figure 4. Bottom air bubble frequency along the chute (**a**): S2; (**b**): S3; (**c**): S4; (**d**): S5.

Overall, the present data could be described by

$$\frac{f_b h_0}{V_0} = 2.81F_0 \times \exp\left(-\frac{x}{L}F_0{}^{-0.27}\right) \text{ for } x > L,\ R^2 = 0.94 \tag{3}$$

where R was the correlation coefficient.

4.2. Bottom Bubble Chord Length Distributions

The chord length distribution of the air bubbles shows microscopic variations passing a measurement point in the stream-wise direction. For all the flow conditions, the experimental results exhibited a broad range of the chord length, varying from less than 0.1 mm to more than 30.0 mm. In this study, the chord length interval was 0.1 mm. The chord lengths are not the bubble diameters, but are the characteristic stream-wise bubble sizes.

Garrett et al. [24] and Deane et al. [25] indicated that the bubble chord length distribution N_a depended on the dissipation rate ε and bubble diameter d_{ab}, and the scaling law followed as

$$N_{\mathrm{a}} \propto A\varepsilon^{1/3} d_{ab}{}^{-5/3} \tag{4}$$

where A was a constant, $N_a = N_t/t$, N_t was the number of air bubbles detected in the scan period t (t = 40 s).

The distributions of the bottom bubble chord length in the impact and equilibrium zones are depicted in Figure 5. In the impact zone, the air concentration was within the range of $0.10 < C < 0.90$, such that the intermediate region neither corresponded to the high-air concentration region nor to the bubbly flow region. In the equilibrium zone, the maximum air concentration was less than 0.3, so that the zone corresponded to the bubbly flow region. Note the following: (1) the distributions of bottom bubble chord length were skewed with a large proportion of small bubbles. Compared with the impact region, the range of bubble chord length distributions decreased. In the impact zone, the largest probability of bubble chord lengths was concentrated in the range $0.1 \leq d_{ab} \leq 30$ mm, while in the equilibrium zone, the highest probability of bubble chord lengths was concentrated in the range $0.1 \leq d_{ab} \leq 10$ mm. (2) The curve of distribution of the bottom bubble chord sizes was also sharper and narrower along the chute. The power-law scaling of bubble density on diameter is about $\beta = -5/3$ for larger than about 1–2 mm bubbles in the impact and equilibrium zones. (3) In the impact zone, β showed a slight increase from -1.2 at $x/L = 1.06$ to $-5/3$ at $x/L = 1.15$ due to the air detrainment. In the equilibrium zone, β showed a slightly increase from $-5/3$ at $x/L = 1.21$ to -2.4 at $x/L = 1.64$ due to the buoyancy effect.

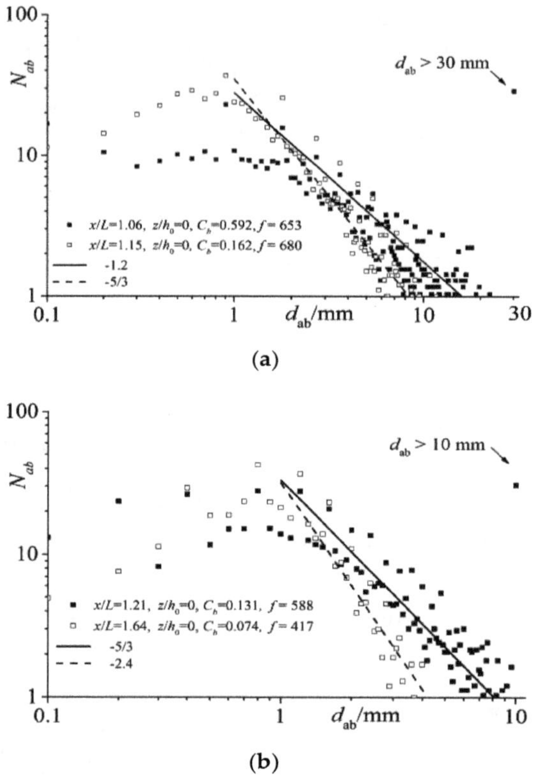

Figure 5. Bottom bubble chord length distributions in the impact zone (**a**) and equilibrium zone (**b**) (S5, V_0 = 8 m/s, F_0 =6.6).

4.3. Bottom Turbulent Dissipation Rate

Xu et al. [22,23] and Brujan et al. proved that the size of air bubbles near the bottom was a significant parameter for the collapse process of cavitation bubbles. There was little information to investigate the air bubble variations on the chute bottom. The equilibrium zone belonged to the bubbly flow region and the dynamic behavior of the flow was considered to be a steady state air-water two-phase flow.

Hinze [26] indicated that any fluid gas bubbles or droplets were likely to fragment if the pressure forced on its surface exceeds the restoring forces associated with surface tension.

$$\rho \frac{u^2}{\sigma} d_c = \mathrm{W_c} \tag{5}$$

where $\mathrm{W_c}$ = the critical Weber numbers, u = the turbulent velocity field on the scale of the bubble. If the fluctuating velocity field was assumed to be described by Kolmogorov's inertial subrange, so that $u^2 = 2\varepsilon^{2/3}{d_{ab}}^{2/3}$, then only bubbles with diameters larger than the Hinze scale would fragment [17–19].

$$d_c = \left(\frac{\mathrm{W_c}\sigma}{2\rho}\right)^{3/5} \varepsilon^{-2/5} \tag{6}$$

where d_c = the Hinze scale, ε = turbulent dissipation rate, which was estimated rather than measured directly. Recent experiments suggest that $\mathrm{W_c}$ lied in the range 0.585–4.7, and we have used the value $\mathrm{W_c}$ = 1.0 (Table 3). For this to exceed a critical value, $\mathrm{W_c}$ required that $d > d_c$.

Table 3. Critical Weber numbers for the splitting of air bubbles in water flows.

Reference	W_c	Fluid	Flow Situation	Comments
Hinze [26]	0.585		Two co-axial cylinders, the inner one rotating	Dimensional analysis.
Killen [27]	1.017	Air bubbles in water	Turbulent boundary layer	Experimental data. V in the range 3.66 to 18.3 m/s.
Lewis and Davidson [28]	2.35	Air and Helium bubbles in water and Fluorisol	Circular jet	Experimental data. V in the range 0.9 to 2.2 m/s.
Evans et al. [29]	0.60	Air bubbles in water	Confined plunging water jet	Experimental data. V in the range 7.8 to 15 m/s.
Chanson [3]	1.00	Air bubbles in water	Self-aerated flow	Experimental data. V in the range 3.2 to 5.3m/s.
Martínez-Bazán et al. [30,31]	1.00	Air bubbles in water	Submerged water jet	Experimental data. V = 17.0 m/s.
Deane and Stokes [25]	4.7	Air bubbles in water	Breaking waves	Experimental data.
Bai et al. [9]	1.00	Air bubbles in water	Chute aerator flow	Experimental data. V in the range 4.0 to 9.0 m/s.

If the Hinze scale was known, the turbulent energy dissipation rate ε on the chute bottom could be estimated from Equation (7). Hinze [26] used the definition that 95% of the air is contained in bubbles with a diameter less than d_c and the calculated turbulent energy dissipation rate ε for Deane's [18] data was much smaller. Garrett et al. [24] used the results of Martínez-Bazán et al. [30,31] and established a one-to-one relationship between the local dissipation rate and bubble size.

$$\varepsilon = (1/2)^{5/2}\left(\frac{\sigma}{\rho}\right)^{3/2}\int_0^{\infty}\left(\frac{d_{ab}}{2}\right)^{1/2} N_a dd_{ab} \Big/ \int_0^{\infty}\left(\frac{d_{ab}}{2}\right)^{3} N_a dd_{ab} \qquad (7)$$

The bottom dissipation rate ε calculated from Equation (7) is shown in Figure 6 and tends to increase in the measured distance. From Equation (6), a change in dissipation rate ε from 0.10 to 0.75 $m^2\,s^{-3}$ corresponded to a change in Hinze scale from 5.5 to 2.4 mm. Thus, the estimate of 2.4 mm for the Hinze scale was in good agreement with the observed change in power-law scaling of the bubble distribution at around 1–2 mm (Figure 5b). Because bubbles smaller than the Hinze scale are stabilized by surface tension forces, a reasonable estimate for this scale is the diameter of the smallest bubbles observed to fragment. In the present study, the bubble chord lengths were smaller than the bubble diameters, due to the measurement technique utilizing conductivity probes. A challenge remains to observe the bubble diameter in a model facility, or even a prototype facility.

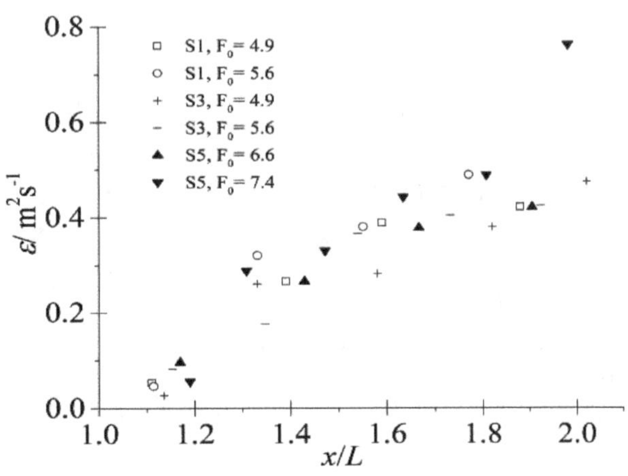

Figure 6. Calculated values of turbulent energy dissipation rate ε on the chute bottom.

5. Conclusions

This study presented new information on the bottom air bubble characteristics in the downstream of chute offset-aerators, which were measured with an accurate measurement technique utilizing conductivity probes. For eliminating the upper aeration, the bottom concentration, bubble frequency, bubble chord length, and bottom dissipation rate were analyzed along the chute longitudinal direction. The present study complemented the existing studies on the bubble properties of chute aerator flows.

The bottom air concentration was found to decrease sharply in the impact zone and to decrease slightly in the equilibrium zone. The bottom air bubble frequency was found to decrease slightly along the chute bottom. A formula to predict the bottom bubble frequency in the impact and equilibrium zones was proposed.

Observations showed that the range of probability of bottom bubble chord lengths tended to decrease sharply in the impact zone and to a lesser extent in the equilibrium zone. The bubbles larger than the Hinze scale were subject to fragmentation and showed $-5/3$ power-law scaling with diameter. Using the relationship between the local dissipation rate and bubble size [24], the bottom dissipation rate was found to increase along the chute bottom. The Hinze scale estimated from Equation (6) showed a good agreement with the observed change in power-law scaling of the bubble distribution.

Author Contributions: Conceptualization, R.B., F.Z. and S.L.; Methodology, R.B. and F.Z.; Software, C.L.; Validation, R.B.; Formal Analysis, R.B. and F.Z.; Investigation, R.B.; Resources, S.L.; D.C., B.F.; Writing-Original Draft Preparation, R.B.; Writing-Review & Editing, R.B.; Visualization, B.F.; Supervision, F.Z.; Project Administration, R.B., F.Z. and S.L.; Funding Acquisition, R.B., F.Z. and S.L.

References

1. Pan, S.; Shao, Y. Scale effects in modeling air demand by a ramp slot. *Scale Eff. Model. Hydraul. Struct.* **1984**, *4*, 1–5.
2. Rutschmann, P.; Hager, H.W. Air entrainment by spillway aerators. *J. Hydraul. Eng.* **1990**, *116*, 765–782. [CrossRef]
3. Chanson, H. Air bubble entrainment in turbulent water jets discharging into the atmosphere. *Aust. Civ. Eng. Trans.* **1996**, *39*, 39–48.
4. Kramer, K.; Hager, W.H. Air transport in chute flows. *Int. J. Multiph. Flow* **2005**, *31*, 1181–1197. [CrossRef]
5. Kramer, K.; Hager, W.H.; Minor, H.-E. Development of air concentration on chute spillways. *J. Hydraul. Eng.* **2006**, *132*, 908–915. [CrossRef]
6. Pfister, M.; Hager, W.H. Chute aerators I: Air transport characteristics. *J. Hydraul. Eng.* **2010**, *136*, 352–359. [CrossRef]
7. Pfister, M.; Hager, W.H. Chute aerators II: Hydraulic design. *J. Hydraul. Eng.* **2010**, *136*, 360–367. [CrossRef]
8. Pfister, M.; Hager, W.H. Chute aerators: pre aerated approach flow. *J. Hydraul. Eng.* **2011**, *137*, 1452–1461. [CrossRef]
9. Bai, R.; Zhang, F.; Liu, S.; Wang, W. Air concentration and bubble characteristics sownstream of a chute aerator. *Int. J. Multiph. Flow* **2016**, *87*, 156–166. [CrossRef]
10. Bai, R.; Liu, S.; Tian, Z.; Wang, W.; Zhang, F. Experimental investigations on air-water flow properties of offse-aerator. *J. Hydraul. Eng.* **2018**, *144*, 04017059. [CrossRef]
11. Li, S.; Zhang, J.M.; Xu, W.L.; Chen, J.G.; Peng, Y. Evolution of pressure and cavitation on side walls affected by lateral divergence angle and opening of radial gate. *J. Hydraul. Eng.* **2016**, *142*, 05016003. [CrossRef]
12. Bai, Z.L.; Peng, Y.; Zhang, J.M. Three-dimensional turbulence simulation of flow in a V-shaped stepped spillway. *J. Hydraul. Eng.* **2017**, *143*, 06017011. [CrossRef]
13. Zhang, J.M.; Chen, J.G.; Wang, Y.R. Experimental study on time-averaged pressures in stepped spillway. *J. Hydraul. Res.* **2012**, *50*, 236–240. [CrossRef]
14. Shi, Q.; Pan, S.; Shao, Y.; Yuan, X. Experimental investigation of flow aeration to prevent cavitation erosion by a deflector. *J. Hydraul. Eng.* **1983**, *3*, 1–13.
15. Pinto, N.L.S.; Neidert, S.H.; Ota, J.J. Aeration at high velocity flows—Part one. *Int. Water Power Dam Constr.* **1982**, *34*, 34–38.
16. Pinto, N.L.S.; Neidert, S.H.; Ota, J.J. Aeration at high velocity flows—Part two. *Int. Water Power Dam Constr.* **1982**, *34*, 42–44.

17. Low, H.S. Model Studies of Clyde Dam Spillway Aerators. Ph.D. Thesis, University of Canterbury, Christchurch, New Zealand, 1986.
18. Peterka, A.J. The effect of entrained air on cavitation pitting. In *Proceedings of the Minnesota International Hydraulic Convention*; ASCE: Reston, VA, USA, 1953; pp. 507–518.
19. Russell, S.O.; Sheenan, G.J. Effect of entrained air on cavitation damage. *Can. J. Civ. Eng.* **1974**, *1*, 97–107. [CrossRef]
20. Semenkov, V.M.; Lentyaev, L.D. Spillway with nappe aeration. *Hydrotech. Constr.* **1973**, *7*, 437–441. [CrossRef]
21. Robinson, P.B.; Blake, J.R.; Kodama, T.; Shima, A.; Tomita, Y. Interaction of cavitation bubbles with a free surface. *J. Appl. Phys.* **2001**, *89*, 8225–8237. [CrossRef]
22. Xu, W.; Bai, L.; Zhang, F. Interaction of a cavitation bubble and an air bubble with a rigid boundary. *J. Hydrodyn.* **2010**, *22*, 503–512. [CrossRef]
23. Brujan, E.A.; Matsumoto, Y. Collapse of micrometer–sized cavitation bubbles near a rigid boundary. *Microfluid. Nanofluid.* **2012**, *13*, 957–966. [CrossRef]
24. Garrett, C.; Li, M.; Farmer, D. The connection between bubble size spectra and energy dissipation rates in the upper ocean. *J. Phys. Oceanogr.* **2000**, *30*, 2163–2171. [CrossRef]
25. Deane, G.B.; Stokes, M.D. Scale dependence of bubble creation mechanisms in breaking waves. *Nature* **2002**, *418*, 839–844. [CrossRef] [PubMed]
26. Hinze, J.O. Fundamentals of the hydrodynamic mechanism of splitting in dispersion processes. *AIChE J.* **1955**, *1*, 289–295. [CrossRef]
27. Killen, J.M. Maximum stable bubble size and associated noise spectra in a turbulent boundary layer. In *Proceedings of the Cavitation and Polyphase Flow Forum*; ASME: New York, NY, USA, 1982; pp. 1–3.
28. Lewis, D.A.; Davidson, J.F. Bubble Splitting in Shear Flow. *Trans. I. Chem. Eng.* **1982**, *60*, 283–291.
29. Evans, G.M.; Jameson, G.J.; Atkinson, B.W. Prediction of the bubble size generated by a plunging liquid jet bubble column. *Chem. Eng. Sci.* **1992**, *47*, 3265–3272. [CrossRef]
30. Martínez-Bazán, C.; Montañés, J.L.; Lasheras, J.C. On the break up of an air bubble injected into a fully developed turbulent flow. Part 1. Breakup frequency. *J. Fluid Mech.* **1999**, *401*, 157–182. [CrossRef]
31. Martínez-Bazán, C.; Montañés, J.L.; Lasheras, J.C. On the break up of an air bubble injected into a fully developed turbulent flow. Part 2. Size PDF of the resulting daughter bubbles. *J. Fluid Mech.* **1999**, *401*, 183–207. [CrossRef]

10

Numerical Study of the Velocity Decay of Offset Jet in a Narrow and Deep Pool

Xin Li, Maolin Zhou *, Jianmin Zhang * and Weilin Xu

State Key Laboratory of Hydraulics and Mountain River Engineering, Sichuan University, Chengdu 610065, China; bixizhen@foxmail.com (X.L.); xuwl@scu.edu.cn (W.X.)

* Correspondence: water636@126.com (M.Z.); zhangjianmin@scu.edu.cn (J.Z.)

Abstract: The present study examines the configuration of an offset jet issuing into a narrow and deep pool. The standard k-ε model with volume-of-fluid (VOF) method was used to simulate the offset jet for three exit offset ratios (OR = 1, 2 and 3), three expansion ratios (ER = 3, 4 and 4.8), and different jet exits (circular and rectangular). The results clearly show significant effects of the circumference of jet exits (L_{exit}) in the early region of flow development, and a fitted formula is presented to estimate the length of the potential core zone (L_{PC}). Analysis of the flow field for OR = 1 showed that the decay of cross-sectional streamwise maximum mean velocity (U_m) in the transition zone could be fitted by power law with the decay rate n decreased from 1.768 to 1.197 as the ER increased, while the decay of U_m for OR = 2 or 3 was observed accurately estimated by linear fit. Analysis of the flow field of circular offset jet showed that U_m for OR = 2 decayed fastest due to the fact that the main flow could be spread evenly in floor-normal direction. For circular jets, the offset ratio and expansion ratio do not affect the spread of streamwise velocity in the early region of flow development. It was also observed that the absence of sudden expansion of offset jet is analogous to that of a plane offset jet, and the flow pattern is different.

Keywords: offset jet; potential core; decay rate; k-ε model

---

1. Introduction

Offset jets are common in drainage systems [1], slot fishways [2], and hydraulics engineering [3]. The offset jet is formed when a fluid jet discharges into an ambient medium above the floor and parallel to the axis of the jet exit but which is offset by a certain height. The submerged offset jets in the narrow and deep pool are more complex flow because it can easily be affected by side walls.

The offset jet flow was widely used as energy dissipation downstream of hydraulic structures—for example, the submerged hydraulic jump at an abrupt drop could be considered as an offset jet flow. From an engineering point-of-view, the length of the stilling basin should be as short as possible. The focus of this paper is to find the factors affect the efficiency of energy dissipation of the offset jet flow. Figure 1 shows a sketch of the submerged offset jet in a pool. The jet is mainly divided into three regions according to the relationship with the floor, viz., recirculation region, impingement region, and wall jet region. The Cartesian coordinate system is used with x, y, and z representing the streamwise, lateral direction, and floor-normal direction, respectively. The exit of the jet is at $x = 0$, and the symmetry plane is at $y = 0$, the pool bottom is at $z = 0$. The symbols L_x, L_y, L_z, S, d, h_t, U_j, and U_m represent the length, width, and height of jet pool, offset height, the diameter of jet exit, depth of tailwater, bulk velocity, and cross-sectional streamwise maximum mean velocity, respectively. Likewise, if the jet exit is rectangular, a_0, b_0 represents the height and width of the rectangle, respectively. The flow field in jet pool can be divided into three zones by the decay of U_m: (a) the potential core zone, where U_m was equal to the jet exit velocity, U_j; (b) transition zone, where the decay of U_m was rapidly in this zone; and (c) fully development zone, where the decay of U_m became very slow.

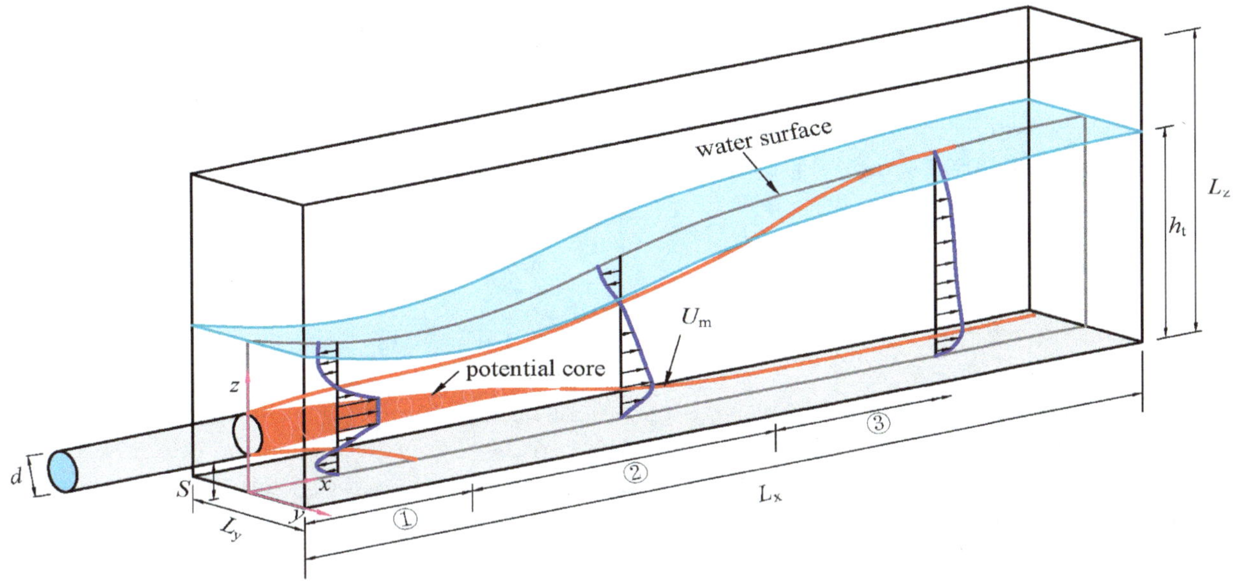

Figure 1. the sketch of the submerged offset jet in a chamber. Flow divided by a relationship with bottom floor: ① recirculation region, ② impingement region, and ③ wall jet region.

The recent studies on submerged offset jets were carried out by Subhasish Dey [4] to investigate the vertical profiles of time-averaged velocity components, and Reynolds stresses in jet flows for offset ratios OR = 2.8–5.1, submergence ratio SR = 2.4–5.6, expansion ratio ER = 1, and jet Reynold number R_0 = 28,475–80,730, where OR = S/a_0 (for circular jet exit $a_0 = d$), SR = h_t/S, ER = L_y/b_0, and $R_0 = (U_0a_0)/v$, v is the coefficient of kinematic viscosity of water. They concluded important characteristic length of submerged offset jets, such as the length of the recirculation region and impingement region, are expressed as a function of the R_0, OR, and SR. In another attempt, Bhuiyan [5] analyzed the characteristics of submerged turbulent plane offset jets (OR = 0.5–3.6) in channels with rough beds and shallow tailwater depths. The results indicated that for an offset height larger than the jet thickness, the peak velocity, the flow momentum decay faster in the downstream direction in an offset jet than in a turbulent plane wall jet. Durand et al. [6] studied the effect of Reynolds number on 3D offset jets both experimentally and numerically for R_0 = 34,000, 53,000 and 86,000. The results indicated the floor-normal location of maximum mean velocity and jet spread to be independent of Reynolds number. The investigated by Nyantekyi-Kwakye [7] on a 3D rectangular offset jet, performed at three offset ratios (OR = 0, 2, and 4) and revealed that large-scale structures dominate the inner layer of the wall jet region. For 3D circular offset jets (with OR ranging from 0–3.5), Agelin-chaab and Tachie [8] using a planar particle image velocimetry (PIV) system conducted the experiments to study the velocity. They observed that OR influenced both the decay of U_m and growth of the shear layer within the developing region. Besides, there have been other experimental studies on 3D offset jets [9–11].

Given the brief overview, it can be concluded that the mechanics of offset jet has been studied extensively. However, the hydraulic properties of submerged offset jets in a narrow and deep pool have not been completely understood, although it was widely used as an efficient energy dissipator in hydraulic engineering located in narrow canyons.

This paper addresses the decay of velocity of offset jets in the narrow and deep pool with various offset ratios, expansion ratios and jet exit shapes, through numerical simulations using a 3D computational fluid dynamics (CFD) model with the k-ε turbulence model coupled with VOF method, which was confirmed perform well in jet flow [10,12–17].

2. Numerical Simulation

2.1. Mathematical Model

An RNG k-ε model has been compared to a standard k-ε model for computing transient jets by Abraham and Maji [18] and concluded that the RNG k-ε model results in predictions of greater mixing in the jets relative to the standard model. The study carried by Nasr and Lai [19] indicated that the standard k-ε turbulence model predicts better than RNG k-ε and Reynolds stress turbulence model. In this study, the standard k-ε model [20] was used as the turbulence closure

$$\frac{\partial(\rho k)}{\partial t} + \frac{\partial(\rho u_i k)}{\partial x_i} = \frac{\partial}{\partial x_i}\left[\left(\mu + \frac{\mu_t}{\sigma_k}\right)\frac{\partial k}{\partial x_i}\right] + G_k - \rho\varepsilon \tag{1}$$

$$\frac{\partial(\rho \varepsilon)}{\partial t} + \frac{\partial(\rho u_i \varepsilon)}{\partial x_i} = \frac{\partial}{\partial x_i}\left[\left(\mu + \frac{\mu_t}{\sigma_\varepsilon}\right)\frac{\partial \varepsilon}{\partial x_i}\right] + C_{1\varepsilon}\frac{\varepsilon}{k}G_k - C_{2\varepsilon}\rho\frac{\varepsilon^2}{k} \tag{2}$$

where ρ is the density of mixture; t is time; k is turbulent kinetic energy; μ is the dynamic viscosity of fluid; u_i is component of velocity in the x_i direction; ε is turbulent energy dissipation rate; μ_t is dynamic turbulent viscosity; σ_k, σ_ε is turbulent Prandtl number for k and ε, respectively; G_k is generation of turbulent kinetic energy due to mean velocity gradients; and μ_t and G_k can be determined as

$$\mu_t = \rho C_\mu \frac{k^2}{\varepsilon} \tag{3}$$

$$G_k = \mu_t\left(\frac{\partial u_i}{\partial x_j} + \frac{\partial u_j}{\partial x_i}\right)\frac{\partial u_i}{\partial x_j} \tag{4}$$

The values of the empirical constants in the turbulence model are $C_\mu = 0.09$, $\sigma_k = 1.0$, $\sigma_\varepsilon = 1.3$, $C_{1\varepsilon} = 1.44$, and $C_{2\varepsilon} = 1.92$.

For this study, the volume-of-fluid (VOF) method was used as the two-phase flow model to track the water surface in the domain. The VOF method is widely used to determine the position of the interface of two or more immiscible flows [21,22]. Air and water were the primary and secondary phases, respectively. In the calculation, all fluids share the turbulence model. For air-water two-phase flow, α_a and α_w are the volume fraction of air and water, respectively. For each control cell,

$$\alpha_w + \alpha_a = 1 \tag{5}$$

If the cell contains only air, the value of $\alpha_w = 0$; if the cell is full of water, the value of $\alpha_w = 1$; and if the interface cuts the cell, then $0 < \alpha_w < 1$. The volume fraction of water, α_w, is calculated from [23]

$$\frac{\partial \alpha_w}{\partial t} + V\cdot\nabla\alpha_w = 0 \tag{6}$$

where the V is the fluid velocity. The fluid properties, such as density ρ and molecular viscosity μ, are adjusted according to the volume fraction. It should be noted that other numerical methods can also be used to study the detailed flow properties in complex contexts [24–29], the choice of numerical methods depends on the research focus.

2.2. Simulation Setup

ANSYS ICEM 16.0 (ANSYS®, Canonsburg, PA, USA) was utilized to develop the numerical models for 13 types of offset jet. Hexahedral grids were used throughout the computational domain. The grid meshing is shown in Figure 2. A Cartesian coordinate system is used so that the origin is at the center of the intersection of the offset wall and floor. The computational domain consists of the pipe and jet pool. The jet is discharged from offset wall offset by a height S above the floor.

The total pipe length is 1 m, and the type of pipe can be divided into the circular pipe and rectangular pipe. The length, L_x, and height, L_z, of the pool are 8 m and 0.8 m, respectively. For the circular pipe, the offset ratio (OR) varied from 1 to 3, the expansion ratio (ER) varied from 3 to 4.8 as the width of the jet pool increased from 0.3 m to 0.48 m. For the rectangular pipe, the aspect ratio (AR) of the exit varied from 0.33 to 11.46. Moreover, the area of exit remains constant for all shape of exits. The parameters of all cases are given in Table 1, and the jet exits and the offset wall are shown in Figure 3.

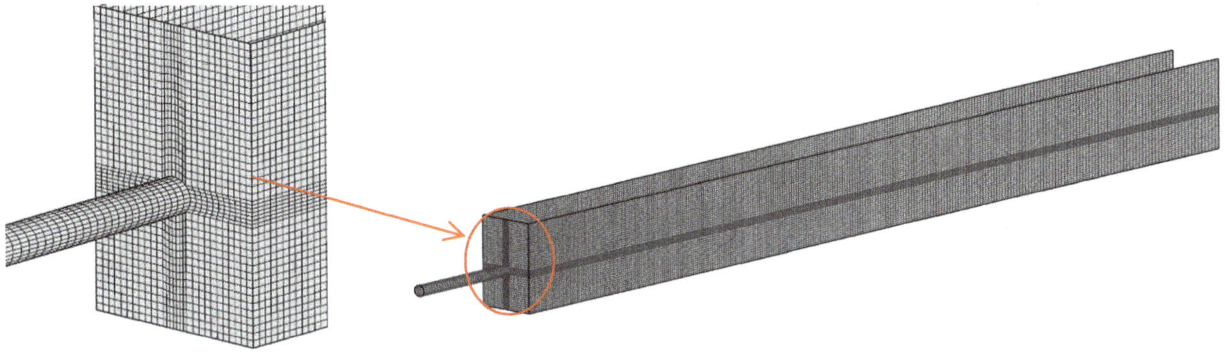

Figure 2. Layout of the calculation (C-O3-E5).

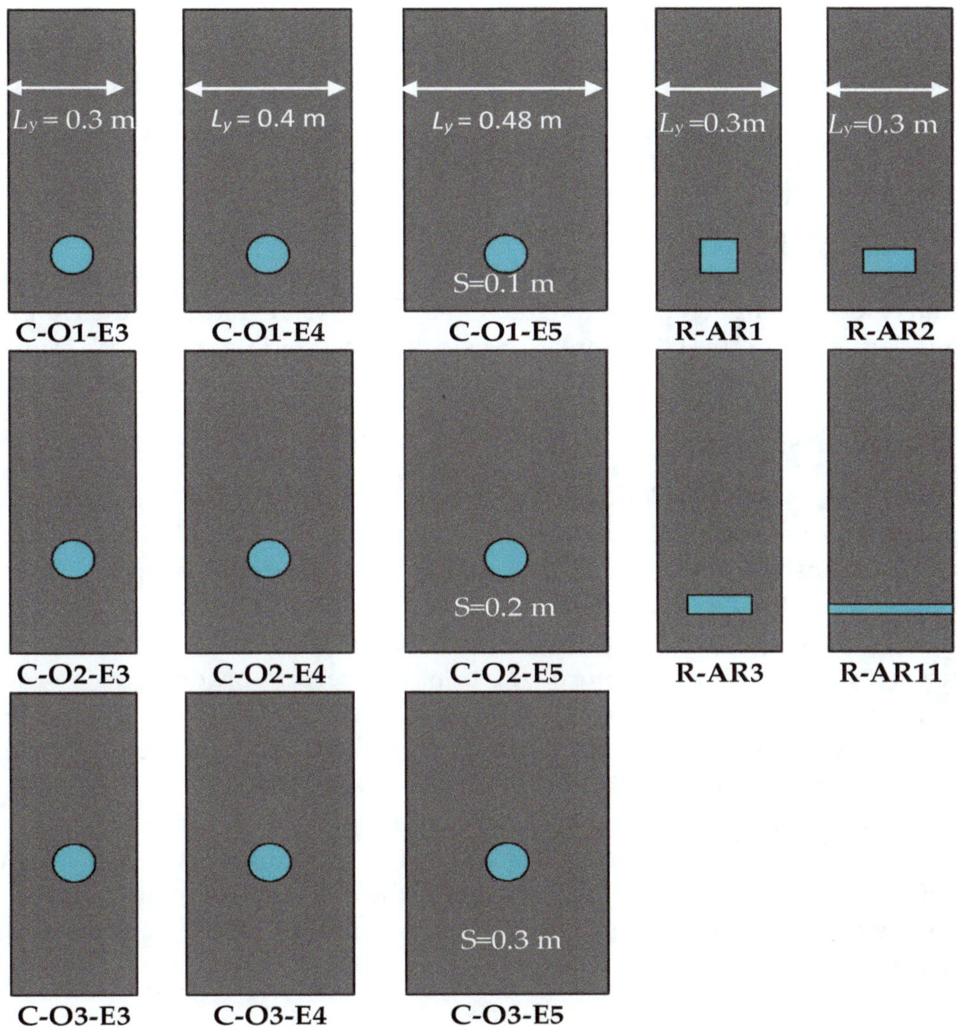

Figure 3. Sketch of jet exits.

Table 1. Parameters of computational cases.

Case Name	Jet Exit Shape	d (m)	a_0 (m)	b_0 (m)	S (m)	L_y (m)	Offset Ratio (OR)	Expansion Ratio (ER)	AR (b_0/a_0)
C-O1-E3	circular	0.1	-	-	0.1	0.3	1	3	-
C-O1-E4	circular	0.1	-	-	0.1	0.4	1	4	-
C-O1-E5	circular	0.1	-	-	0.1	0.48	1	4.8	-
C-O2-E3	circular	0.1	-	-	0.2	0.3	2	3	-
C-O2-E4	circular	0.1	-	-	0.2	0.4	2	4	-
C-O2-E5	circular	0.1	-	-	0.2	0.48	2	4.8	-
C-O3-E3	circular	0.1	-	-	0.3	0.3	3	3	-
C-O3-E4	circular	0.1	-	-	0.3	0.4	3	4	-
C-O3-E5	circular	0.1	-	-	0.3	0.48	3	4.8	-
R-AR1	rectangular	-	0.089	0.089	0.1	0.3	1.13	3.39	1
R-AR2	rectangular	-	0.125	0.063	0.1	0.3	1.60	2.39	2
R-AR3	rectangular	-	0.153	0.051	0.1	0.3	1.95	1.95	3
R-AR11	rectangular	-	0.1	0.026	0.1	0.3	3.82	1.00	11.46

2.3. Initial Conditions and Boundary Conditions

All the variables except at boundaries (such as u_i, P, k, and ε) are initialized at zero. The computed data are stored at every alternate time step for post-processing. All computations are conducted on an Inter core i7 3.60 GHz Windows machine.

The boundary conditions were set as follows.

- Inflow boundary: the inlet was treated as an inlet velocity boundary with the velocity was set as 5 m/s;
- Outflow boundary: pressure outlet boundary was selected at the outlet, the depth of tailwater was fixed at 0.55 m with the help of user-defined function (UDF);
- Free surface: pressure inlet was employed, and its value was the standard atmospheric pressure, the operating pressure and density were selected as 101,325 Pa and 1.225 kg/m^3, respectively.
- Wall boundary: for the parameters investigated in this study, the data near the wall was ineffective. No slip boundary condition is considered for velocity. To avoid the fine mesh required to resolve the viscous sub-layer near the boundary, so standard wall function method has been used.

2.4. Numerical Discretizations

ANSYS Fluent 16.0 (ANSYS®, Canonsburg, PA, USA) was utilized to perform the simulation. The governing equations are discretized using the implicit Finite Volume Method (FVM). The SIMPLE algorithm, using a relationship between velocity and pressure corrections to enforce mass conservation and to obtain the pressure field, was applied to couple the velocity and pressure. The least-squares cell-based method was used to calculate the gradient. PRESTO! was used to discretize the pressure and Geo-Reconstruct was used for the volume fraction. The second-order upwind scheme was used for the momentum and the first-order upwind for the turbulent kinetic energy and the dissipation rate with ANSYS Fluent's default under relaxation values for all parameters. The time step was Δt = 0.0001~0.001 s and the iteration number was always less than 2.

The computational results are considered to be converged when the residual becomes smaller than 0.001 for all equations. Here, the residual is defined as the square root of the summation of the squares of the difference between right and lefts sides of the discretized equations for a single control volume [30]. The results analyzed and presented in this study were taken from the simulation when quasi-steady state was reached.

2.5. Grids Sensitivity and Model Validation

Grid independence was examined to ensure the reliability of the numerical simulation results. The Case C-O3-E5 was chosen to test mesh sensitivity. Three sets of grids had 1,065,575 (grid 1), 438,221 (grid 2) and 181,036 (grid 3) cells, respectively. The grid convergence index (GCI) is widely used to estimate discretization uncertainty [31]. The GCI is given by

$$GCI = \frac{1.25|(\varnothing_3 - \varnothing_2)/\varnothing_3|}{(h_2/h_3)^P - 1} \tag{7}$$

$$P = \frac{1}{\ln(h_2/h_3)}\left|\ln|(\varnothing_1 - \varnothing_3)/(\varnothing_2 - \varnothing_3)| + \ln\left|\frac{(h_2/h_3)^P - \mathrm{sgn}[(\varnothing_2 - \varnothing_3)/(\varnothing_2 - \varnothing_3)]}{(h_1/h_2)^P - \mathrm{sgn}[(\varnothing_1 - \varnothing_3)/(\varnothing_2 - \varnothing_3)]}\right|\right| \tag{8}$$

where $\varnothing_i$ is the solution on the ith grid and h_i is the average grid size on the ith grid and $h_1 > h_2 > h_3$.

Figure 4 presents an axial velocity profile along the jet exit axis. In this figure, 80% out of 40 points exhibited oscillatory convergence. The maximum uncertainties in velocity were approximately 12.6%, which corresponds to a maximum uncertainty in velocity of about ±0.25 m/s. Considering the simulation accuracy and the computation efficiency, the final grid number in this study was taken as 438,221.

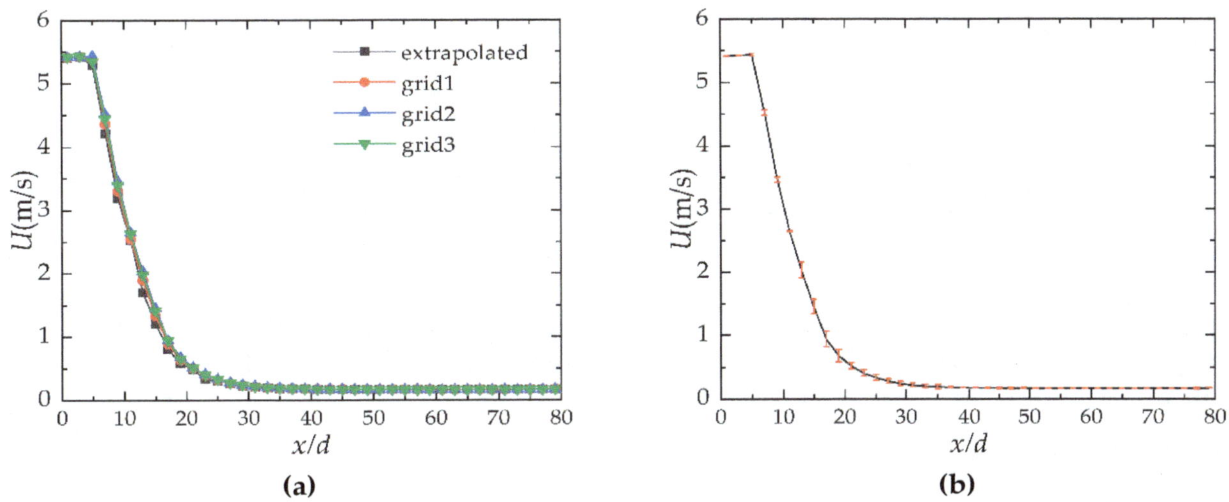

Figure 4. (**a**)Velocity profiles on jet exit axis with three grid sizes; (**b**) Results from grid 2, with discretization error bars computed using GCI.

To examine simulation accuracy, the physical model experiment of Case C-O3-E5 was performed at the State Key Laboratory of Hydraulics and Mountain River Engineering, Sichuan University, Chengdu. The experimental model is shown in Figure 5. It can be seen from Figure 6 that the calculated results of the height of water surface and velocity distribution were fairly consistent with that of laboratory tests. The results are indicating that the numerical simulation produced reliable and acceptable results.

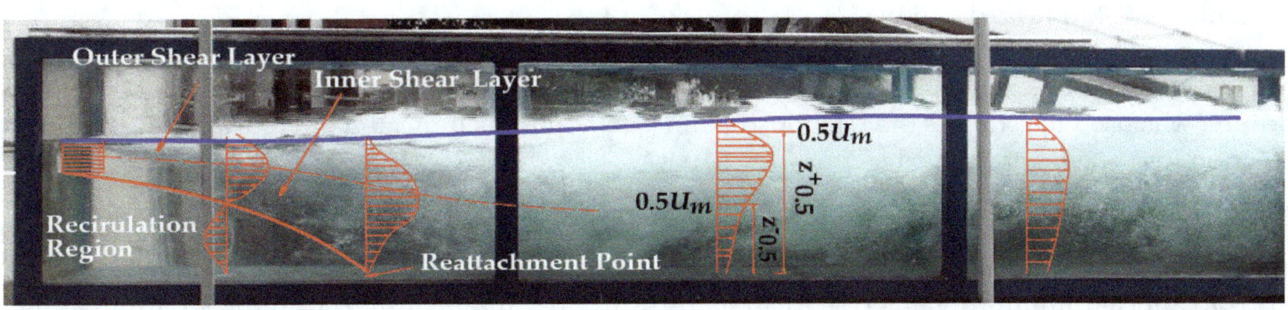

Figure 5. Experiment model.

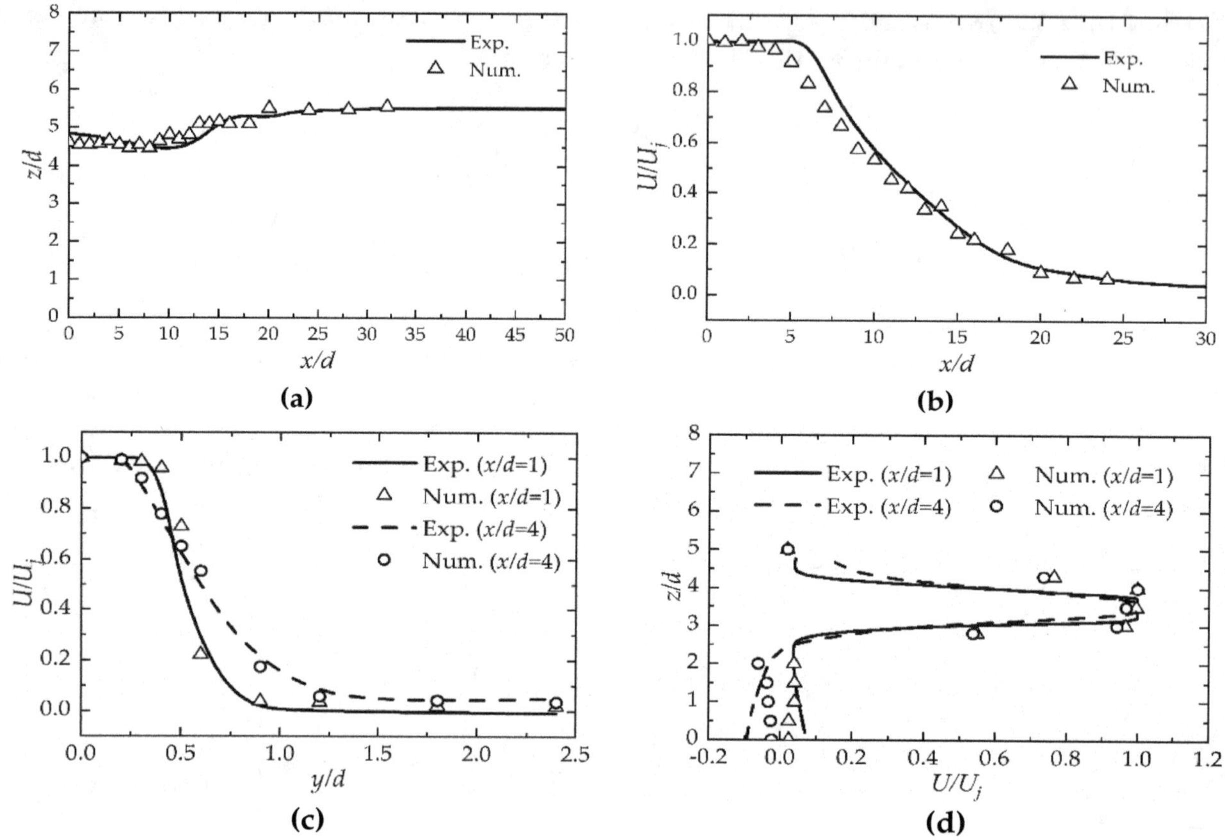

Figure 6. Model validation: (**a**) Heights of the water surface in streamwise direction; (**b**) Velocity distribution in streamwise direction (central line of exit); (**c**) Velocity distribution in lateral direction; (**d**) Velocity distribution in floor-normal direction.

3. Results and Analysis

3.1. Velocity Attenuation

The attenuation of the jet flow in the narrow and deep pool was characterized using the decay of the cross-sectional maximum streamwise mean velocity (U_m). Previous studies indicated that U_m is constant within the potential core region, followed by a rapid decay with streamwise distance in transition zone [32,33]. The entrained ambient fluid gets momentum since the jet enters the pool, the section of jet flow continues to expand, and the velocity is decreasing as the results of the mixing of jet and ambient fluid. Figure 7a shows the distribution of U_m normalized by the jet exit velocity, U_j. Normalized values of U_m were observed decayed sharply within the region $x/d > L_{PC}$ and the trend of decay of U_m varies greatly in the transition zone for different circular offset jet. It can be seen from Figure 7b that the length of the potential core zone (L_{PC}) varied from 5.0 to 5.5 with changes in ER, and a higher ER indicated a higher L_{PC}.

As can be observed in Figure 7c–d, the L_{PC} varies greatly with AR of the rectangular jet exit due to the circumference of exit (L_{exit}) were changed. When L_{exit} increased, the L_{PC} decreased sharply, a fitting formula was used to estimate the relationship of L_{exit} and L_{PC} as indicated in Figure 7d, the empirical equation for L_{PC} was employed as $L_{PC}/d = 2100e^{-2(L_{exit}/d)} + 0.5$.

The decay of the U_m over distance x (scaled by d) for Case R-AR11 is shown in Figure 7c. The most striking feature of Case R-AR11 is that there is no sudden expansion (ER = 1). Thus, the Case R-AR11 is analogous to that of a plane offset jet. The decay of U_m in the recirculation region is quite sharp, the U_m drops to a local minimum value due to an increase in pressure resulting from the jet impingement after the reattachment of the jet. The decay of U_m is rather gradual in the wall jet region. A similar observation was reported by Gu [16] and Dey [4].

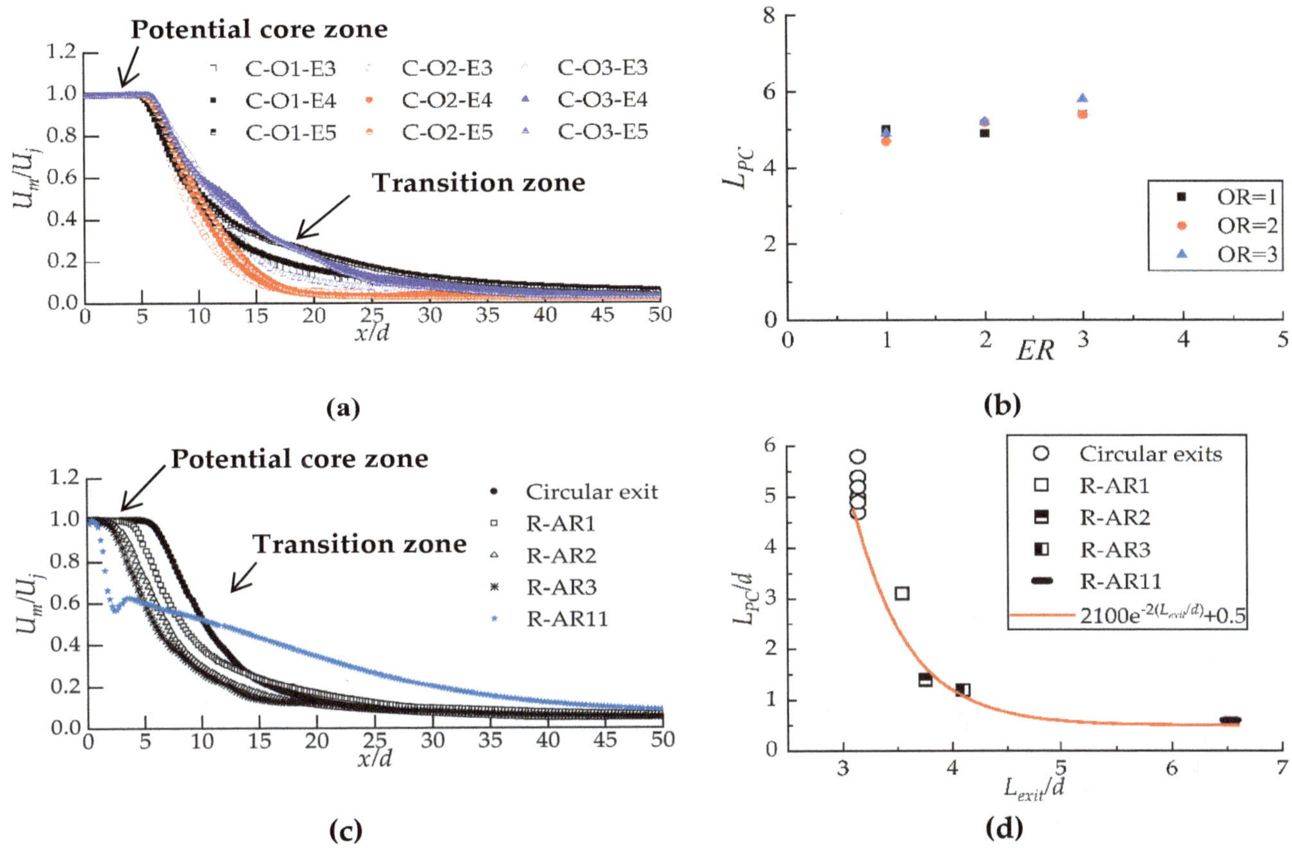

Figure 7. (**a**) U_m decay for circular offset jets; (**b**) the relationship of L_{PC}, ER, and OR for circular offset jets; (**c**) U_m decay for rectangular offset jets (AR $\leq$ 3); (**d**) the relationship of L_{PC} and L_{exit}.

The U_m/U_j decay profiles can be grouped based on their characteristics. The power law of the form $U_m/U_j = C(x/d)^{-n}$ is typically used to describe the velocity decay in the transition zone [6,32,34,35], where C and n are a proportionality symbol and decay rate, respectively. The decay rate indicated the extent of entrainment and boundary effects on the jet. Figure 8 shows the variation of U_m/U_j for all cases with the normalized longitudinal coordinate x/d within the transition zone. Rajaratman reported $n = 0.5$ for a plane free jet by considering simplified conservation laws and entrainment hypotheses [36]. The wall jet was observed to be accurately estimated by using the power law. However, the distribution of U_m/U_j for the offset jets are not accurately described by the power law, which should be replaced by a linear fit [32].

For circular offset jet, using the power law fit, the decay rate for the lower offset ratio (OR = 1) circular jet was observed to be accurately estimated with the decay rate n varies from 1.197 to 1.768 as shown in Figure 8a. As can be seen in Figure 8b–c, the distributions of U_m/U_j, for the higher offset ratio (OR = 2 or 3) circular jet are not accurately described by the power law. Therefore, a linear fit was used to estimate the decay rates for higher offset ratio jet, the slope κ the linear equation was considered as the decay rate. Decay rate κ values of 0.110, 0.096 and 0.090 with R^2 above 0.94 were obtained for ER = 3, 4 and 4.8, respectively, when OR = 2. Decay rate values of 0.081, 0.048 and 0.049 with R^2 above 0.968 were obtained for ER = 3, 4 and 4.8, respectively, when OR = 3. Above analysis and as shown in Figure 8a–c indicated that when the offset ratio is moderate (OR = 2) produces a faster decay than that of higher or lower offset ratio (OR = 1 or 3). The reason is that two circulating vortices were developed below and above jet, respectively. The circulating vortices develop reverse flows against the inflow-jet direction. Negative momentum of the reverse flows reduced the inflow momentum slowing the jet rapidly down.

The decay rates for the various circular jet cases are reported in Table 2. The decay rate value is the largest as the expansion ratio (ER) is 3. There is a stronger jet interaction with the ambient fluid due to the more confined environment.

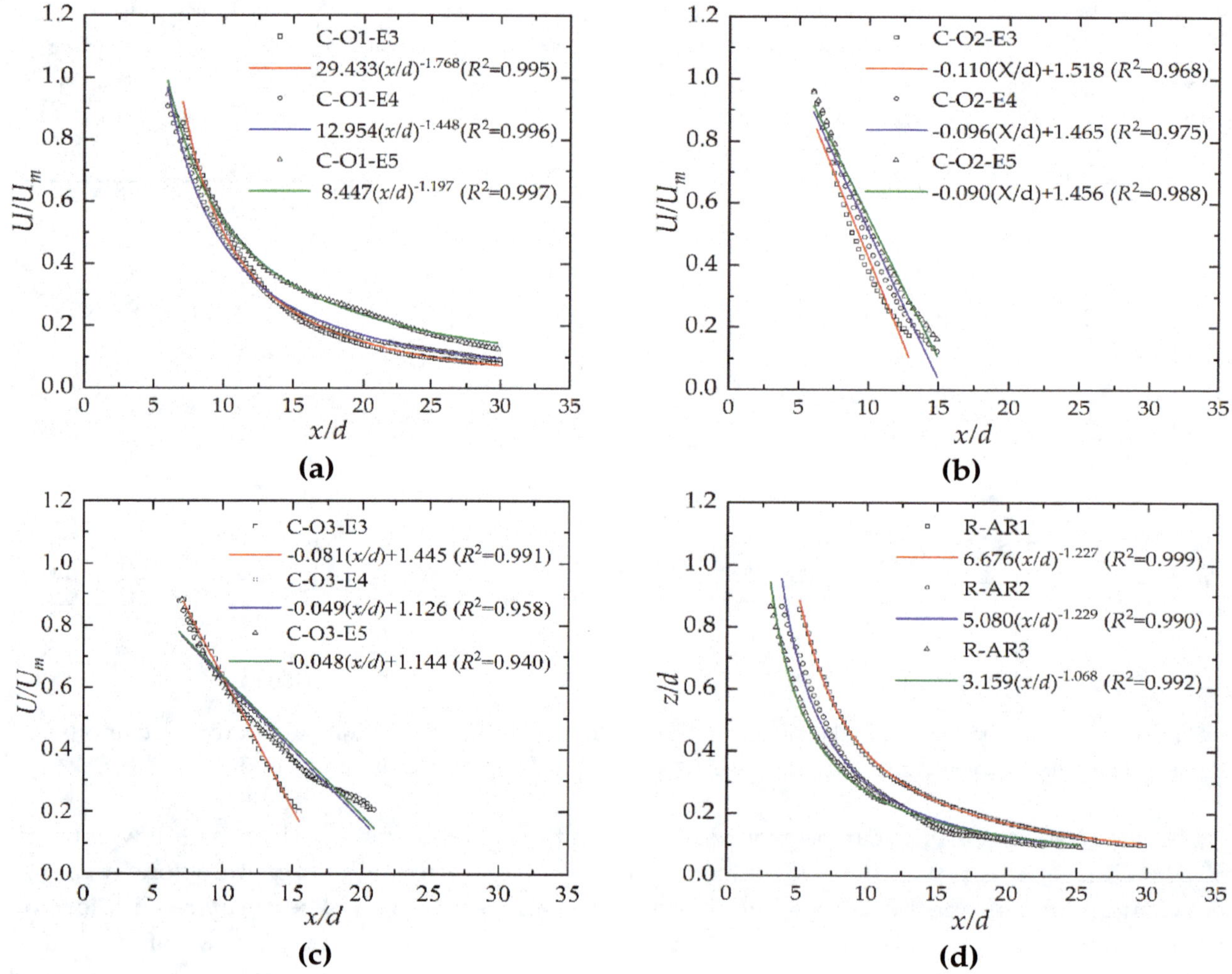

Figure 8. (**a**) power law fit to U_m for circular offset jets as OR = 1; (**b**) linear fit to U_m for circular offset jets as OR = 2; (**c**) linear fit to U_m for circular offset jets as OR = 3; (**d**) power law or linear fit to U_m for rectangular offset jets.

Table 2. The decay rate of the circular jet.

Decay Rate	**n (OR = 1)**	**κ (OR = 2)**	**κ (OR = 3)**
Fit Equation	**Power Law**	**Linear Equation**	
ER = 3	1.768	0.110	0.081
ER = 4	1.448	0.096	0.049
ER = 4.8	1.197	0.090	0.048

For rectangular offset jet, the U_m/U_j for $1 \leq \text{AR} \leq 3$ is expressed well by power law form with the decay rate n varies from 1.068 to 1.228 (Table 3), and the U_m decay in transition zone does not show substantial differences (Figure 8d). Therefore, it appears reasonable to assume the effect of the aspect ratio of the rectangular jet on the mainstream attenuation is mainly to influence the length of the potential core zone.

Table 3. The decay rate of the rectangular jet.

Decay Rate	n (OR = 1)
Fit Equation	**Power Law**
AR = 1	1.227
AR = 2	1.229
AR = 3	1.068

The algorithm for calculating the length of the potential core (L_{PC}) and decay profiles in the transition zone could be used to roughly calculate the length of the energy dissipation region (L_{ED}). For example, a square offset jet is issued to a pool. The side length of the square is 2.67 m (equivalent diameter is 3 m), the offset height is 3 m, the width of the pool is 12 m, and the height of tail water is 16.5 m. If the $U/U_m = 0.2$ as the end of the length of the energy dissipation region, the distance U/U_m decreases from 1 to 0.2 is called L_{tr}. The L_{PC} can be described by:

$$L_{ED} = L_{PC} + L_{tr} \tag{9}$$

The $L_{PC} = 6.60$ m calculated by $L_{PC}/d = 2100e^{-2(L_{exit}/d)} + 0.5$. The $L_{tr} = 39.53$ m calculated by $0.2 = -0.096(x/d) + 1.465$. As the result, the length of energy dissipation region is 43.13 m.

3.2. Vertical Velocity Spread

Figure 9 shows the flow field of the offset jet with different offset ratios. There is a large range of vortices formed near the jet flow, the location and size of the vortex on the streamwise section have notable differences. When OR = 1, in the range of *x/d* = 0–25, a large vortex was formed above the jet flow; When OR = 3, in the range of *x/d* = 0–25, a large vortex was formed below the jet flow. However, there is a smaller vortex above and below the jet flow in the range of *x/d* = 0–15.

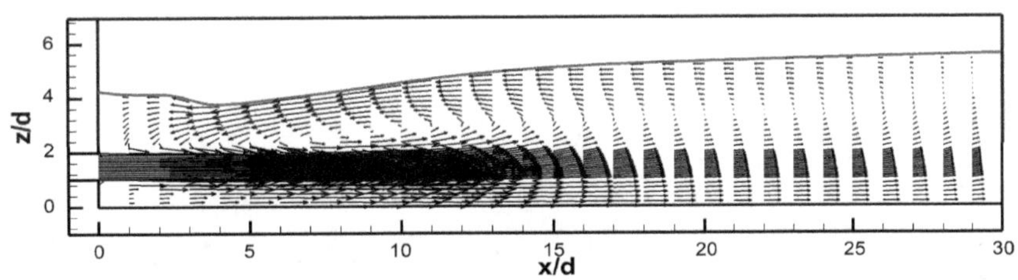

(a)

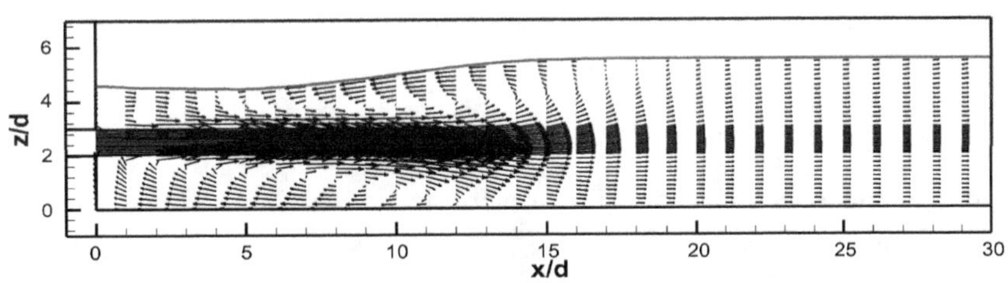

(b)

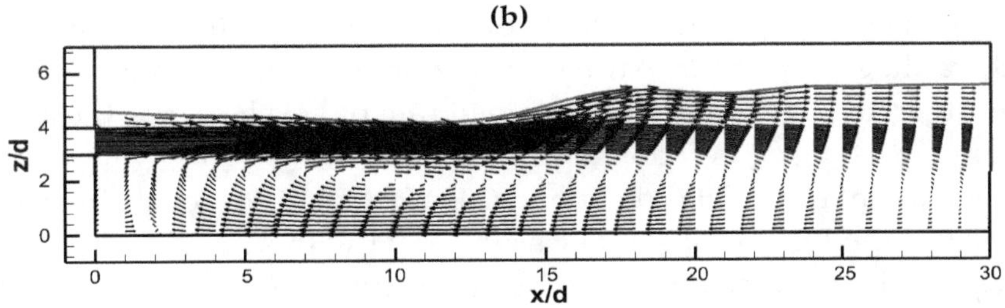

(c)

Figure 9. Velocity vectors of offset jet with different offset ratio: (**a**) OR = 1 (C-O1-E4); (**b**) OR = 2 (C-O2-E4); (**c**) OR = 3 (C-O3-E4).

Profiles of the streamwise mean velocities (U) extracted at x/d = 1, 4, 9, 13 and 17 are shown in Figure 10. It may be noted that to cover a range of ER = 4, Cases C-O1-E4, C-O2-E4, and C-O3-E4 were selected; otherwise, using the data of all Cases would make the plots clumsy. The velocity and length scale used was U_m and d, respectively. Here, z_c is the distance from the center of exit to the floor, that is, $z - z_c = 0$, where is the axis of the jet exits. Vertical profiles of U are shown in Figure 10a for ER is 4. The effect of offset ratio (OR) on the vertical velocity spread was not evident within the region $x/d < 9$. It can be noted that the velocity distribution in the floor-normal direction conforms with the Gaussian distribution within the region of $5 \leq x/d \leq 9$. However, as the jet travels away from the jet exits ($x/d > 9$), the data of jets at different offset height are no longer overlapped. The location of the maximum of U/U_m for OR = 2 is almost at the axis of the jet exits. The location of the maximum of U/U_m is above and below the axis of the jet exits for OR = 3 and OR = 1, respectively. The results reveal offset height changes travel direction of the bulk of the jet.

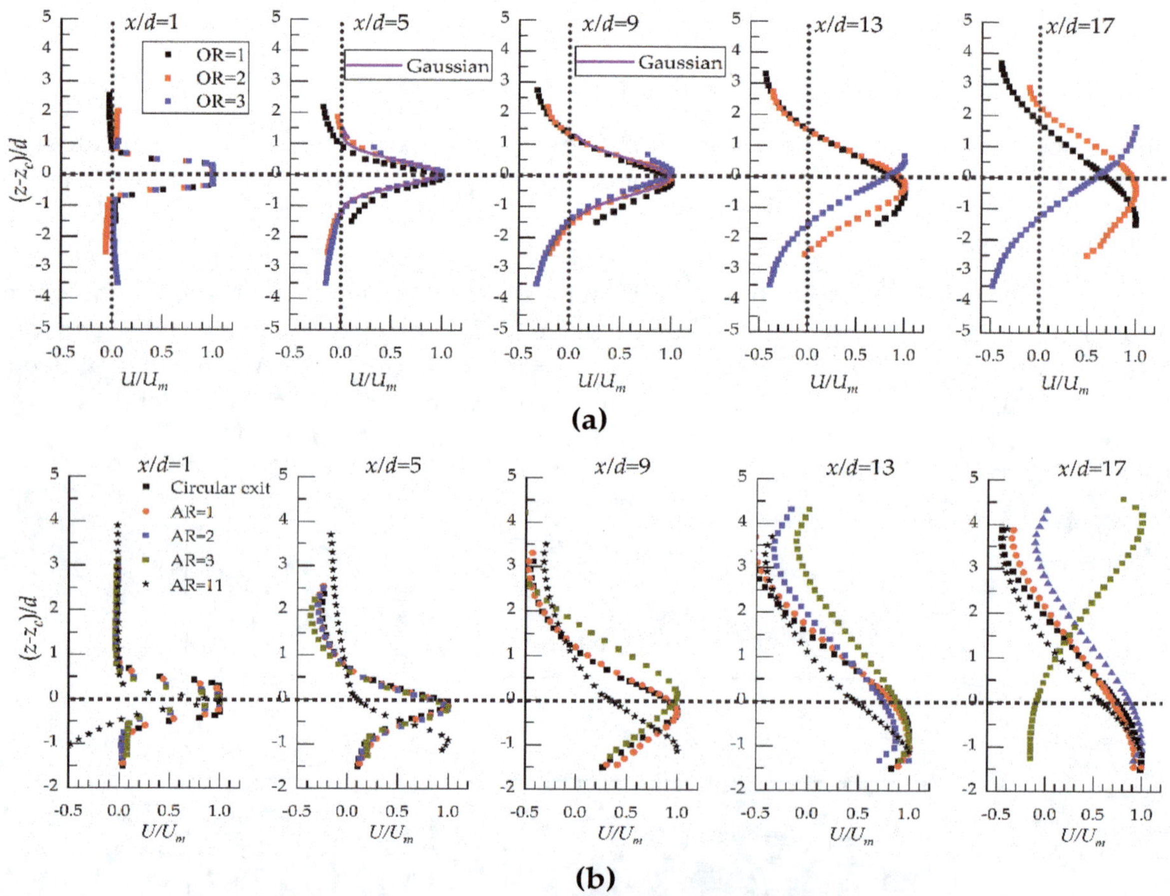

Figure 10. Streamwise mean velocity distribution in symmetry (x–z) plane: (**a**) circular offset jets (ER = 4); (**b**) rectangular offset jets.

Figure 10b shows the development of the U of the offset jet with rectangular exits. The results demonstrate that the bulk of the jet turns travel direction and then attaches to the floor within the region of $x/d < 13$. Further, the figure shows the spread rate of U is larger as the aspect ratio increases, and the flow direction for AR = 3 was changed at x/d = 17.

The U distribution in floor-normal direction in Case R-AR11 is different from those in other cases, the difference can be attributed to the jet exit of Case R-AR11 without sudden expand (ER = 1), there is an obviously lower pressure zone below the jet at the jet exit, then the main jet bends to the floor in short distance ($x/d < 2.5$). Thus, the profile is close to that of a wall jet.

Figure 11 shows contours of U/U_m in the symmetry plane for the offset jets, and the lower and upper lines represent the loci of $z_{0.5}^{-}$ ($U = 0.5\ U_m$, the inner separation line), and $z_{0.5}^{+}$ ($U = 0.5\ U_m$, the outer separation line), respectively. The jet between $z^-_{0.5}$ and $z^+_{0.5}$ was considered as the main flow in the present paper. The $L^-_{Z0.5}$ is the distance from the exit to where the $z^-_{0.5}$ reaches the floor, and $L^+_{Z0.5}$ is the distance from the exit to where the $z^+_{0.5}$ reaches the water surface. For circular exits, the main flow for OR = 1 bends to the floor at $x/d = 12$ ($L^-_{Z0.5} < 12$ and $L^+_{Z0.5} > 46$), then the flow spreads from floor to water surface (Figure 11a). The main flow for OR = 2 spreads evenly in the floor-normal direction at $x/d > 20$, the values of $L^-_{Z0.5}$ and $L^+_{Z0.5}$ are roughly equivalent (Figure 11b). Furthermore, the main flow for OR = 3 bends to the water surface, with the $L^+_{Z0.5} < 20$ and $L^-_{Z0.5} > 42$, which means the flow in the pool is spreading from water surface to floor (Figure 11c). For rectangular exits (Figure 11d), the main flow for AR = 1 is identical to that of a circular jet. The main flow for AR = 2 spreads from floor to water surface. The main flow for AR = 3 bends sharply to water surface as the jet enters the pool, the reason is that jet is too thin to effectively maintain a stable flow pattern. As the expansion ratio increases to 11.48, the spread range of the main flow in floor-normal direction is small, and the high velocity region is close to the floor.

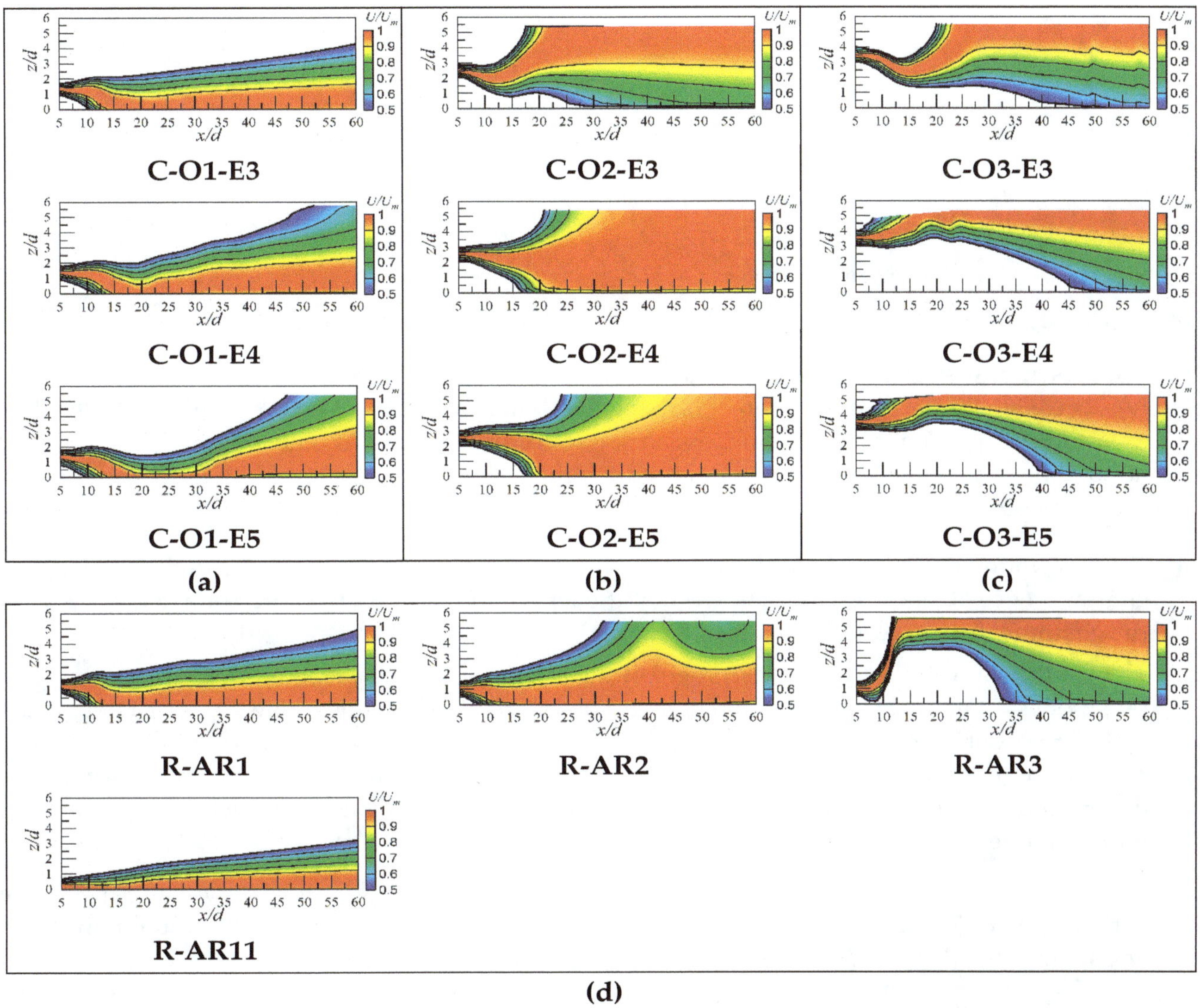

Figure 11. Contour of U/U_m in symmetry (x–z) plane: (**a**) circular offset jets (ER = 3); (**b**) circular offset jets (ER = 4); (**c**) circular offset jets (ER = 4.8); (**d**) rectangular offset jets.

3.3. Lateral Velocity Spread

As observed in Figure 12, there is a negligible effect of the expansion ratio on the lateral velocity spread within the region $x/d < 9$. This observation reveals that the profiles of U were independent of the expansion ratios, the Gaussian distribution is overlapped on the present data plots within the region $5 \leq x/d \leq 9$. The results indicated that the offset height did not greatly alter the flow within early development regions of the offset jet. However, at $x/d = 13$ and $x/d = 17$, U/U_m are lower for the expansion ratios, ER, are lower within the region $y/d > 0.5$. Figure 10 indicates that a more confining enclosure enhances the diffusion of the velocity than a less confining condition. The velocity profiles in the x-y plane of the jet with rectangular exits are not obtained. The flow patterns are not stable as the expansion ratio was changed [37].

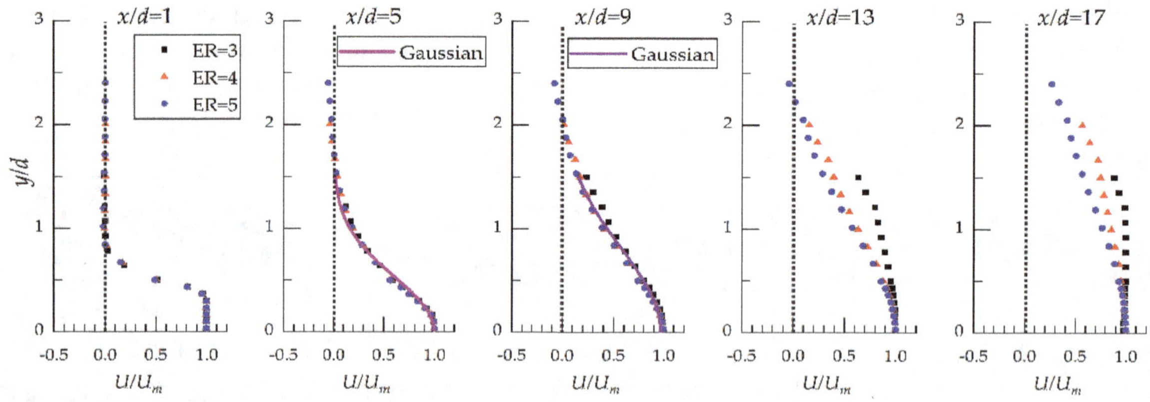

Figure 12. Profiles of U/U_m in (x–y) plane of circular jet.

4. Conclusions

The present investigation sought to elucidate the velocity development of a submerged 3D offset jet flow occurred in the narrow and deep pool using the standard k-ε model with VOF method. The flow field was studied for different offset ratio, expansion ratio, and jet exits. The numerical results showed that the solid boundaries of the narrow pool produce circulation in the surrounding fluid opposite the jet flow, which produces a rapid jet decay. The following conclusions can be drawn:

1. The development of the jet within the potential core zone was observed to be dependent on the circumference of jet exits. It was noted that the length of the potential core zone decreases with the circumference of exit.
2. The results showed that the development of the U_m in the transition zone could be divided into two modes. That is, (a) the decay of U_m could be estimated from power law fit with a decay rate n of 1.089–1.451 as the offset ratio is lower (OR = 1). (b) A linear fit was used to estimate the decay rates for higher offset ratio jet (OR = 2 or 3). It was observed that U_m decay is the fastest as the offset ratio (OR) is 2. Further, the results indicate that a more confining enclosure produces a more rapid jet decay than a less confining condition.
3. The spread rate of the circular jet is not affected by the expansion ratios and offset ratio in the early region of jet decay. The main flow for OR = 2 was traveled straightly, and the main flow was toward the floor and the water surface as the OR = 1 and OR = 3, respectively.
4. The absence of the sudden expansion coupled with the high aspect ratio of the exit contributed to form a unique flow pattern compared with other cases.

Author Contributions: Conceptualization and Methodology, X.L. and M.Z.; Software, X.L.; Formal analysis, X.L. and M.Z.; Investigation; X.L. and M.Z.; Writing—original draft, X.L.; Writing—Review & Editing, J.Z. and M.Z.; Supervision, J.Z. and W.X.; Project administration, J.Z. and W.X.; Funding acquisition, J.Z.

Abbreviations

x	Stream wise direction
y	Lateral direction
z	Floor direction
L_x	Length of the jet pool in the streamwise direction (m)
L_y	The width of the jet pool in the lateral direction (m)
L_z	The height of the jet pool in the floor-normal direction (m)
d	The diameter of circular jet exit (m) or equivalent diameter of rectangular exit (m)
a_0	The height of the rectangular exit (m)
b_0	Width of rectangular exit (m)
h_t	The height of tailwater (m)
S	Offset height of jet exit (m)
U	Mean velocity in streamwise (m/s)
U_m	Cross-Sectional streamwise maximum mean velocity (m/s)
U_j	Velocity in jet exit (m/s)
L_{PC}	Length of potential core
L_{exit}	The circumference of jet exit
z_c	Distance from center of exit to floor, $z_c = S + 0.5d$ or $z_c = S + 0.5a_0$
$z^+_{0.5}$	The loci of where $U = 0.5\ U_m$ above the jet
$z^-_{0.5}$	The loci of where $U = 0.5\ U_m$ below the jet
$L^-_{Z0.5}$	Distance from the exit to where the $z^-_{0.5}$ reaches floor
$L^+_{Z0.5}$	Distance from the exit to where the $z^+_{0.5}$ reaches water surface
n	Decay rate used in power law
κ	Decay rate used in the linear fit
R_0	Reynolds number
OR	Offset ratio of jet exit (S/d or S/a_0)
ER	The expansion ratio of jet exit (L_y/d or L_y/b_0)
AR	Aspect ratio of jet exit (b_0/a_0)

References

1. SoaresFrazão, S. *Wastewater Hydraulics*; Springer: Berlin/Heidelberg, Germany, 2010; pp. 842–843.
2. Liu, M.; Rajaratnam, N.; Zhu David, Z. Mean Flow and Turbulence Structure in Vertical Slot Fishways. *J. Hydraul. Eng.* **2006**, *132*, 765–777. [CrossRef]
3. Chen, J.-G.; Zhang, J.-M.; Xu, W.-L.; Li, S.; He, X.-L. Particle image velocimetry measurements of vortex structures in stilling basin of multi-horizontal submerged jets. *J. Hydrodyn.* **2013**, *25*, 556–563. [CrossRef]
4. Dey, S.; Ravi Kishore, G.; Castro-Orgaz, O.; Ali, S.Z. Hydrodynamics of submerged turbulent plane offset jets. *Phys. Fluids* **2017**, *29*, 065112. [CrossRef]
5. Bhuiyan, F.; Habibzadeh, A.; Rajaratnam, N.; Zhu David, Z. Reattached Turbulent Submerged Offset Jets on Rough Beds with Shallow Tailwater. *J. Hydraul. Eng.* **2011**, *137*, 1636–1648. [CrossRef]
6. Durand, Z.M.J.; Clark, S.P.; Tachie, M.F.; Malenchak, J.; Muluye, G. Experimental Study of Reynolds Number Effects on Three-Dimensional Offset Jets. In Proceedings of the 12th International Conference on Nanochannels, Microchannels, and Minichannels, Chicago, IL, USA, 3–7 August 2014. [CrossRef]
7. Nyantekyi-Kwakye, B.; Tachie, M.F.; Clark, S.P.; Malenchak, J.; Muluye, G.Y. Experimental study of the flow structures of 3D turbulent offset jets. *J. Hydraul. Res.* **2015**, *53*, 773–786. [CrossRef]
8. Agelin-Chaab, M.; Tachie, M.F. Characteristics and structure of turbulent 3D offset jets. *Int. J. Heat Fluid Flow* **2011**, *32*, 608–620. [CrossRef]
9. Li, J.-N.; Zhang, J.-M.; Peng, Y. Characterization of the mean velocity of a circular jet in a bounded basin. *J. Zhejiang Univ. Sci. A* **2017**, *18*, 807–818. [CrossRef]
10. Kumar Rathore, S.; Kumar Das, M. Effect of Freestream Motion on Heat Transfer Characteristics of Turbulent Offset Jet. *J. Therm. Sci. Eng. Appl.* **2015**, *8*, 011021. [CrossRef]
11. Camino, G.A.; Zhu, D.Z.; Rajaratnam, N. Jet diffusion inside a confined chamber. *J. Hydraul. Res.* **2012**, *50*, 121–128. [CrossRef]

12. Zhang, Z.; Guo, Y.; Zeng, J.; Zheng, J.; Wu, X. Numerical Simulation of Vertical Buoyant Wall Jet Discharged into a Linearly Stratified Environment. *J. Hydraul. Eng.* **2018**, *144*, 06018009. [CrossRef]
13. Assoudi, A.; Habli, S.; Mahjoub Saïd, N.; Bournot, H.; Le Palec, G. Three-dimensional study of turbulent flow characteristics of an offset plane jet with variable density. *Heat Mass Transf.* **2016**, *52*, 2327–2343. [CrossRef]
14. Mondal, T.; Guha, A.; Das, M.K. Computational study of periodically unsteady interaction between a wall jet and an offset jet for various velocity ratios. *Comput. Fluids* **2015**, *123*, 146–161. [CrossRef]
15. Vishnuvardhanarao, E.; Das, M.K. Computation of Mean Flow and Thermal Characteristics of Incompressible Turbulent Offset Jet Flows. *Numer. Heat Transf. A* **2007**, *53*, 843–869. [CrossRef]
16. Gu, R. Modeling Two-Dimensional Turbulent Offset Jets. *J. Hydraul. Eng.* **1996**, *122*, 617–624. [CrossRef]
17. Hnaien, N.; Marzouk, S.; Ben Aissia, H.; Jay, J. CFD investigation on the offset ratio effect on thermal characteristics of a combined wall and offset jets flow. *Heat Mass Transf.* **2017**, *53*, 2531–2549. [CrossRef]
18. Abraham, J.; Magi, V. *Computations of Transient Jets: RNG k-e Model Versus Standard k-e Model*; SAE: Warrendale, PA, USA, 1997.
19. Nasr, A.; Lai, J.C.S. A turbulent plane offset jet with small offset ratio. *Exp. Fluids* **1998**, *24*, 47–57. [CrossRef]
20. Launder, B.E.; Sharma, B.I. Application of the energy-dissipation model of turbulence to the calculation of flow near a spinning disc. *Lett. Heat Mass Transf.* **1974**, *1*, 131–137. [CrossRef]
21. Bai, Z.; Wang, Y.; Zhang, J. Pressure distributions of stepped spillways with different horizontal face angles. In *Proceedings of the Institution of Civil Engineers-Water Management*; Thomas Telford Ltd.: London, UK; pp. 1–12.
22. Li, S.; Zhang, J. Numerical Investigation on the Hydraulic Properties of the Skimming Flow over Pooled Stepped Spillway. *Water* **2018**, *10*, 1478. [CrossRef]
23. Hirt, C.W.; Nichols, B.D. Volume of Fluid (VOF) Method for the Dynamics of Free Boundaries. *J. Comput. Phys.* **1981**, *39*, 201–225. [CrossRef]
24. Peng, Y.; Mao, Y.F.; Wang, B.; Xie, B. Study on C–S and P–R EOS in pseudo-potential lattice Boltzmann model for two-phase flows. *Int. J. Mod. Phys. C* **2017**, *28*, 1750120. [CrossRef]
25. Peng, Y.; Wang, B.; Mao, Y. Study on Force Schemes in Pseudopotential Lattice Boltzmann Model for Two-Phase Flows. *J. Math. Probl. Eng.* **2018**, *2018*, 9. [CrossRef]
26. Peng, Y.; Zhang, J.; Meng, J. Second-order force scheme for lattice Boltzmann model of shallow water flows. *J. Hydraul. Res.* **2017**, *55*, 592–597. [CrossRef]
27. Peng, Y.; Zhang, J.M.; Zhou, J.G. Lattice Boltzmann Model Using Two Relaxation Times for Shallow-Water Equations. *J. Hydraul. Eng.* **2016**, *142*, 06015017. [CrossRef]
28. Peng, Y.; Zhou, J.G.; Burrows, R. Modeling Free-Surface Flow in Rectangular Shallow Basins by Using Lattice Boltzmann Method. *J. Hydraul. Eng.* **2011**, *137*, 1680–1685. [CrossRef]
29. Peng, Y.; Zhou, J.G.; Burrows, R. Modelling solute transport in shallow water with the lattice Boltzmann method. *Comput. Fluids* **2011**, *50*, 181–188. [CrossRef]
30. Van Doormaal, J.P.; Raithby, G.D. Enhancements of the Simple Method for Predicting Incompressible Fluid Flows. *Numer. Heat Transf.* **1984**, *7*, 147–163. [CrossRef]
31. Celik, I.; Ghia, U.; Roache, P.; Freitas, C.J. Procedure for estimation and reporting of uncertainty due to discretization in CFD applications. *J. Fluids Eng.* **2008**, *130*, 078001. [CrossRef]
32. Nyantekyi-Kwakye, B.; Clark, S.P.; Tachie, M.F.; Malenchak, J.; Muluye, G. Flow characteristics within the recirculation region of three-dimensional turbulent offset jet. *J. Hydraul. Res.* **2015**, *53*, 230–242. [CrossRef]
33. Li, X.; Wang, Y.; Zhang, J. Numerical Simulation of an Offset Jet in Bounded Pool with Deflection Wall. *Math. Probl. Eng.* **2017**, *2017*, 11. [CrossRef]
34. Nyantekyi-Kwakye, B.; Tachie, M.F.; Clark, S.P.; Malenchak, J.; Muluye, G.Y. Acoustic Doppler velocimeter measurements of a submerged three-dimensional offset jet flow over rough surfaces. *J. Hydraul. Res.* **2017**, *55*, 40–49. [CrossRef]
35. Padmanabham, G.; Lakshmana Gowda, B.H. Mean and Turbulence Characteristics of a Class of Three-Dimensional Wall Jets—Part 1: Mean Flow Characteristics. *J. Fluids Eng.* **1991**, *113*, 620–628. [CrossRef]
36. Rajaratnam, N. *Turbulents Jets*; Elsevier: New York, NY, USA, 1976.
37. Ohtsu, I.; Yasuda, Y.; Ishikawa, M. Submerged Hydraulic Jumps below Abrupt Expansions. *J. Hydraul. Eng.* **1999**, *125*, 492–499. [CrossRef]

11

Effect of Particle Size and Shape on Separation in a Hydrocyclone

Zhaojia Tang [1], Liming Yu [1,*], Fenghua Wang [1], Na Li [1], Liuhong Chang [2] and Ningbo Cui [3]

1 Faculty of Modern Agriculture Engineering, Kunming University of Science and Technology, Kunming 650500, China; tangzhaojia01@sina.com (Z.T.); wangfenghua018@163.com (F.W.); kjclina@163.com (N.L.)

2 School of Hydraulic Engineering, Changsha University of Science and Technology, Changsha 410114, China; claire886@163.com

3 State Key Laboratory of Hydraulics and Mountain River Engineering and College of Water Resource and Hydropower, Sichuan University, Chengdu 610065, China; cuiningbo@scu.edu.cn

* Correspondence: liming16900@sina.com

Abstract: Given the complex separation mechanisms of the particulate mixture in a hydrocyclone and the uncertain effects of particle size and shape on separation, this study explored the influence of the maximum projected area of particles on the separation effect as well as single and mixed separations based on CFD–DEM (Computational Fluid Dynamics and Discrete Element Method) coupling and experimental test methods. The results showed that spherical particles flowed out more easily from the downstream as their sizes increased. Furthermore, with the enlargement of maximum projected area, the running space of the particles with the same volume got closer to the upward flow and particles tended to be separated from the upstream. The axial velocity of the combined separation of 60 μm particles and 120 μm particles increased by 25.74% compared with that of single separation of 60 μm particles near the transition section from a cylinder to a cone. The concentration of 60 μm particles near the running space of 120 μm particles increased by 20.73% and those separated from the downstream increased by 4.1%. This study showed the influence of particle size and maximum projected area on the separation effect and the separation mechanism of mixed sand particles in a hydrocyclone, thereby providing a theoretical basis for later studies on the effect of particle size and shape on sedimentation under the cyclone action in a hydrocyclone.

Keywords: CFD–DEM coupling; hydrocyclone; particle shape; particle size; water and sediment separation

---

1. Introduction

A hydrocyclone is widely used for various separation purposes in different fields, such as mining, chemical industry, and agriculture. In the engineering field, it is often adopted to separate and classify the sand in water. The specific working process is as follows. As the water and sand mixture enters from the inlet, the water flows out from the upstream and downstream. The sand particles with strong water-following ability are carried out from the upstream and those with poor water-following ability are carried out from the downstream under the internal swirling flow.

In recent years, the separation mechanism of hydrocyclone has been mainly studied from the following aspects:

(1) The separation effect of various solid particle properties and different external conditions: The equation of particle motion in a hydrocyclone is established and the formula for calculating the optimum separation medium density is deduced according to the force analysis on different particles in the separation [1]. The calculation method and equation are put forward based on the relationship between particle size and fluid drag force [2]. The separation effect of submicron particles in a microclone at different temperatures and water pressures [3], the effect of internal flow field on the separation of different-sized particles [4], and the effect of inlet velocity and particle shape on the separation effect [5] are also studied.

(2) The effect of a hydrocyclone structure on separation: The optimal particle separation effect is determined by the orthogonal analysis on the optimal insertion depth, wall thickness, and optimal inlet flow [6]. The influence of the inlet structure in the performance of a hydrocyclone is explored [7]. Three kinds of hydrocyclones are designed to improve the particle separation efficiency, and the optimum separation structure is determined by studying the pressure and velocity distribution inside a hydrocyclone [8]. The effect of cone angle structure on the classification of fine particles [9] and of the inlet angle, cylinder height, and inner cone on separation [10,11] are discussed. In addition, studies are conducted on the separation results with hydrocyclones of diverse structures and different working conditions. Factors such as feed flow rate, size of the downstream, isolation tank, suspension concentration, viscosity, and pressure drop are included [12].

The separation effect of solid particles has been investigated by observing changes in the external conditions, optimization of hydrocyclone structure, and so on. Only a few basic studies have been conducted on the shape and size of sand particles. Thus, more efforts are called for on the study of the interaction between particles and the complex motion of different-shaped particles in a hydrocyclone. Due to the differences in particle sizes and shapes in nature, the separation becomes more complex, with unpredictable separation effects. In this study, the CFD–DEM coupling method and experimental testing method were employed to study the effect of maximum projected area on the separation performance with the same volume of sand. The mechanism underlying the separation of water and sand in a hydrocyclone was proposed by comparing the separation of single 60 μm spherical sand particles and 60 and 120 μm mixed particles.

2. Experimental Modes and Methods

2.1. Hydrocyclone Model

The parameters of the hydrocyclone used in this study were as follows. The diameters of the inlet, upstream, and downstream are 2, 3.4, and 2.2 mm, respectively. The cylinder section was 40 mm in height and 10 mm in diameter. The cone angle was 4 degrees. The shape and mesh conditions are shown in Figure 1. Based on the separation characteristics of hydrocyclones, it is generally believed that the sand particles moving against the wall where the cylinder and cone connect can be separated from the downstream. Therefore, the D-D section was set on the model to divide the area between the radial center and the wall into five regions. Figure 1a shows the model segmentation and meshing, and Figure 1b provides a magnified view of the D-D section 1:5. The three-dimensional structure was drawn using the UG10.0 (Unigraphics NX10.0) software, the meshing was done using the GAMBIT 2.4.6 software, and the numerical simulation was based on the ANSYS 16.0 and DEM 2.6.1 coupling method.

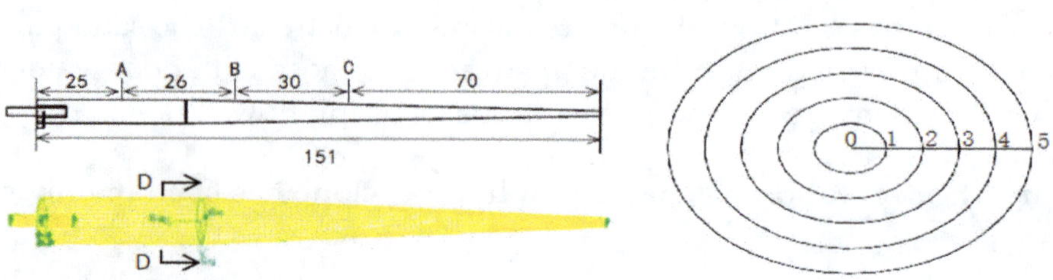

Figure 1. The axial length of Z axis and cross section of the cylinder of the hydrocyclone.

2.2. Numerical Simulation

2.2.1. Particle Characteristics

In this study, a single separation simulation of five spherical sand particles with diameters of 60, 70, 90, 106, and 120 μm was performed. A single separation simulation was conducted on three plate-like particles with the same volume as spherical sand particles of 60 μm (in Figure 2, the maximum projected area is S), and their maximum projected area was 3.11 S (type A), 2.24 S (type B), and 1.36S (type C), respectively. A mixed separation simulation was conducted on spherical sand particles with diameters of 60 and 120 μm. Table 1 shows the deformation amount, thickness, maximum projected area, and equivalent diameters of particles are shown in Table 1. The type C, B, and A particles share the maximum projected area with particles of 70 μm, 90 μm, and 106 μm, respectively.

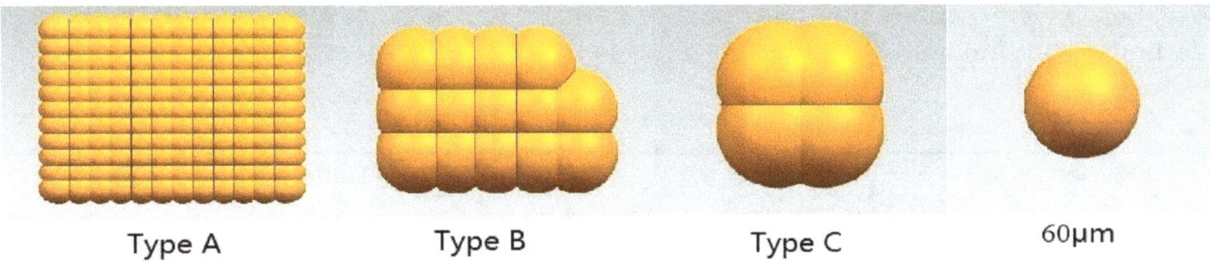

Figure 2. Particle shapes in numerical simulation.

Table 1. Parameters of particles in the numerical simulation.

Title 1	Type A	Type B	Type C	60 μm	70 μm	90 μm	106 μm	120 μm
Mass (μg)	0.28	0.28	0.28	0.28	0.44	0.95	1.54	2.23
Thickness (mm)	0.015	0.030	0.045	0.060	0.070	0.090	0.106	0.120
Maximum projected area (S)	3.11	2.24	1.36	1.00	1.36	2.24	3.11	4.00

In simulation experiments, the recorded residence time of sand particles in the cyclone equaled the period from their entering the inlet to their leaving the downstream or upstream. The data of particle velocity, position, and time were exported from the software DEM and edited by the MATLAB 2017b (Matrix Laboratory) postprocessor program. The values between the third and fourth seconds were collected as the analysis data.

2.2.2. Mathematical Model and Simulation Method

Flow in the hydrocyclone is considered to be an incompressible viscous fluid. The effect of gravity and the roughness of the hydrocyclone walls were taken into account, though surface tension was neglected. In the present study, the inlet velocity of the hydrocyclone was 2.00 m s^{-1}, The continuous medium flow is calculated from the continuity and the Navier–Stokes equations based on the local mean variables over a computational cell, which are given by:

$$\frac{\partial \varepsilon}{\partial t} + \nabla\cdot(\varepsilon u) = 0 \tag{1}$$

$$\frac{\partial(\rho_f \varepsilon u)}{\partial t} + \nabla\cdot(\rho_f \varepsilon u u) = -\nabla P - F_{p-f} + \nabla\cdot(\varepsilon \tau) + \rho_f \varepsilon g \tag{2}$$

F_{p-f} is Interaction forces between fluid and solid phases, equal to $\sum_{i=1}^{k_c} f_{P-f,i}/\Delta V_c$, N/$m^3$; the flow solved in Equations (1) and (2) represents the mixture flow of medium and sand, and was obtained by use of the models in a commercial ANSYS software package, i.e., Fluent. The details of the medium flow calculation and its validation can be found elsewhere [13,14]. In this work, we only give an overall description of the ANSYS model for the hydrocyclone.

The governing equations that are described above were discretized by the control volume numerical technique, and then the SIMPLE pressure–velocity coupling technique, with a second-order upwind scheme for the convection terms that were employed to solve the discretized equations over the computational domain.

The maximum particle volume fraction for the sand used in this study was less than 1%, which means the mixture of water and sand belonged to dilute phase flow. The Lagrangian coupling method was employed and Table 2 summarizes the parameter settings for the sand [15], Table 3 summarizes the boundary condition in the model. Ansys16.0 and DEM2.6.1 were employed during this study. The discrete approach was used to simulate sand movement, collisions among sand particles and between the sand particle and the hydrocyclone wall, and the effects of sand movement on the surrounding continuous phase, energy and momentum exchange. Collisions among sand particles and between sand particles and the wall did not lead to significant plastic deformation, which in consequence were attributed to hard particle contact, which is a wet grain contact model. The 'Hertz-Mindlin (no slip) built-in' model was utilized in this work [16].

Table 2. Parameters used in the model.

Phase	Parameter	Symbol	Units	Value
Solid	Density distribution	ρ	$kg{\cdot}m^{-3}$	2500
	Rolling friction coefficient	μ_r	-	0.01
	Sliding friction coefficient	μ_s	-	0.30
	Poisson's ratio	V	-	0.40
	Young's modulus	E	$N{\cdot}m^{-2}$	2×10^{-7}
	Coefficient of Restitution	c_r	-	0.55
fluid	Density	ρ	$kg{\cdot}m^{-3}$	998.20
	Viscosity	μ	$kg{\cdot}m^{-1}{\cdot}s^{-1}$	0.001

Table 3. Boundary condition.

	Symbol	Units	Value
Particle velocity at inlet	v	$m{\cdot}s^{-1}$	2.00
Viscosity of Water Phase	v	$m{\cdot}s^{-1}$	2.00
Turbulent intensity	I	-	5%
Hydraulic radius	D	mm	2.00
Pressure at upstream	-	Pa	0.00
Pressure at downstream	-	Pa	0.00
Back-flow turbulence intensity	I_h	-	5%
Number of particle	-	$N{\cdot}s^{-1}$	1000
Particle diameter	d_i	μm	-

In the simulation, the contributions of viscous drag force, pressure gradient force and gravity were considered. Other additional forces such as virtual mass force and Saffman force were not considered, as they are an order of magnitude smaller compared with the foregoing [17].

The CFD-DEM coupling process is described as follows: the continuous phase is resolved by ANSYS to acquire fluid drag force with sand, which was transformed from flow field information through the drag force model. The stress state of sands was measured by DEM to obtain new information such as the position and velocity of sands and to the flow field. ANSYS was used to model the flow and the most representative stress condition of sediments. The two approaches were coupled with a compound model that included the particle–fluid interaction force [16]. The relevant equations can be seen in Table 4.

Table 4. Equations and specifications.

S/N	Name	Formula	Description
1	The normal force (F_n)	$F_n = \frac{4}{3}E^*(R^*)^{\frac{1}{2}}\alpha^{\frac{3}{2}}$	R* is the equivalent radius α is the normal overlap
2	The equivalent elastic modulus (E*)	$\frac{1}{E^*} = \frac{1-V_1^2}{E_1} + \frac{1-V_2^2}{E_2}$	E_1, v_1 and E_1, v_1 are elastic modulus and Poisson's ratio of sand 1 and sand 2
3	The damping force ($F_n{}^d$)	$F_n^d = -2\sqrt{\frac{5}{6}}\beta\sqrt{S_n m^*}V_n^{rel}$	V_n^{rel} is the normal relative velocity; S_n is the normal stiffness β is coefficient
4	The equivalent mass (m*)	$m^* = \frac{m_1 m_2}{m_1+m_2}$	m_1 and m_2 are the mass of sand 1 and sand 2
5	The tangential force among the sands (F_t)	$F_t = -S_t\delta$	δ is the tangential overlap
6	The tangential stiffness (S_t)	$S_t = 8G^*\sqrt{R^*\alpha}$	
7	The equivalent shear modulus (G*)	$G^* = \frac{2-V_1^2}{G_1} + \frac{2-V_2^2}{G_2}$	G_1 and G_2 are shear modulus of sand 1 and sand 2
		-	V_1 and V_2 are velocity of sand 1 and sand 2
8	The tangential damping force among sand particles (F_t)	$F_t = -2\sqrt{\frac{5}{6}}\beta\sqrt{S_t m^*}V_t^{rel}$	V_t^{rel} is the tangential relative velocity
9	The rolling friction (T_i)	$T_i = -\mu_r F_n R_i \omega_i$	μ_r is coefficient of rolling friction R_i is the distance between the center of mass to the point of contact; ω_i is unit angular velocity vector of object at the contact point

Particle–fluid interaction, force ($f_{p-f,\ i}$):

$$f_{D,i} = \left(0.63 + \frac{4.8}{Re_{p,i}^{0.5}}\right)^2 \frac{\rho_f|u_i - v_i|(|u_i - v_i|)}{2}\frac{\pi d_i^2}{4}\varepsilon_i^{-\beta}$$
$$Re_{p,i} = \frac{d_i\rho_f\varepsilon_i|u_i - v_i|}{\mu_f},\ \beta = 3.7 - 0.65\exp\left[-\frac{\left(1.5 - \log Re_{p,i}\right)^2}{2}\right] \tag{3}$$
$$\varepsilon = 1 - \frac{\sum_{i=1}^{k_C} V_i}{\Delta V_C}$$

$$f_{pg,i} = V_{p,i}\nabla P \tag{4}$$

The free settling velocity (u_t)

$$u_t = \frac{gd^2}{18\upsilon}\left(\frac{\rho_p - \rho}{\rho}\right) \tag{5}$$

Here, the drag coefficient (C) is

$$C = \frac{(\rho_P - \rho)Vg}{A\rho\left(\frac{u^2}{2}\right)} \tag{6}$$

And the particle Reynolds number (Re) is

$$Re = \frac{D_{1p}u\rho}{\mu} \tag{7}$$

The drag coefficient varies depending on the inverse of the particle Reynolds number, thus the approximated drag coefficient was calculated as a function of the particle Reynolds number. The approximated drag coefficient Ca is expressed as

$$C_a = C_0 + C_1\left(\frac{1}{Re}\right) \tag{8}$$

and C_0 and C_1 were determined from the correlation of the particle Reynolds numbers and measured drag coefficients using the ratio of the particle diameter D_2 to thickness T as the particle shape factor

$$C_0 = 0.4555Ln\left(\frac{D_2}{T}\right) + 0.4687 \tag{9}$$

$$C_1 = 19.285 \tag{10}$$

Finally, the approximated drag coefficient was expressed as [18]:

$$C_a = 19.285\left(\frac{1}{Re}\right) + 0.4555Ln\left(\frac{D_2}{T}\right) + 0.4687 \tag{11}$$

Pressure gradient force [19]: Due to the uneven distribution of hydraulic pressure on the surfaces of particles, the force is given by:

$$F_p = -\frac{\nabla p}{\rho_p} \tag{12}$$

The total drag force of fluid on sand particles (F_D)

$$F_D = \xi A_p \frac{\rho\mu^2}{2} \tag{13}$$

When a particle accelerates in the viscous fluid, it will cause the surrounding fluid to accelerate, too. Due to the inertia effect, the fluid has a reaction force to the particle, which makes the inertia of the particle virtually increase. The related force is given by [2]:

$$F_A = \frac{1}{2}\frac{\rho}{\rho_p}\left(\frac{du}{dt} - \frac{du_p}{dt}\right) \tag{14}$$

When a particle moves in a flow field with velocity gradient, it will experience a lateral force from the low flow rate side to the high flow rate side, given by:

$$F_S = \frac{9.69}{\pi d_p \rho_p}(\mu\rho|\gamma|)^{\frac{1}{2}} C_{LS}(u - u_p) \tag{15}$$

when a particle rotates at an angular velocity of ω, a lateral force perpendicular to the relative velocity and the rotation axis of the particle is generated from the upstream side to the downstream side, given by:

$$F_M = \frac{3}{4}\frac{\rho}{\rho_p}\frac{(u - u_p)^2}{d_p} C_{LM} \frac{\omega_r \times u_r}{|\omega_r||u_r|},\ \omega_r = \omega - \Omega \tag{16}$$

When a particle moves in a viscous fluid, the boundary layer on the surface of the particle will carry a certain amount of fluid. Due to the inertia of the fluid, the fluid cannot immediately follow the acceleration or deceleration of the particle, so this boundary layer is unstable, and the particle will be affected by a time-dependent force, given by:

$$F_B = \frac{9}{d_p\rho_p}\sqrt{\frac{\rho\mu}{\pi}}\int_{t_0}^{t}\frac{(du/d\tau)-(du_p/d\tau)}{\sqrt{t-\tau}}d\tau \quad (17)$$

In the Lagrangian coordinate system, the motions of particles are predicted based on the forces acting on them. In our studied system, the governing equation of particles can be given by

$$\frac{du_p}{dt} = \left(1-\frac{\rho}{\rho_p}\right)g + F_D + F_P + F_A + F_S + F_M + F_B \quad (18)$$

The formula for the separation effect of the hydrocyclone:

$$\begin{aligned} P_{downstream} &= \frac{N_{downstream}}{N_{downstream}+N_{upstream}}\% \\ P_{upstream} &= \frac{N_{upstream}}{N_{downstream}+N_{upstream}}\% \end{aligned} \quad (19)$$

2.3. Separation Experiments

Figure 3 provides the flow chart of separation experiment in the hydrocyclone. The water pressure was controlled at 20 kPa by adjusting the feed valve (consistent with the pressure set in numerical simulation).

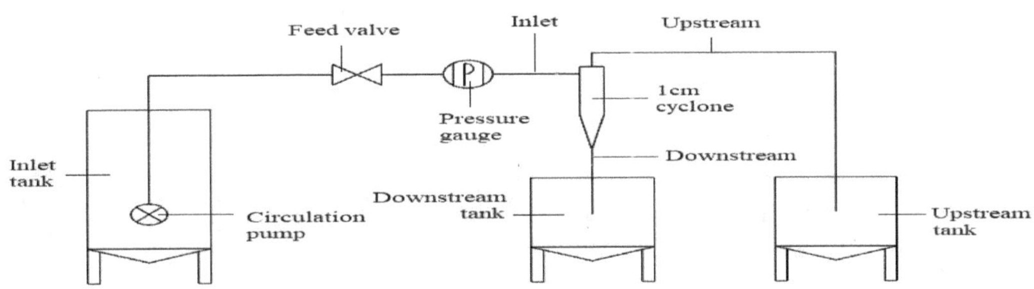

Figure 3. Schematic diagram of the hydrocyclone test.

Sand particles with D50 of 120 µm were screened using a 115 to 125 mesh screen to substitute the 120 µm particles in numerical simulation, and particles with D50 of 60 µm were screened using a 230 to 250 mesh screen to substitute the 60 µm particles in numerical simulation.

The volume of 120 µm particles was eight times that of 60 µm particles. The volume of 120 µm particles put into the hydrocyclone was eight times that of 60 µm particles to ensure an equal number of both particles.

Table 5 shows the standard volume concentration adopted. The concentrations for 60 and 120 µm particles were 5 kg·m^{-3} and 40 kg·m^{-3}, respectively.

Table 5. Experiments for particle separation.

Particles	Particle Size (µm)	Concentration (kg·m^{-3})	Separation Time (s)
Singe	60	5	16
	120	40	16
Mixed	60 + 120	5 + 40	16

Experiment 1. *Sand particles (61–106 µm; 9 kg) were screened out using a 150 to 240 mesh screen to substitute the 60 to 106 µm irregular particles in numerical simulation, and a separation simulation was performed on them.*

Experiment 2. *Three experiments were carried out, including two independent single separations of 60 and 120 µm particles and one mixed separation of both particles.*

3. Analysis of Experimental Results

3.1. Flow Field, Flow Rate and Particle Distribution

The separation process of water and sand in a hydrocyclone is as follows: the turbulent flow of water drives the sediment particles to move, as a result of which the latter eventually flow out from the upstream and downstream separately. It can be seen that turbulent flow of water plays an important role in the separation of sediment particles. After it flows in from the inlet, the water flows into a swirling turbulent flow in the cavity, and finally carries the sediment particles out of the upstream or downstream separately.

Figure 4 shows the velocity of water flow in the hydrocyclone along the *X*-, *Y*- and *Z*-axes, the resultant velocity of water flow (scalar quantity obtained by calculating the values of three coordinate axis), and the simulated result of water flow pressure distribution.

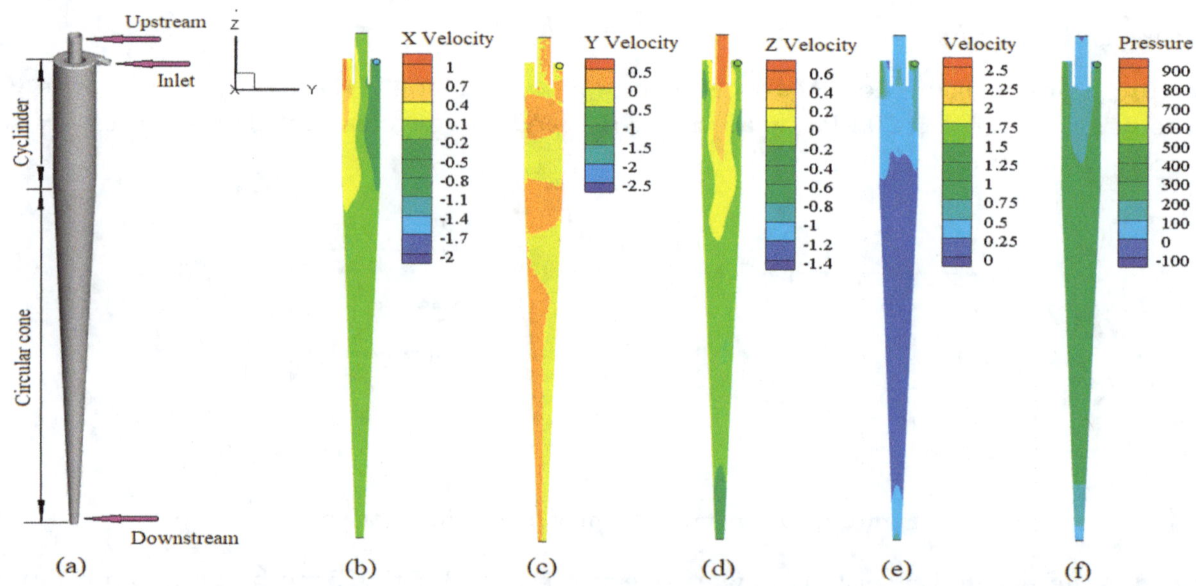

Figure 4. Structure of the cyclone, water flow velocity (m·s^{-1}) and pressure (Pa) distribution: (**a**) structure of hydrocyclone; (**b**) flow velocity in X-axis; (**c**) flow velocity in Y-axis; (**d**) flow velocity in Z-axis; (**e**) water flow velocity distribution; (**f**) water flow pressure distribution.

Figure 4a shows the structure of the hydrocyclone, Figure 4b, 4c and 4d are the water flow velocity (m/s) distributions along the *X*-, *Y*- and *Z*-axes, respectively. Figure 4e and 4f presents the water flow velocity distribution and the water flow pressure distribution, separately. The velocity distribution along the *X*-axis is symmetrically distributed on both sides centered with the *Z* axis. The velocity distribution along the *Y*-axis exhibits periodic positive and negative phases. With regard to the speed distribution along the *Z*-axis, the velocity near the wall of the cylindrical section is low, and particles there flow from the downstream. The water flows through a zero-speed buffer zone when approaching the axis center, and eventually flows to the upstream. The general trend is that the water flow velocity in the cavity changes significantly at the intersection between cylinders, and the flow velocity increases near the downstream (as shown in Figure 4e). The pressure distribution in the cavity did not show obvious changes, but changes significantly when near the outlet. By observing the water flow velocity and pressure distribution, it can be known that the water flow carries the sediment particles from the inlet, and being blocked by the wall, particles rotate in circles in the cyclone chamber, and finally flow out from the upstream or the downstream. Since the water carrying capacity is influenced by the shape and size of the particles themselves, the hydrocyclone is usually used for sediment particle separation.

Figure 5 shows the position distribution of the particles in the hydrocyclone during the simulation. Figure 5a shows the separation simulation of 60 and 120 μm particles, and Figure 5b shows the simulation of the effect of the cross-section change of sand particles in the separation. Figure 5c shows the distribution of sand particles along the Z-axis and Figure 5d shows the particle distribution at the D-D section in Figure 5b.

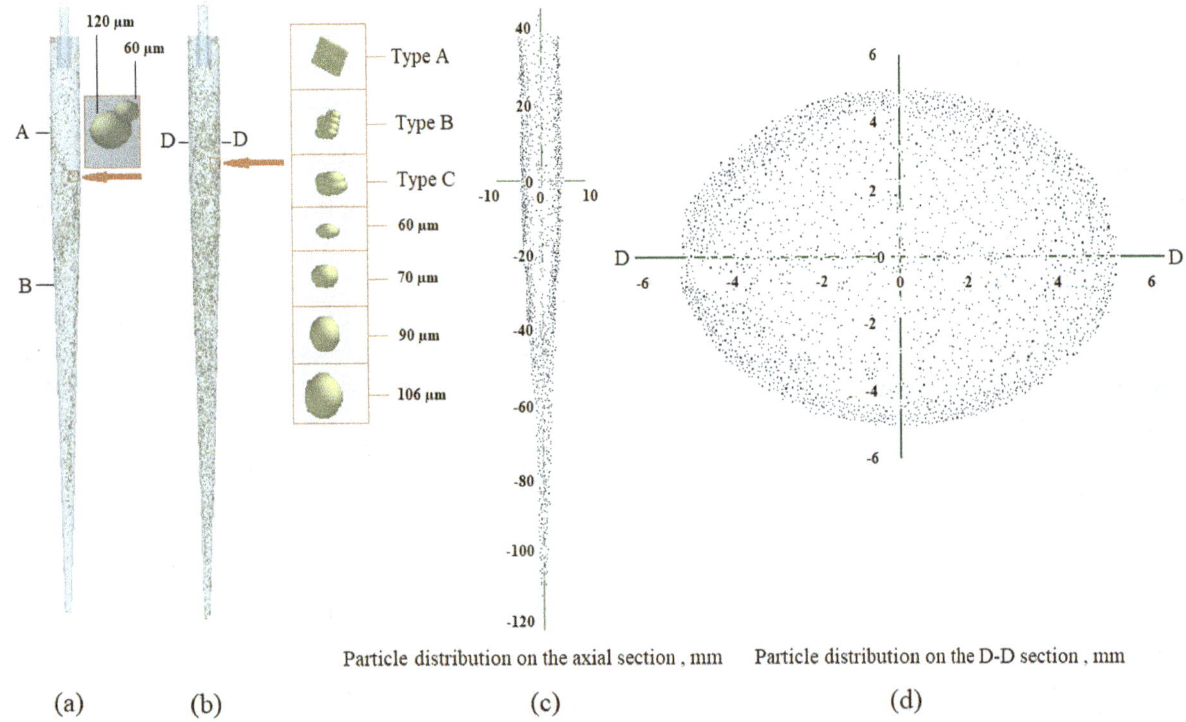

Figure 5. Sand particle distribution in simulation: (**a**) the separation simulation of 60 and 120 μm particles; (**b**) the simulation of the effect of the cross-section change of sand particles in the separation; (**c**) the distribution of sand particles along the Z-axis; (**d**) the particle distribution at the D-D section in (**b**).

From Figure 5c,d, the overall distribution of sand particle in the cylindrical section and part of the conical section can be obtained, and the closer of particles are to the wall, the higher the density is. The radial (*X*-axis) concentration distribution, as well as the velocity and the separation results in Figure 5b are shown in Tables 6 and 7, respectively. The axial velocity of the *Z*-axis, the radial (*X*-axis) distribution percentage, and the separation results in Figure 5a are shown in Figure 6, Tables 8 and 9, respectively.

3.2. Relationship between Maximum Projected Area and Separation Results

3.2.1. Radial Concentration Distribution of Particles

Table 6 shows the concentration distribution of different sand particles in the radial area of Figure 1. The general trend of sand particle distribution was that the higher the concentration distribution and the nearer the sands to the wall, the more concentrated the large sand particles, especially the 106 μm sand particles, which accounted for 86.93% of the total area.

Table 6. Concentration distribution of different sand particles in the cylindrical section.

Radial Distance	Type A	Type B	Type C	60 μm	70 μm	90 μm	106 μm
	Percentage of Content (%)						
0–1	8.10	6.82	3.96	2.65	2.11	1.81	1.01
1–2	9.01	7.12	5.39	5.14	3.23	2.34	1.96
2–3	12.87	11.12	9.01	8.63	7.05	5.91	3.21
3–4	18.59	17.36	16.20	13.85	11.61	7.32	6.89
4–5	51.43	57.58	65.44	69.73	76.00	82.62	86.93

The volume of 60 μm particles was the same as that of particle types C, B, and A. Their concentrations were 69.73%, 65.44%, 57.58%, and 51.43%, respectively, in the area with 4 to 5 mm radial distance near the wall. Therefore, the concentration of sand particles with the same volume tended to decrease gradually near the wall of the hydrocyclone with the increase in the maximum projected area.

The concentrations of 60, 70, 90, and 106 μm particles in the area with 4 to 5 mm radial distance were 69.73%, 76.00%, 82.62%, and 86.93%, respectively. That is to say, the concentration of spherical particles tended to increase in the direction of the hydrocyclone wall with the increase in the volume.

Taking the 2-3 area as the center, the concentrations of types C, B, and A in the area with 0 to 2 mm radial distance were 4.01%, 9.79%, and 14.14% higher than those of 70, 90, and 106 μm particles, respectively. Furthermore, their concentrations in the area with 3 to 5 mm radial distance were 5.97%, 15.00%, and 23.80% lower, respectively. Therefore, for the particles with the same volume and maximum projected area, the higher the concentration in the center of the hydrocyclone, the more likely the central part (the upwelling) flowing out from the upstream; for the particles with the same maximum projected area, the bigger the volume, the higher the concentration near the hydrocyclone wall, and the closer to the side wall, the more likely they might flow out from the downstream.

3.2.2. Separation Results in Numerical Simulation

Table 7 shows the separation results of the upstream and downstream with changes in the volume and maximum area.

Table 7. Separation results of particles from the upstream.

Radial Distance	Maximum Projected Area (S)	$t_{upstream}$ (s)	$t_{downstream}$ (s)	$V_{upstream}$ (m s^{-1})	$V_{downstream}$ (m s^{-1})	$L_{upstream}$ (m)	$L_{downstream}$ (m)	p (%)
Type A	3.11	0.63	2.21	0.41	0.59	0.26	1.30	24.0
Type B	2.24	0.66	2.15	0.41	0.59	0.27	1.27	17.8
Type C	1.36	0.69	2.11	0.41	0.59	0.28	1.25	14.9
60 μm	1.00	0.72	2.08	0.40	0.58	0.29	1.21	9.4
70 μm	1.36	0.75	2.02	0.39	0.59	0.29	1.19	7.4
90 μm	2.24	0.79	1.98	0.39	0.60	0.31	1.19	3.2
106 μm	3.11	0.81	1.92	0.39	0.60	0.32	1.15	0.9

$t_{upstream}$, the average residence time of particles crossing the upstream; $t_{downstream}$ is the average residence time of particles crossing the downstream; $V_{upstream}$ is the average absolute velocity of particles crossing the upstream; $V_{downstream}$ is the average absolute velocity of particles crossing the downstream; $L_{upstream}$ is the average path length of particles crossing the upstream; $L_{downstream}$ is the average path length of particles crossing the downstream; and p is the percentage of the particles crossing the upstream in the separated particles, which was obtained using Equation (19).

As shown in Table 7, 9.4%, 7.4%, 3.2%, and 0.9% of spherical particles with diameters of 60, 70, 90, and 106 μm flowed out from the upstream, respectively. Therefore, it was harder for larger spherical particles to be separated from the upstream. The 60 μm particles had the same volume as types C, B, and A particles. As the maximum projected area increased gradually, 9.4%, 14.9%, 17.8%, and 24% of them flowed out from the upstream, accordingly. Thus, it was easier for the particles with the same volume to be separated from the upstream as the maximum projected area increased.

The 60 μm particles and types C, B, and A particles flowed out from the upstream in less time and traveled a shorter distance with the increase in the maximum projected area. For instance, the particles of type A took 0.09 s less compared with 60 μm particles and the average distance decrreased by 0.03 m.

However, the 60, 70, 90, and 106 μm particles needed more time to flow out from the upstream and travel a longer distance with the increase in their size. Among them, the 106 μm particles took the most time, 0.09 s more compared with 60 μm particles on average, and the average distance increased by 0.03 m.

For particles obtained from the upstream, the average velocity of 60 μm spherical particles and types A, B, and C particles was greater than that of 70, 90, and 106 μm particles. However, the average velocity of particles obtained from the downstream was in contrast to those from the upstream. The absolute velocity of different particles was less than that of water flow. Therefore, the smaller the particles or the larger the projected area, the stronger the ability to flow.

3.3. Separation Mechanism of Single and Mixed Particles

3.3.1. Distribution of Axial Velocity and Radial Concentration

Figure 6 reflects the radial velocity of 60 and 120 μm particles. Three curves are seen from the top downward. The first one is about the average velocity of 120 μm particles in a positive direction, the direction of downstream. (The average velocity of 120 μm particles in single and mixed separations is shown by the same curve as it is the same in different separations, and 60 μm particles have no influence on 120 μm particles.) The second curve shows the average velocity of 60 μm particles in mixed separation in a positive direction (the direction of downstream). The third one is about the average velocity of 60 μm particles in single separation in a positive direction of Z axis. Figure 6b shows the partially enlarged image of Figure 6a (the former is five times bigger than the latter).

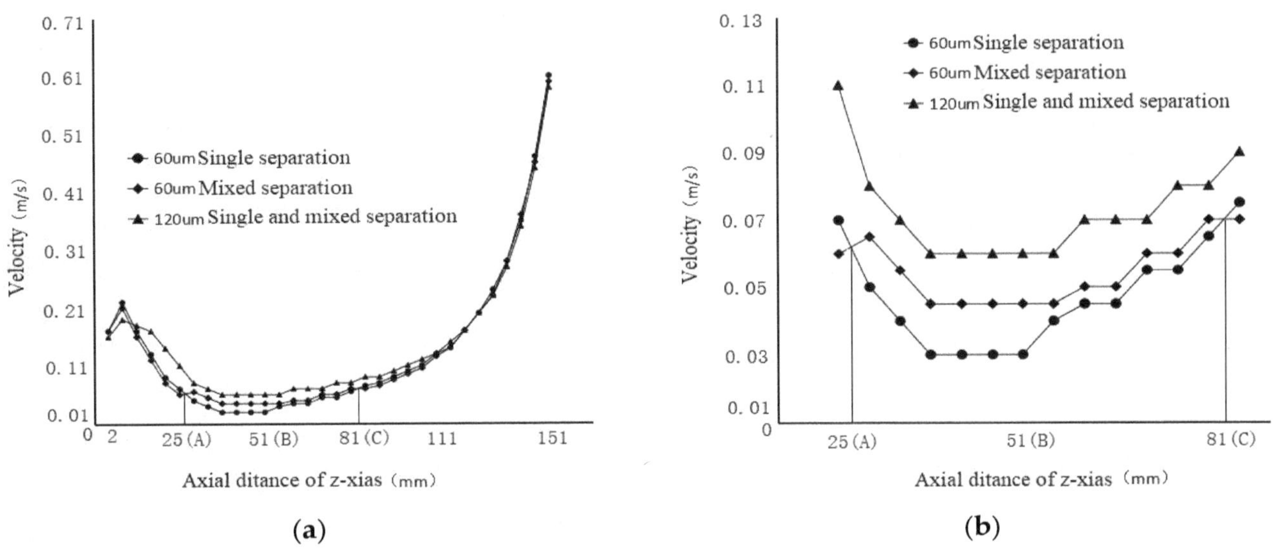

Figure 6. Average velocity distribution in a positive direction of Z axis: (**a**) radial velocity of 60 and 120 μm particles; (**b**) partially enlarged image of (**a**).

The velocity and direction of water flow significantly changed at 25–81 mm in Figure 6, which is the transition from a cylinder to a cone. Particles slowed down in a positive direction on the Z axis due to a lower water velocity, and the average velocity of different-sized particles also decreased. Thus, particles in this section mostly flow slowly in the whole separation process.

The velocity of 60 and 120 μm particles on Z axis had four changing stages: rising, falling, flattening, and rising again. Compared with a single separation, the average velocity of 60 μm particles increased by 25.74% at 25–81 mm in mixed separation, which was quite obvious at 25–51 mm. However, the velocity tended to be consistent at 0–25 mm and 81–151 mm.

The velocity was the lowest at 25–81 mm in the whole process. As shown in Figure 1, it was the border of cylinder and cone sections, 25–51 mm in the cylinder section and the rest in the cone section.

Table 8 shows the distribution of particle concentration at A to B in Figure 4. It was found that 93.23% of 120 μm particles in single separation were concentrated 4–5 mm away from the center of the circle in the radial direction. The percentage of particles nearest to the center of the circle was 1.33%. However, 69.73% of 60 μm particles in single separation were concentrated at 4–5 mm, and the percentage of particles nearest to the center of the circles was 2.65%. When the particles of these two sizes were mixed, the radial concentration distribution of 120 μm particles was in accordance with that of single separation generally. However, the radial concentration distribution of 60 μm particles in single separation was quite different from that of mixed separation. The difference was reflected in the section of 4–5 mm, and the percentage reached 20.73%.

In other words, influenced by 120 μm particles, 20.73% of 60 μm particles were concentrated in the aforementioned section in mixed separation.

Table 8. Concentration distribution of 60 and 120 μm particles in the cylinder section.

		0–1 mm	1–2 mm	2–3 mm	3–4 mm	4–5 mm
60 μm particles	Single	2.65	5.14	8.63	13.85	69.73
	Mixed	1.41	1.16	1.19	5.78	90.46
	Difference	−1.24	−3.98	−7.44	−8.07	20.73
120 μm particles	Single	1.33	1.50	1.64	2.30	93.23
	Mixed	1.21	1.32	1.79	2.95	92.73
	Difference	−0.12	−0.18	0.15	0.65	−0.5

3.3.2. Particle Separation Results

The calculation of particle separation took 4 s (see Table 9). The number of particles that flowed into the hydrocyclone was more than the sum of particles separated from upstream and desilting port in 0–3 s; the separation was unstable. However, it reached a dynamic balance because the number of particles flowing into the hydrocyclone was close to that of the particles separated from it in 3–4 s. When 60 μm particles were mixed with others, the average time spent in separation from the downstream was 0.2 s less than that in single separation while the separation increased by 4.1%. Furthermore, 120 μm particles were not separated in either single separation or mixed separation from the upstream.

Table 9. Separation of 60 and 120 μm particles.

		$t_{avearge}$	$N_{downstream}$	$N_{upstream}$	p (%)
Single	X	2.00	906	94	90.6
	D	1.70	1000	0	100
Mixed	X	1.80	947	53	94.7
	D	1.70	1000	0	100

X and D, the number of X (60 μm particles) and D (120 μm particles) was 1000 per second; $t_{avearge}$ is the average time duration in which particles were separated from the downstream staying in the hydrocyclone in 3–4 s; $N_{downstream}$ is the number of particles passing the downstream; $N_{upstream}$ is the number of particles passing the upstream in 3–4 s; and p is the percentage of particles in the downstream, which was calculated using Formula (19).

4. Experimental Tests

4.1. Test on the Effect of Cross-Section Change on Separation

Figure 7 depicts the observation results of sand samples in the experiment. Figure 7a shows the representative sand particles screened using a vibrating screen. Both particles 1 and 2 could pass the 110 μm hydrocyclone screen, but they were not typical spherical particles, and their volume was less than that of spherical particles of the same size. Figure 7b shows the typical comparison of thickness. Obviously, particles 1 and 3 were thicker than particle 2 in terms of light transmittance. Figure 7c shows the particles with different projected area and thickness. Among them, the volumes of particles 1, 2, 3, and 4 might be the same. Figure 7d shows statistics on the maximum projected area of typical lamelliform particles separated from the upstream. The maximum projected area of particles 1, 2, 3, and 4 in Figure 4 was 7968.39, 8346.47, 8596.55, 3826.37, and 3949.87 μm^2, respectively. Accordingly, the size of spherical particles was 100.75, 103.11, 104.65, 69.81, and 70.93 μm, whose volume was less than that of spherical particles with the same projected area. Smaller particles were more than larger particles. Moreover, most were lamelliform particles with a similar size.

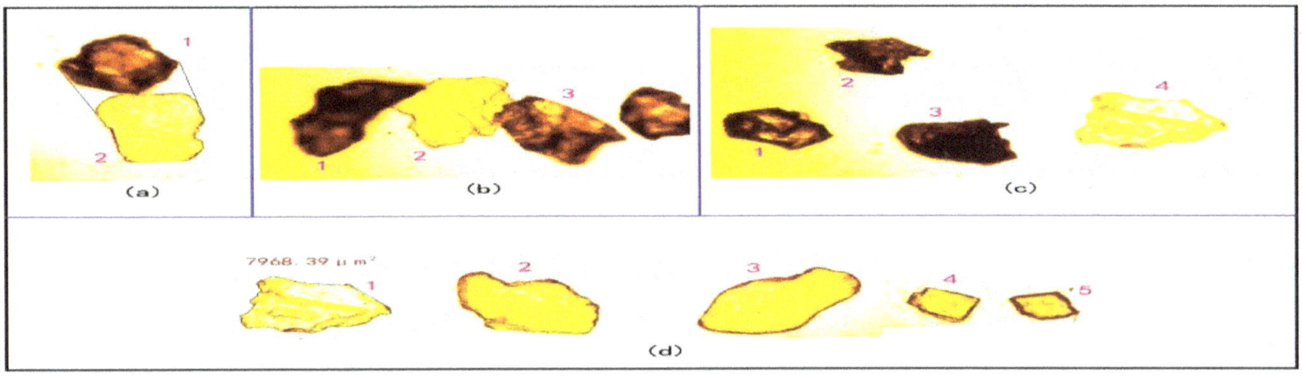

Figure 7. Distribution of particle characteristics (×1 k).

Different particles (10 groups) were observed with a transflective polarized microscope whose number was 59XC-PC before the experiment. It was found that most of 500 particles had a bad light transmittance and only a small number of them (10–30 particles) had a good light transmittance. However, about 160–210 among 500 particles obtained from the upstream were good in light transmittance (because they were relatively thin) after the experiment.

The results indicated that the larger the size of particles, the less the particles were separated from the upstream. As shown in Figure 8, the percentage of 60 and 106 μm particles was 27.75 and 1.91%, respectively.

Therefore, smaller and thinner particles were separated from the upstream more easily. In other words, the process of separating sand particles in the hydrocyclone led to the classification of particles of different sizes, which was equivalent to the filtering of particles of different shapes (lamelliform particles). The experimental result was consistent with the regularities reflected in numerical simulation.

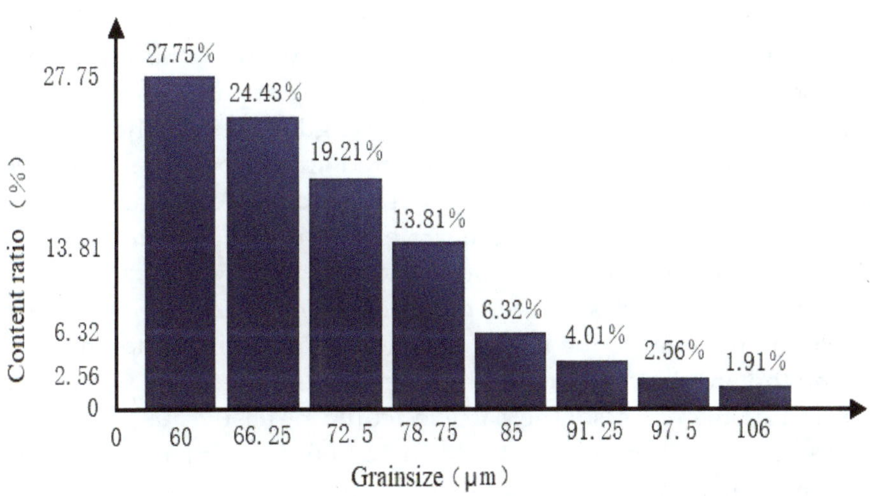

Figure 8. Percentage of particles with different sizes on the upstream.

4.2. Separation Test of Sand Particles 60 and 120 μm

The concentration of 60 and 120 μm sand particles was set as 5 kg·m^{-3} and 40 kg·m^{-3}, respectively, to ensure an equal number of sand particles put into the hydrocyclone. The capacity of the container was 1 m^3. The separation experiment lasted for 16 min.

Table 10 shows the results: 120 μm sand particles were separated from the downstream in both single and mixed separations. However, 312 g of 60 μm sand particles were separated from the upstream in single separation, and 255 g were separated in mixed separation. It was concluded that larger particles improved the separation of smaller ones from the upstream in mixed separation, which was consistent with the regularities reflected in numerical simulation.

Table 10. Separation experiments for 60 and 120 μm sand particles.

	Single		Mixed	
	60 μm	120 μm	60 μm	120 μm
Upstream	0.312	0	0.255	0
Downstream	28.85		29.25	

5. Discussion

5.1. Influence of Maximum Projected Area and Volume Changes on Separation Results

Several parameters affect the performance of hydrocyclones, including the inlet velocity, solid content, liquid-phase viscosity, particle size and shape, and so forth. Saber et al. [20] believed that the separation effect would improve with the increase in the size of spherical particles, which was obviously better than that of plate-like particles. Also, the separation effect declined with the increase in the size of plate-like particles with diameters more than 60 μm. Abdollahzadeh et al. [21] thought that the separation effect decreased along with the increase in flatness when the particle diameter was more than 10 μm. Endoh et al. [22] found that particles with a large size and high proportion were separated from the upstream. Wills et al. [23] held the view that plate-like particles, such as mica, were often discharged upstream even though they were relatively coarse. Kashiwaya et al. [17] believed that the shape of particles would influence the drag coefficient and separation effect. The drag coefficient would increase when the particles became flatter. Then, the increased drag force would increase the influence of water movement on particles. Wang et al. [24] found that the change in particle velocity fluctuated evidently as the size of particles became smaller and the drag force increased. Some particles passed through the locus of zero vertical velocity, joined the upward flow, and then moved out from the upstream. In this study, particles of types C, B, and A, corresponding to 70, 90, and 106 μm particles, respectively, could enter the upward flow and then flow out of the upstream easily as particles became smaller and drag force increased. Therefore, the corresponding percentage dropped sharply in the area with 3 to 5 mm radial distance and increased in the area with 1 to 3 mm radial distance.

Drag force is proportional to the maximum projected area of particles along the direction of fluid flow (Formula (19)), which means that the drag force is strengthened with the increase in the maximum projected area of particles. Hence, the maximum projected area of particles has a certain influence on the separation results of the hydrocyclone. Kashiwaya et al. considered that when Reynolds number reached a certain level (above 50), the maximum projected area of particles was a major influencing factor. They also claimed that the radial variation in the tangential velocity in the hydrocyclone gave rise to a moment on plate-like particles and the particles settled, maintaining the orientation as their broad planes became perpendicular to the radial direction. In this study, the pressure gradient force in the upwelling direction toward the center of the hydrocyclone in the same position increased with the increase in the cross-sectional area. Also, the particles could be easily pushed into the upwelling and then out of the upstream. Figure 9 shows that the volume of 60 μm spherical particles was the same as that of particles of types C, B, and A, but the cross-sectional area was larger and hence the track was simpler in the hydrocyclone.

Figure 9 shows the representative motion tracks of sand particles with different cross-sectional areas when they moved from the upstream. The figure shows that 60 μm spherical particles had the same volume as that of particles of types C, B, and A. With the increase in the maximum projected area, the tracks became simpler and the time needed for separation became shorter. Furthermore, 60, 70, 90, and 106 μm particles were all spherical. The tracks separated from the upstream become more complex with the increase in the volume of particles. That is, it was harder for particles to separate from the upstream as the time taken for separation increased. Tracks (2) and (8), (3) and (7), and (4) and (6) belong to the particles with the same maximum projected area. The smaller the volume of particles, the simpler the tracks when separated from the upstream, indicating that the time taken for separation was shorter, and it was harder for particles to separate from the upstream. Hence, the maximum projected area might have an obvious and a stable influence on the motion tracks of particles, hence affecting the separation effect.

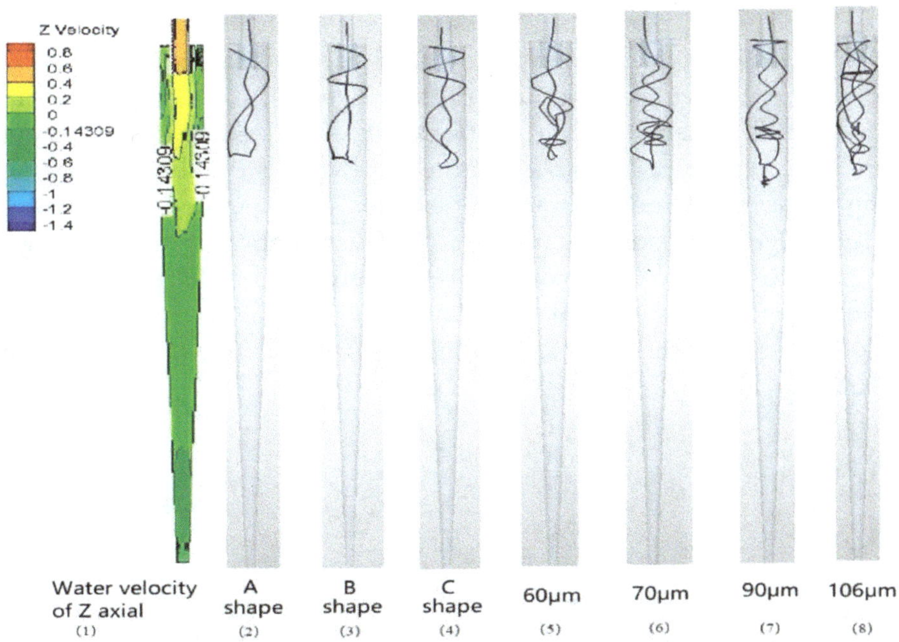

Figure 9. Representative motion tracks of sand particles.

5.2. Analysis of the Following Phenomenon of Fine Particles

Zhang et al. [19] believed that the movement of particles was controlled by the combined effect of the radial fluid drag force, pressure gradient force, and centrifugal force. Of these, the fluid drag force of the inward direction increased exponentially with decreasing size of small particles. A part of the small particles was pushed into the center of the hydrocyclone and then flowed out of upstream. Xu et al. [25] held the view that large particles moved toward the wall under the effect of inertial force and gravity. In this test, a part of 60 μm particles flowed out of the upstream, while the 120 μm particles did not. However, the following phenomenon enhanced the density of the 60 μm particles near the wall, where most 120 μm sand particles were present, thereby reducing the possibility of 60 μm particles flowing into the center of the hydrocyclone and out of the upstream. Therefore, the force on particles in the hydrocyclone was complicated. Zhu et al. [26] noted that the comprehensive effect of the fluid drag force, pressure gradient force, virtual mass force, and gravity on particles should be taken into consideration. Peng et al. used the lattice Boltzmann method to successfully simulate the shallow water flows and river bed evolution [27–34] as well as two-phase flows [35,36].

Tsuji et al. [37] believed that the large leading particles had a huge drag force on small trailing particles, which had little or even no drag force on leading particles in turn. The experiment showed that 120 μm particles had a great impact on 60 μm particles in terms of distribution, velocity, or separation effect. However, no 120 μm particles separated from the upstream. Furthermore, 120 μm sand particles could influence 60 μm particles to a great degree, but the latter's influence on the former could be neglected, which was consistent with the findings of other studies. Meanwhile, Zhu et al. [38] claimed that Re could affect drag, and the increase in Re could obviously weaken the influence of leading particles on small ones. Zhang et al. [18] held the view that the following drag of small particles toward large ones would decrease with the increasing distance, and it would be affected by particle's ratio and the local wake flow compared with that between the particles and the incoming fluid stream. Schubert et al. [39] considered that this phenomenon was due to particle interaction, while the finer particles were captured by the larger particles and the effect was more prominent with the increase in the fraction of large particles. In this study, Re did not change at all under the force of inlet pressure, which was the only one used in this experiment. However, the following phenomenon was quite obvious from the perspectives of process and results.

Under the ideal circumstances, the deposition velocity of spherical particles is proportional to particle diameter (Formula (13)). The velocity is high with the increase in the particle diameter. The numerical calculation and experimental results in this study showed that larger particles were faster than smaller particles in terms of deposition velocity in the hydrocyclone. CFD–DEM coupling was adopted to capture the following phenomena of the movement of small particles toward large ones (see Figure 10). Figure 10a,b presents the motion tracks of 120 and 60 μm particles in the downstream in mixed separation, whereas Figure 10c is the enlarged drawing of the complex segment in Figure 10b. In the figure, among particles around the transition section from a cylinder to a cone, 60 μm particles firstly entered the hydrocyclone, and their motion track was chaotic. After meeting with the 120 μm particles, the 60 μm particles changed their direction and were separated from the downstream together with the 120 μm ones. This phenomenon also explains why the average velocity of mixed separation of 60 and 120 μm particles near the boundary of cylinder and cone was higher than that of single separation (Figure 6).

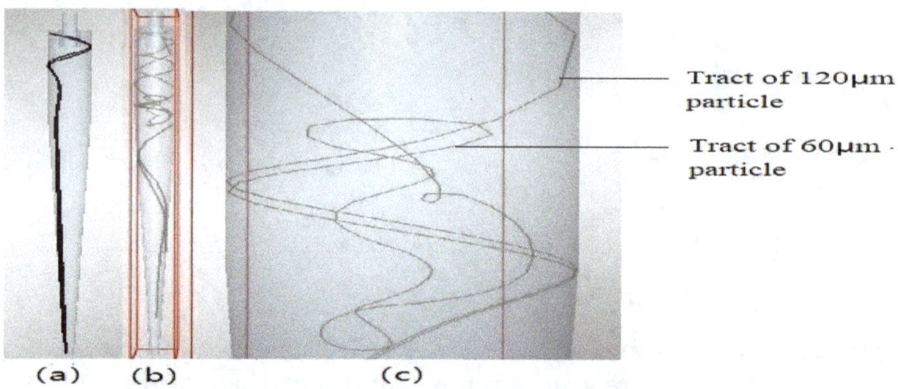

Figure 10. Representative motion tracks of particles.

6. Conclusions

In this study, numerical simulation and experimental methods were applied to reveal the separation mechanism of the hydrocyclone for different-sized particles and the influence of maximum projected area on the separation effect. The conclusions were as follows:

The percentage of 60, 70, 90, 106, and 120 μm spherical particles separated from the upstream was 9.4%, 7.4%, 3.2%, 0.9%, and 0%, respectively, indicating that spherical particles were more easily separated from the downstream as their size increased.

For 60 μm spherical particles and types C, B, and A particles with the same volume, the separation percentages from the upstream were 9.4%, 14.9%, 17.8%, and 24.0%, respectively, implying that the running space was closer to the upward flow in the hydrocyclone as the maximum projected area increased. Hence, particles with the same volume were easier to be separated from the upstream.

The axial velocity of 60 μm particles mixed with 120 μm particles in separation increased by 25.74% compared with that in the single separation of 60 μm particles near the transition section from a cylinder to a cone. The concentration near the wall, where mainly 120 μm particles were present, increased by 20.73%, and the separation effect in the downstream improved by 4.1%. Therefore, large particles could optimize the separation effect of small particles from the downstream.

Author Contributions: Software, Z.T.; methodology, N.L. and F.W.; writing—original draft preparation, Z.T.; writing—review and editing, L.Y.; funding acquisition, L.Y., L.C. and N.C.

Acknowledgments: The authors appreciate the financial support received from the National Natural Science Foundation of China (NSFC) (No. 51769009, 51379024, 51809022) and Funding Program for University Engineering Research Center of the Yunnan Province in China.

Nomenclature

A	measure of area
A_P	maximum projected area of sand particles in the direction of fluid flow, m^2
C_{LS}	Saffman lit coefficient
C_{LM}	magnus lift coefficient
D_1	drag
D_2	particle diameter
d_p	particle diameter, m
d	damping
d_i	diameter of sand particles, m
f	fluid phase
F_A	added mass force per unit mass, $m \cdot s^{-2}$
F_B	basset force per unit mass, $m \cdot s^{-2}$
F_D	total drag force;
F_M	Magus force per unit mass, $m \cdot s^{-2}$
F_P	pressure gradient force per unit mass, $m \cdot s^{-2}$
$F_{p\text{-}f}$	interaction forces between fluid and solids phases
F_S	Saffman lift force per unit mass, $m \cdot s^{-2}$
g	gravity acceleration vector, 9.81 $m \cdot s^{-2}$
i(j)	corresponding to i(j)th particle
k_c	number of particles in a computational cell, dimensionless
$N_{downstream}$	number of sand particles separated from the downstream
$N_{upstream}$	number of sand particles separated from the upstream
$P_{downstream}$	fractional flow of the downstream
$P_{upstream}$	fractional flow of the upstream
P	pressure, Pa
pg	pressure gradient
p	particle phase
Re	Reynolds number, dimensionless
T	particle shape factor (thickness)
t	time, s
u	fluid velocity vector, $m \cdot s^{-1}$
u_t	settling velocity, $m \cdot s^{-1}$
u_r	relative velocity, $m \cdot s^{-1}$
V	volume, m^3
v	fluid kinematic viscosity, $kg \cdot m^{-1} s^{-1}$
ΔV_c	volume of a computational cell, m^3
ΔP	pressure drop, Pa
∇p	pressure gradient, $kg \cdot m^{-2} s^{-2}$

Greek letters

ε	porosity, dimensionless
μ	fluid viscosity, $kg \cdot m^{-1} \cdot s^{-1}$
τ	viscous stress tensor, $N \cdot m^{-3}$
ρ	density, $kg \cdot m^{-3}$
β	Empirical coefficient defined in Equation (3), dimensionless
ρ_p	particle density, $kg \cdot m^{-3}$
γ	fluid strain rate, s^{-1}
ω_r	relative angular velocity, $rad \cdot s^{-1}$
ω	particle rotation angular velocity, $rad \cdot s^{-1}$
ξ	drag force coefficient
Ω	fluid rotation angular velocity, $rad \cdot s^{-1}$
f	force

References

1. Fu, S.; Yong, F.; Yuan, H.; Tan, W.; Dong, Y. Effect of the medium's density on the hydrocyclonic separation of waste plastics with different densities. *Waste Manag.* **2017**, *67*, 27–31. [CrossRef] [PubMed]
2. Zhang, Y.; Cai, P.; Jiang, F.; Dong, K.; Jiang, Y.; Wang, B. Understanding the separation of particles in a hydrocyclone by force analysis. *Powder Technol.* **2017**, *322*, 471–489. [CrossRef]
3. Neesse, T.; Dueck, J.; Schwemmer, H.; Farghaly, M. Using a high pressure hydrocyclone for solids classification in the submicron range. *Miner. Eng.* **2015**, *71*, 85–88. [CrossRef]
4. Yu, L.; Zou, X.; Hong, T.; Yan, W.; Chen, L.; Xiong, Z. 3D numerical simulation of water and sediment flow in hydrocyclone based on coupled cfd-dem. *Trans. Chin. Soc. Agric. Mach.* **2016**, *01*, 126–132.
5. Zhu, G.; Liow, J. Experimental study of particle separation and the fishhook effect in a mini-hydrocyclone. *Chem. Eng. Sci.* **2014**, *111*, 94–105. [CrossRef]
6. Liu, H.; Wang, Y.; Han, T. Influence of vortex finder configurations on separation of fine particles. *CIESC J.* **2017**, *05*, 1921–1931.
7. Tang, B.; Xu, Y.; Song, X.; Xu, J. Effect of inlet configuration on hydrocyclone performance. *Trans. Nonferrous Met. Soc. China* **2017**, *27*, 1645–1655. [CrossRef]
8. Hwang, K.-J.; Chou, S.-P. Designing vortex finder structure for improving the particle separation efficiency of a hydrocyclone. *Sep. Purif. Technol.* **2017**, *172*, 76–84. [CrossRef]
9. Vakamalla, T.R.; Koruprolu, V.B.R.; Arugonda, R.; Mangadoddy, N. Development of novel hydrocyclone designs for improved fines classification using multiphase cfd model. *Sep. Purif. Technol.* **2017**, *175*, 481–497. [CrossRef]
10. Zhang, C.; Wei, D.Z.; Cui, B.Y.; Li, T.S.; Luo, N. Effects of curvature radius on separation behaviors of the hydrocyclone with a tangent-circle inlet. *Powder Technol.* **2017**, *305*, 156–165. [CrossRef]
11. Zhou, Q.; Wang, C.; Wang, H.; Wang, J. Eulerian–lagrangian study of dense liquid–solid flow in an industrial-scale cylindrical hydrocyclone. *Int. J. Miner. Process.* **2016**, *151*, 40–50. [CrossRef]
12. Zhao, Q.; Li, W.; He, W. Experimental study of pressure drop characteristics of a novel solid/liquid hydrocyclone. *J. Eng. Thermophys.* **2010**, *01*, 61–63.
13. Wang, B.; Chu, K.W.; Yu, A.B. Numerical study of particle–fluid flow in a hydrocyclone. *Ind. Eng. Chem. Res.* **2007**, *46*, 4695–4705. [CrossRef]
14. Wang, B.; Chu, K.W.; Yu, A.B.; Vince, A. Modelling the multiphase flow in a dense medium cyclone. *Ind. Eng. Chem. Res.* **2009**, *48*, 3628–3639. [CrossRef]
15. Asakura, K.; Asari, T.; Nakajima, I. Simulation of solid–liquid flows in a vertical pipe by a collision model. *Powder Technol.* **1997**, *94*, 201–206. [CrossRef]
16. Chu, K.W.; Wang, B.; Yu, A.B.; Vince, A. Cfd-dem modelling of multiphase flow in dense medium cyclones. *Powder Technol.* **2009**, *193*, 235–247. [CrossRef]
17. Niu, W.; Liu, L.; Chen, X. Influence of fine particle size and concentration on the clogging of labyrinth emitters. *Irrig. Sci.* **2013**, *31*, 545–555. [CrossRef]
18. Kashiwaya, K.; Noumachi, T.; Hiroyoshi, N.; Ito, M.; Tsunekawa, M. Effect of particle shape on hydrocyclone classification. *Powder Technol.* **2012**, *226*, 147–156. [CrossRef]

19. Zhang, J.; Fan, L.S. A semianalytical expression for the drag force of an interactive particle due to wake effect. *Ind. Eng. Chem. Res.* **2002**, *41*, 5094–5097. [CrossRef]
20. Niazi, S.; Habibian, M.; Rahimi, M. A comparative study on the separation of different-shape particles using a mini-hydrocyclone. *Chem. Eng. Technol.* **2017**, *40*, 699–708. [CrossRef]
21. Abdollahzadeh, L.; Habibian, M.; Etezazian, R.; Naseri, S. Study of particle's shape factor, inlet velocity and feed concentration on mini-hydrocyclone classification and fishhook effect. *Powder Technol.* **2015**, *283*, 294–301. [CrossRef]
22. Endoh, S. Study of the shape separation of fine particles using fluid fields-dynamic properties of irregular shaped particles in wet cyclones. *J. Soc. Powder Technol. Jpn.* **1992**, *29*, 125–132. [CrossRef]
23. Wills, B.A.; Napier-Munn, T. *Wills' Mineral Processing Technology*, 7th ed.; Elsevier Science & Technology Books: Amsterdam, The Netherlands, 2006; pp. 145–206.
24. Wang, B.; Yu, A.B. Computational investigation of the mechanisms of particle separation and "fish-hook" phenomenon in hydrocyclones. *AIChE J.* **2010**, *56*, 1703–1715. [CrossRef]
25. Xu, P.; Wu, Z.; Mujumdar, A.S.; Yu, B. Innovative hydrocyclone inlet designs to reduce erosion-induced wear in mineral dewatering processes. *Dry. Technol.* **2009**, *27*, 201–211. [CrossRef]
26. Zhu, C.; Liang, S.C.; Fan, L.S. Particle wake effects on the drag force of an interactive particle. *Int. J. Multiph. Flow* **1994**, *20*, 117–129. [CrossRef]
27. Peng, Y.; Zhou, J.G.; Burrows, R. Modelling the free surface flow in rectangular shallow basins by lattice Boltzmann method. *J. Hydraul. Eng. ASCE* **2011**, *137*, 1680–1685. [CrossRef]
28. Peng, Y.; Zhou, J.G.; Burrows, R. Modelling solute transport in shallow water with the lattice Boltzmann method. *Comput. Fluids* **2011**, *50*, 181–188. [CrossRef]
29. Peng, Y.; Zhang, J.M.; Zhou, J.G. Lattice Boltzmann Model Using Two-Relaxation-Time for Shallow Water Equations. *J. Hydraul. Eng. ASCE* **2016**, *142*, 06015017. [CrossRef]
30. Peng, Y.; Zhang, J.M.; Meng, J.P. Second order force scheme for lattice Boltzmann model of shallow water flows. *J. Hydraul. Res.* **2017**, *55*, 592–597. [CrossRef]
31. Peng, Y.; Zhou, J.G.; Zhang, J.M.; Burrows, R. Modeling moving boundary in shallow water by LBM. *Int. J. Mod. Phys. C* **2013**, *24*, 1–17. [CrossRef]
32. Peng, Y.; Zhou, J.G.; Zhang, J.M.; Liu, H.F. Lattice Boltzmann Modelling of Shallow Water Flows over Discontinuous Beds. *Int. J. Numer. Method Fluids* **2014**, *75*, 608–619. [CrossRef]
33. Peng, Y.; Zhou, J.G.; Zhang, J.M. Mixed numerical method for bed evolution. *Proc. Inst. Civ. Eng. Water Manag.* **2015**, *168*, 3–15. [CrossRef]
34. Peng, Y.; Meng, J.P.; Zhang, J.M. Multispeed lattice Boltzmann model with stream-collision scheme for transcritical shallow water flows. *Math. Probl. Eng.* **2017**, *2017*, 8917360. [CrossRef]
35. Peng, Y.; Mao, Y.F.; Wang, B.; Xie, B. Study on C-S and P-R EOS in pseudo-potential lattice Boltzmann model for two-phase flows. *Int. J. Mod. Phys. C* **2017**, *28*, 1750120. [CrossRef]
36. Peng, Y.; Wang, B.; Mao, Y.F. Study on force schemes in pseudopotential lattice Boltzmann model for two-phase flows. *Math. Probl. Eng.* **2018**, *2018*, 6496379. [CrossRef]
37. Tsuji, Y.; Morikawa, Y.; Terashima, K. Fluid-dynamic interaction between two spheres. *Int. J. Multiph. Flow* **1982**, *8*, 71–82. [CrossRef]
38. Zhu, G.; Liow, J.L.; Neely, A. Computational study of the flow characteristics and separation efficiency in a mini-hydrocyclone. *Chem. Eng. Res. Des.* **2012**, *90*, 2135–2147. [CrossRef]
39. Schubert, H. On the origin of "Anomalous" shapes of the separation curve in hydrocyclone separation of fine particles. *Part. Sci. Technol.* **2004**, *22*, 219–234. [CrossRef]

12

Hydraulic Features of Flow through Local Non-Submerged Rigid Vegetation in the Y-Shaped Confluence Channel

Xuneng Tong, Xiaodong Liu *, Ting Yang, Zulin Hua, Zian Wang, Jingjing Liu and Ruoshui Li

Key Laboratory of Integrated Regulation and Resource Development on Shallow Lake of Ministry of Education, 1 Xikang Road, Nanjing 210098, China; txn@hhu.edu.cn (X.T.); yang92005006@163.com (T.Y.); Zulinhua@hhu.edu.cn (Z.H.); wangzian1996@hotmail.com (Z.W.); 15205168389@163.com (J.L.); 1514050118@hhu.edu.cn (R.L.)

* Correspondence: xdliu@hhu.edu.cn

Abstract: A laboratory measurement with acoustic Doppler velocimeter (ADV) was used to investigate the flow through a Y-shaped confluence channel partially covered with rigid vegetation on its inner bank. In this study, the flow velocities in cases with and without vegetation were measured by the ADV in a Y-shaped confluence channel. The results clearly showed that the existence of non-submerged rigid plants has changed the internal flow structure. The velocity in the non-vegetated area is greater than in the vegetated area. There is a large exchange of mass and momentum between the vegetated and non-vegetated areas. In addition, due to the presence of vegetation, the high-velocity area moved rapidly to the middle of the non-vegetated area in the vicinity of tributaries, and the secondary flow phenomenon disappeared. The presence of vegetation made the flow in non-vegetated areas more intense. The turbulent kinetic energy of the non-vegetated area was smaller than that of the vegetated area.

Keywords: Y-shaped confluence channel; non-submerged rigid vegetation; longitudinal velocity; secondary flow; turbulent kinetic energy

---

1. Introduction

Channel confluences are a common occurrence in fluvial networks, where significant changes can occur in hydraulics, sediment transport, water environment, and ecology. Many previous studies have concentrated on hydraulic characteristics in channel confluence areas [1–6]. Lan [7] divided the confluences into two types: (1) the confluence where post-confluence channel forms a linear extension of the upstream main channel (also named asymmetrical confluences), (2) the Y-shaped junction. According to Lan's data the quantities of two types of confluences are about equal in nature. Most previous work has paid more attention to the asymmetrical river confluences, but few studies have been conducted on the Y-shaped confluences [8–13]. Many studies confirmed that there are many differences between the two types of confluence channel. Nadia el al. [14], with their experimental work, found that the higher the junction angle, the wider and longer the retardation zone at the upstream junction corner and the separation zone, and the greater the flow deflection at the entrance of the tributary into the post-confluence channel.

The aquatic vegetation is ubiquitous in natural rivers. The vegetation has many ecological, aesthetic and economic benefits, such as providing a terrestrial wildlife habitat; improving water quality; stabilizing streambanks and floodplains; supplying energy subsidies for aquatic and terrestrial ecosystems [15–17]. However, from an engineering point of view, aquatic vegetation practices have not been frequently encouraged because of increased flow resistance, sediment transport effects and decreased flood discharge efficiency compared to unvegetated regions [18]. Thus, a better understanding of the physical processes governing flow resistances in vegetated areas could help resolve these conflicting engineering and ecological considerations [19].

The studies conducted so far in this field consider straight flumes with different types of vegetation on the bed. Nepf [20] developed a physically based model to predict turbulence intensity and diffusion within rigid and emergent vegetation. Ghisalberti and Nepf [21] emphasized the importance of vegetation in open channels and used a model of vegetation to study the flow features in a straight flume. This model showed that the coherent vortices of vegetated shear layer dominate the vertical transport. Since then, many laboratory and numerical studies have been conducted to discuss the effects of vegetation on moving fluids [22–25]. For example, Nezu and Sanjou [26] investigated the turbulence characteristics and the evolution of coherent structures in a flume with rigid and submerged vegetation. They highlighted that the development of sweep and/or ejection events near the top of vegetation affects the exchange processes between vegetation and the overflow. Zhao and Cheng [27] used an array of rigid cylindrical rods to simulate emergent vegetation stems that were subject to unidirectional open channel flows. Until now, straight open channel flows with vegetation have received enough research attention. To address the new objectives of river restoration and environmental flood management, a better understanding of hydraulic features of flow through local non-submerged rigid vegetation in Y-Shaped confluences channel is required.

In this study, we focus on how local non-submerged vegetation influence hydraulic features in a Y-shaped confluence channel. A physical model was used to simulate changes of flow state, velocity distribution, turbulence structures, and turbulent kinetic energy caused by vegetation by comparing the results with non-vegetated conditions under the same flow regime. The primary objective is to investigate the features of distribution of velocity, secondary flow structure and turbulent kinetic energy with the existence of vegetation in Y-shaped confluence channel.

2. Laboratory Experiments

2.1. Experimental Setup and Measurement Technique

Experiments were conducted in a flat-bottom Plexiglas flume (Figure 1) in the Hydraulics Laboratory of Hohai University. Water was pumped into a stilling cistern and then flowed into the flume. The length of the upstream tributaries was 3 m, and the widths were 0.26 and 0.22 m, respectively. The length of the downstream main stream was 6 m, and the width was 0.4 m. In order to make this study have a certain practical significance, the design of the flume model was combined with the analysis of the morphological characteristics of the Xitiaoxi River Basin (30°23′–31°11′ N, 119°14′–120°29′ E) to select the parameters (Table 1). The average width before and after the intersection of rivers and the intersection angle at the intersection were counted respectively. Based on the analysis of the morphological characteristics of river network and the actual conditions, the convergence angle between the tributaries (the angle between the geometric axes) was 60°. A large number of non-submerged rigid vegetation, such as reeds, existed on both sides of these Y-shaped intersections. Polyvinyl chloride(PVC) baseboards ($1 \times 0.1 \times 0.01$ m) were used to cover the entire bottom of the flume. Rigid cylinders were used to simulate rigid vegetation (8 mm diameter, 0.2 m height). The flexural rigidity of simulated rigid vegetation was calculated using the approach of Łoboda et al. [28], based on Niklas [29] and ASTM [30]. The flexural rigidity of simulated rigid vegetation is 40,960 N· mm^2.

The distance measured between the plants (rigid cylinders) was 0.025 m, and the linear spacing was 0.1 m. Vegetation was distributed on both sides of the flume in 2-m-long bands along the two tributaries, and 5-m-long bands along the main stream were planted perpendicularly with the artificial vegetation. In addition, two transition segments and a tailgate were installed to prevent large-scale disturbances from the inlet, thus enabling the development of a quasi-constant water flow by depth.

The outlet and the inlet, which were both connected to the tank, enabled continuous recirculation of the steady-state discharges.

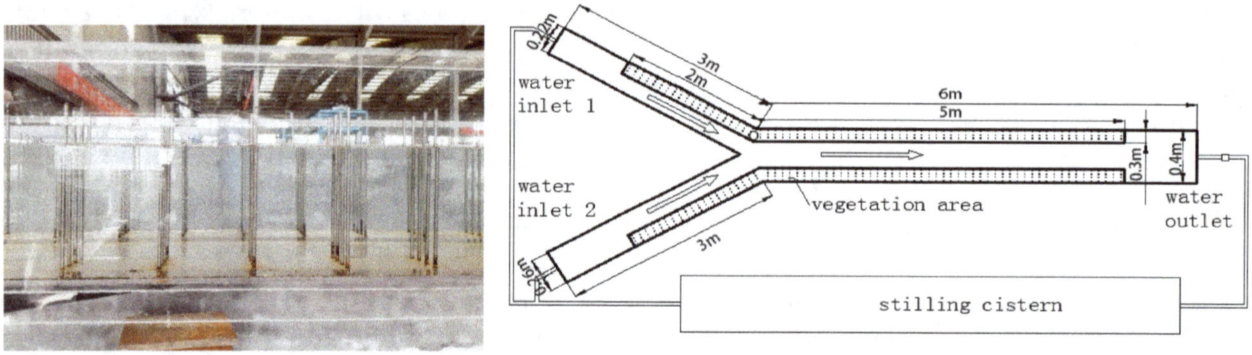

Figure 1. The sketch of the experiment.

Table 1. The morphological characteristics of the Xitiaoxi River Basin.

Y-Shaped Confluence Channel	Average Width of Left River before Confluence (m)	Average Width of Right River before Confluence (m)	Average Width of River after Confluence (m)	The Convergence Angle (°)
1	100.21	107.81	198.81	58.59
2	69.06	80.15	106.15	48.35
3	27.81	30.83	54.65	30.81
4	98.23	100.88	163.15	60.98
5	54.14	73.18	100.88	69.27
6	33.31	48.02	54.15	75.025
7	32.61	38.17	44.63	92.73
8	29.56	40.3	74.14	45.57
Average	55.616	64.918	99.570	60.166

This experiment used two flowmeters to measure the discharge of both tributaries. 3D velocities and velocity fluctuations were measured using a 3D sideways-looking acoustic Doppler velocimeter (ADV) manufactured by SonTek, Inc. (San Diego, CA, USA). The ADV technology is based on the pulse-to-pulse coherent measurement method. The instrument consists of three modules: a measuring probe, a conditioning module, and a processing module [31]. In this study, the ADV was used to sample each measurement point at a frequency of 50 Hz for 30 s. The WinADV-program, a post-processing program, was used to filter and post-process the sampled data. Data with average correlations less than or equal to 80% were filtered out. The result was that each point had a total of 1500 instantaneous data points, which ensured the adequacy and accuracy of the dataset. To achieve three-dimensional movement control of the ADV, a special regulating device was used to realize three-dimensional free positioning movement of the ADV probe in the test flume. The probe could be easily moved between measurement lines and sections. The ADV was mounted in a wood frame across the center section of the test segment and could be easily moved upstream or downstream, so that all sampling points were vertically aligned. The probe could also channel real-time data to the user's computer through a data acquisition program.

2.2. Test Series Description

To conduct comparative experiments and analyses, both the flume with vegetation and one without vegetation were measured while keeping other conditions unchanged. In this experiment, 18 sections were set up (two sections each on the tributaries and 14 sections on the main stream after the intersection). The coordinate origin was set at the bottom of the flume, directly below the intersection point of the two tributaries.

Before the intersection, five perpendiculars were established on each section of the tributaries, with each perpendicular having eight measurement points. In other words, there were 40 points in each section. After the intersection, eight perpendiculars were established on each section of the main stream, with each perpendicular having eight measurement points. This means that there were 64 points for each section. There were five survey lines on the two tributaries and seven survey lines on the mainstream. Figure 2a,b show the sections and the measuring lines.

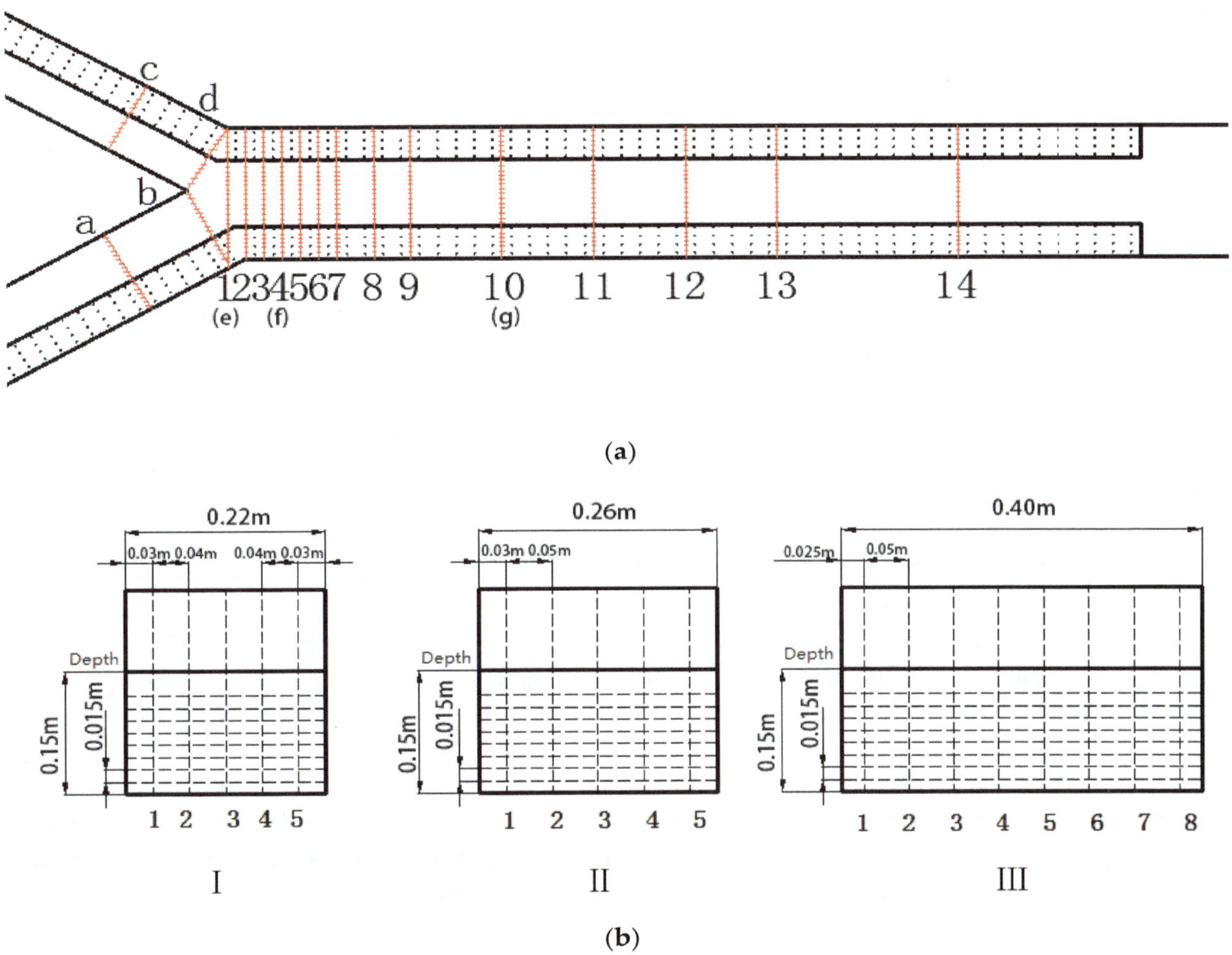

Figure 2. The sections and measuring lines; (**a**) Distribution of measuring cross sections; (**b**) Distribution of measuring lines.

The 3-D Doppler ultrasound anemometer was used to measure the flow velocities on 18 typical cross-sections. By comparing date, seven of those typical cross-sections were chosen to study. The seven sections were the initial section and intersection of both tributaries and sections 1, 4, and 10 of the main streams, as seen in Figure 2a. These seven cross sections were defined as cross Section a, cross Section b, cross Section c, cross Section d, cross Section e, cross Section f, and cross Section g. Section a is 2 m upstream of the right confluence apex, Section b is the right confluence apex, Section c is 2 m upstream of the left confluence apex, Section d is the left confluence apex, Sections e, f, g are 0, 0.2, 0.6 m downstream of the confluence apex, respectively. Vertical measuring lines (Figure 2b) were arranged on each cross-section. All the experimental data were derived from detailed high-resolution measurements on each measuring line.

In the flow characteristics test, the flow rate of the left tributary was Q_1 = 20 m^3/h, and the flow rate of the right tributary was Q_2 = 20 m^3/h. Therefore, the confluence ratio = Q_1/Q_2 = 1.0, and the intersection angle = 60°. Some parameters of the experiment are shown in Table 2, Q denotes to the flow rate; b and h means the width and water section depth, respectively; V points to the velocity of the water section; R_e and F_r is the Reynolds number and Froude number of the flow, respectively.

Table 2. Parameters of the experiment.

Element	Q (m^3/h)	b (m)	h (m)	V(m/s)	R_e (10 °C)	F_r
Left tributary	20	0.22	0.15	0.17	8168	0.14
Right tributary	20	0.26	0.15	0.14	7584	0.12
Main stream	40	0.4	0.15	0.19	12135	0.16

3. Analytical Methods

3.1. Velocity Distribution

Li et al. [32] proposed a method to calculate the velocity of a curved open channel with partially non-submerged rigid vegetation by comparing the integral of the longitudinal depth averaged velocity between vegetated and non-vegetated conditions to analyze the changes in the velocity distribution. The streamwise velocities at different water depths in different sections under vegetated conditions were depth averaged along the depth to obtain the corresponding average streamwise velocity. Due to the limited number of measurement points, the area method of experimental points can be used to circumscribe the area and to calculate the experimental measurement points. By comparing the integral of the streamwise depth velocity between vegetated and non-vegetated conditions, the changes in velocity distribution could be analyzed.

3.2. Secondary Flow Structure

Secondary motions are commonly present in open channel flows. In contrast to the facility in identifying longitudinal bedforms, secondary flows are much more difficult to measure because of their relatively small magnitude [33]. Around the 1950s, the existence of cellular secondary flows in rivers was only inferred from laterally periodic distribution of primary flow and sediment concentration, rather than being confirm through direct velocity measurements, by geologists and river engineers [34]. The precise filed measurement of cellular secondary flows was possible only after the ascent of the electromagnetic current meter (EM) and acoustic Doppler velocimeter (ADV) [35]. The water mixed energetically in the Y-shaped channels, and three-dimensional flow characteristics were prominent. A single vector or contour map was not enough to explain these complex hydrodynamic phenomena. For ease of explanation, this paper provides and explains a longitudinal velocity contour slice map. Each specific section of the diagram is presented and explained. The study presents a comparative analysis of the flow characteristics of the Y-shaped channels with and without vegetation. The velocity data obtained through the experiments were processed using the Tecplot software.

3.3. Turbulent Kinetic Energy

The quantitative analysis of turbulent flow is based on measurements of velocity fluctuations at a single point with local non-submerged rigid vegetation and without any vegetation in a Y-shaped confluence channel. In this study, the mean flow velocity components (u, v, and w) and the velocity fluctuation components in turbulent flow (u', v', and w') correspond to the streamwise, lateral, and vertical directions, respectively. Velocity fluctuations can be defined as deviations from the mean velocity. The turbulent kinetic energy was calculated as noise free [36].

In general, turbulent kinetic energy can be considered a measure of turbulence intensity [37]. In this study, the turbulent kinetic energy E_k was calculated using Equation (1), which was proposed by Li [38] and defines E_k as follows in terms of streamwise, lateral, and vertical directions:

$$E_k = \rho\left(\overline{u'^2} + \overline{v'^2} + \overline{w'^2}\right)/2 \quad (1)$$

where ρ is the density of water and u', v', and w' are the velocity fluctuations of streamwise, lateral, and vertical flow, respectively, and the overbars denote mean values.

4. Results

4.1. Distribution of Stream-Wise Velocity

The depth-averaged streamwise velocity distribution of the seven typical cross sections (Figure 3) showed that under the retarding effect of vegetation, the depth-averaged streamwise velocities (U_{avg}) in the vegetated region were much lower than in the non-vegetated region. A large velocity gradient was apparent near the junction of the vegetated and non-vegetated areas, we can see all the figures show a large velocity gradient was apparent near the junction of the veg and non-veg areas, it not only can be seen from the tributaries, but also from the mainstreams, which showed typical values as reported by Huai et al. [39].

There was a difference in the streamwise velocity distribution between non-vegetated and vegetated cases (Figure 3). The velocities in the vegetated case have a higher gradient than in the non-vegetated case, which can be ascribed to the fact that the retardation caused by vegetation makes the transverse distribution slightly more non-uniform. Along the streamwise direction, the velocities in the vegetated area decreased and those in the non-vegetated area increased when heading downstream, which is consistent with results for open channel flows found by Nikora [40] and Caroppi [41].

Three-dimensional flow characteristics were prominent after the intersection. The depth-averaged streamwise velocity distribution was not enough to explain the complex hydrodynamic phenomena. Clearly, the flow rate gradient was larger with vegetation than without vegetation, which indicates that the presence of vegetation caused drastic changes in the flow structure of the river (Figure 4). With vegetation (Figure 4a), the flow velocity in vegetated zones was smaller than in non-vegetated zones. It can also be observed that in the absence of vegetation (Figure 4b), the streamwise velocity distribution was polarized in the longitudinal direction, with the branch on the left side going downstream having a large flow area, whereas the area close to the right-side tributaries near the bottom was in a low-velocity zone. Downstream of the confluence in non-vegetated cases, a small velocity separation zone appeared on the lateral surface of the left tributary. This is consistent with Wang's results [42]. After vegetation planting (Figure 4a), the polarized character of the streamwise velocity distribution in the longitudinal direction disappeared, the downstream flow velocities in the vegetated areas in the left and right branches obviously diminished, and the flow velocity distribution with depth became insignificant.

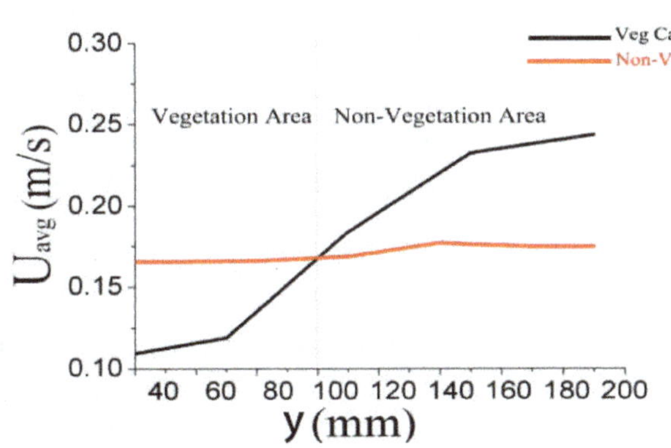

(a)

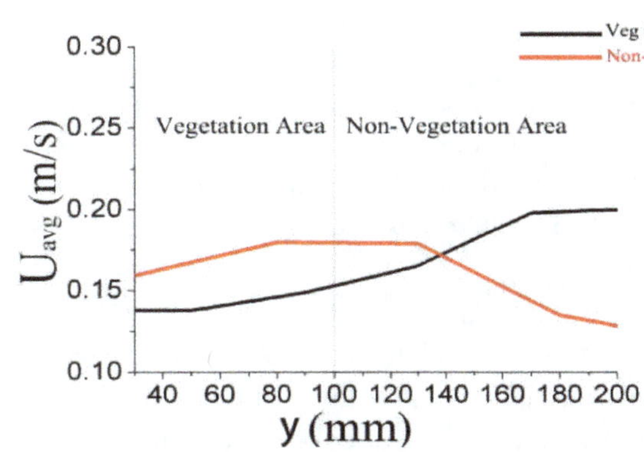

(b)

Figure 3. *Cont.*

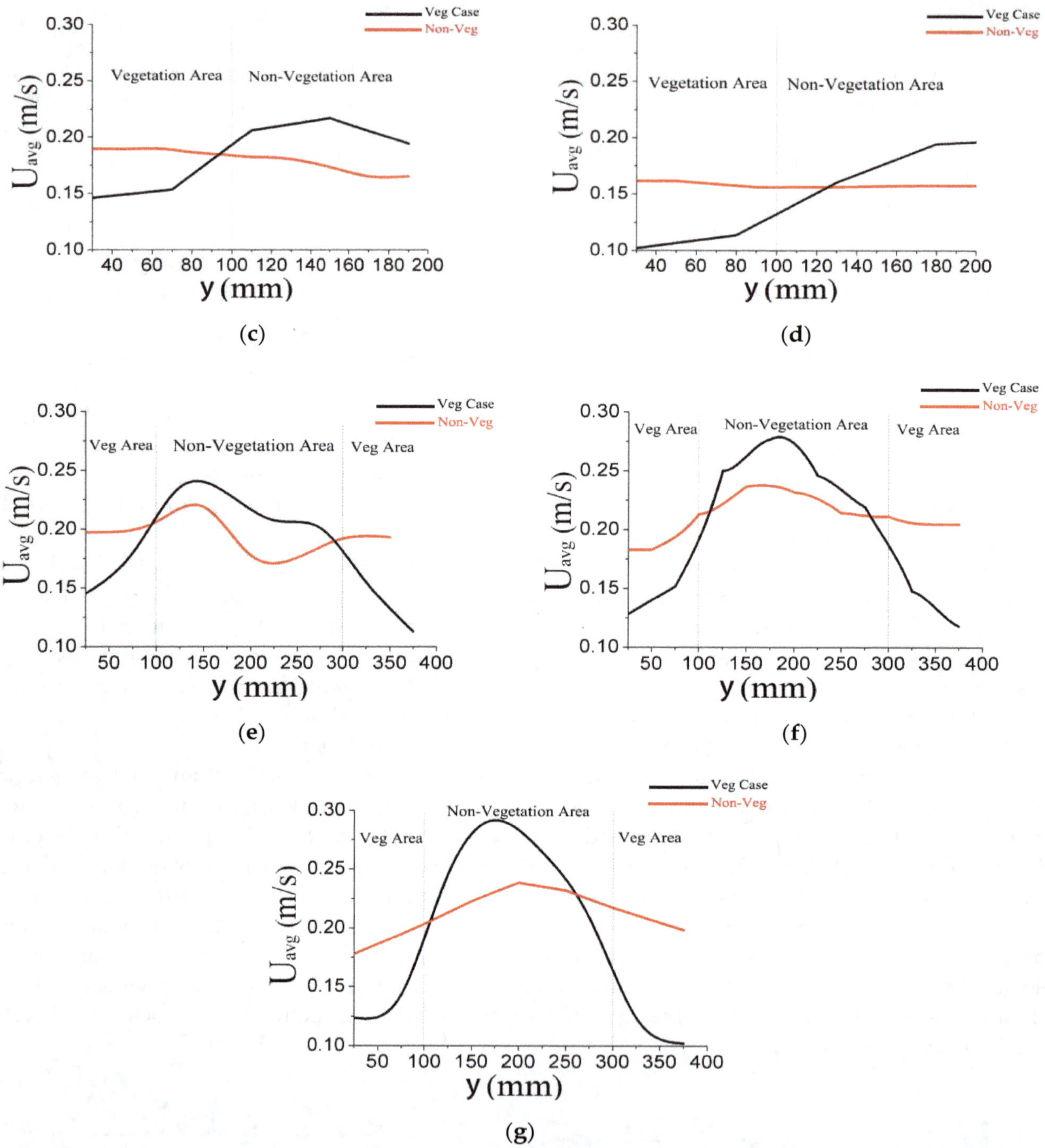

Figure 3. Comparison of depth-averaged streamwise velocity transverse distribution between vegetated and non-vegetated cases, where U_{avg} is the depth-averaged streamwise velocity and y is the flume width. (**a**) Cross Section a; (**b**) Cross Section b; (**c**) Cross Section c; (**d**) Cross Section d; (**e**) Cross Section e; (**f**) Cross Section f; (**g**) Cross Section g.

4.2. Secondary Flow Structure

There was a difference in Secondary Flow Structure by comparing the experimental results for the confluence of the Y-shaped channels with vegetation (Figure 5) and without vegetation (Figure 6). For the mainstream Sections (e), (f), and (g), when vegetation was present, the velocity difference in the confluence was more remarkable. The Section e is the confluence cross section, without vegetation (Figure 5), the left velocity was higher, and the right velocity was lower. A small clockwise circulation appeared at the bottom of the left side. But there was no circulation in the vegetated case (Figure 6). In addition, with the presence of vegetation, the velocity rapidly became partitioned, with the velocity on both sides decreasing with resistance from vegetation and increasing in the middle. Cross sections e and g showed that the higher-velocity area extended to the center in the non-vegetated case, the bottom of the separation zone became larger, the secondary flow phenomena along the left bank slowly disappeared, and a slight counterclockwise motion appeared in the lower right corner in the mainstream cross Section f. This result is similar to the report by Vaghefi et al. [43]. Under the influence of vegetation resistance, the low-velocity zone gradually became wider, the high-speed zone gradually narrowed, the low-speed area on the left changed faster than the low-speed area on the right, and there was no circulation.

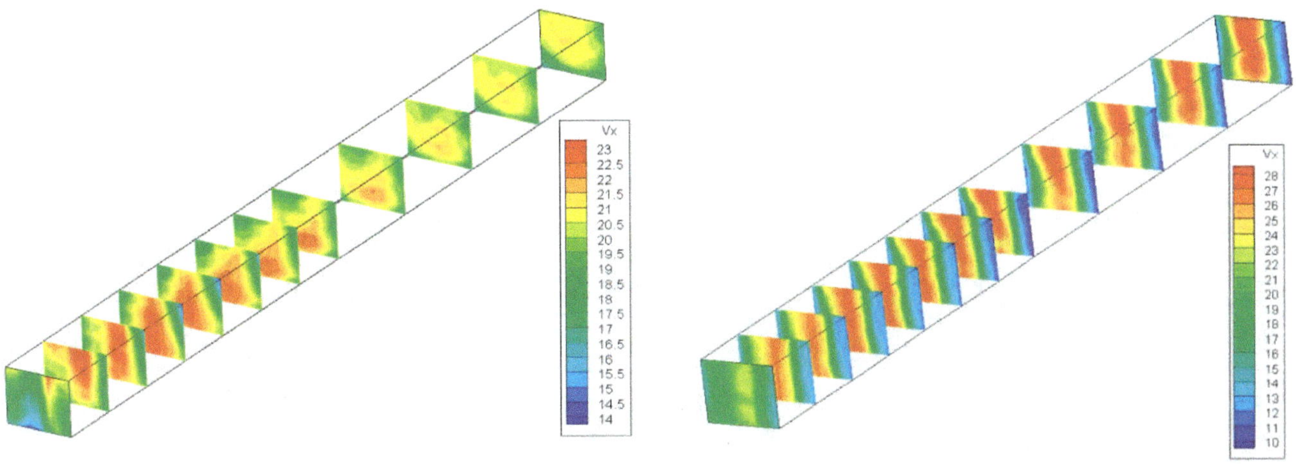

(a) (b)

Figure 4. Slice diagram of streamwise velocity contour after intersection: (**a**) vegetated case; (**b**) non-vegetated case.

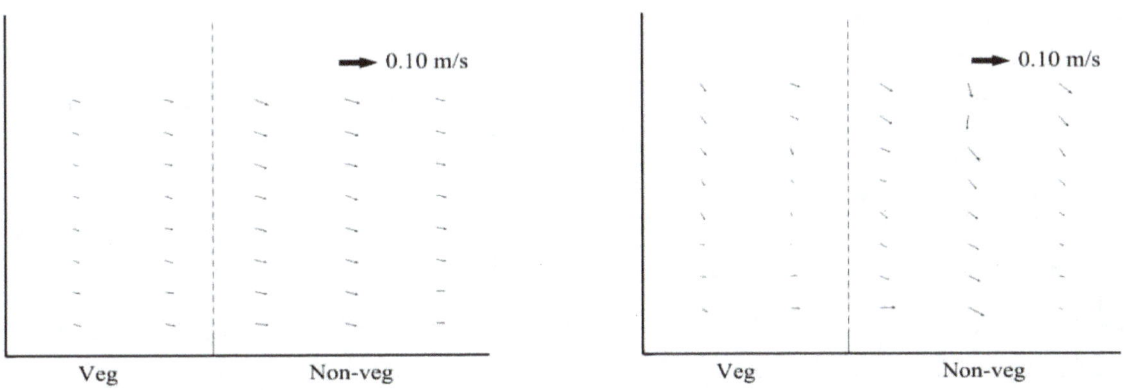

(a) (b)

Figure 5. *Cont.*

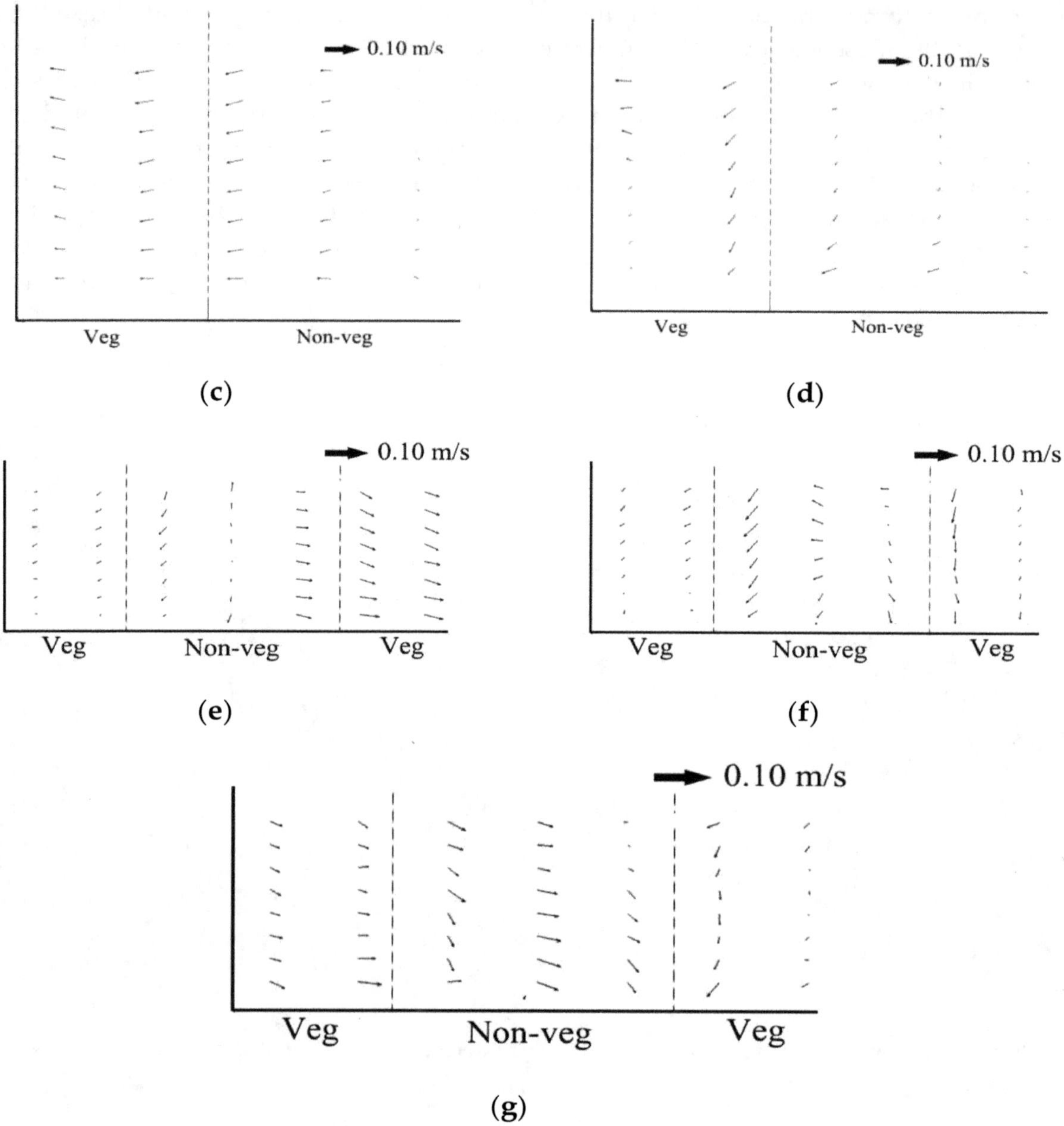

Figure 5. Comparison of secondary flow structures with vegetation. (**a**) Cross Section a; (**b**) Cross Section b; (**c**) Cross Section c; (**d**) Cross Section d; (**e**) Cross Section e; (**f**) Cross Section f; (**g**) Cross Section g.

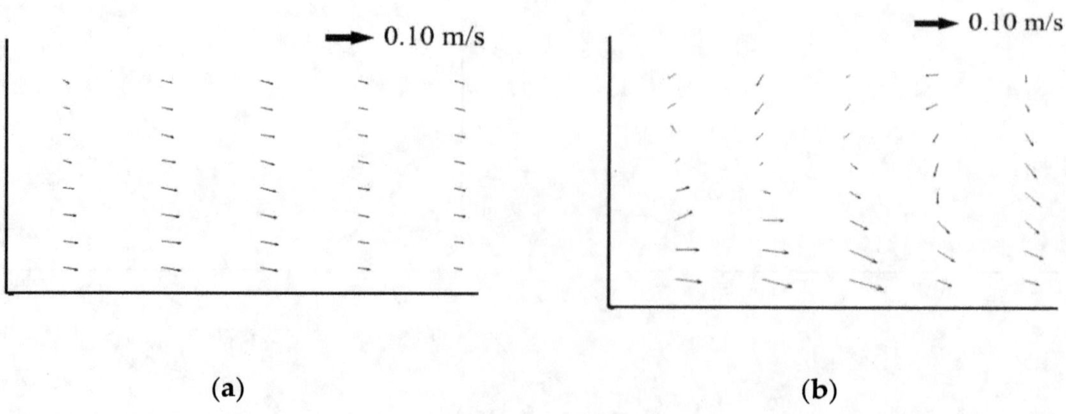

Figure 6. *Cont.*

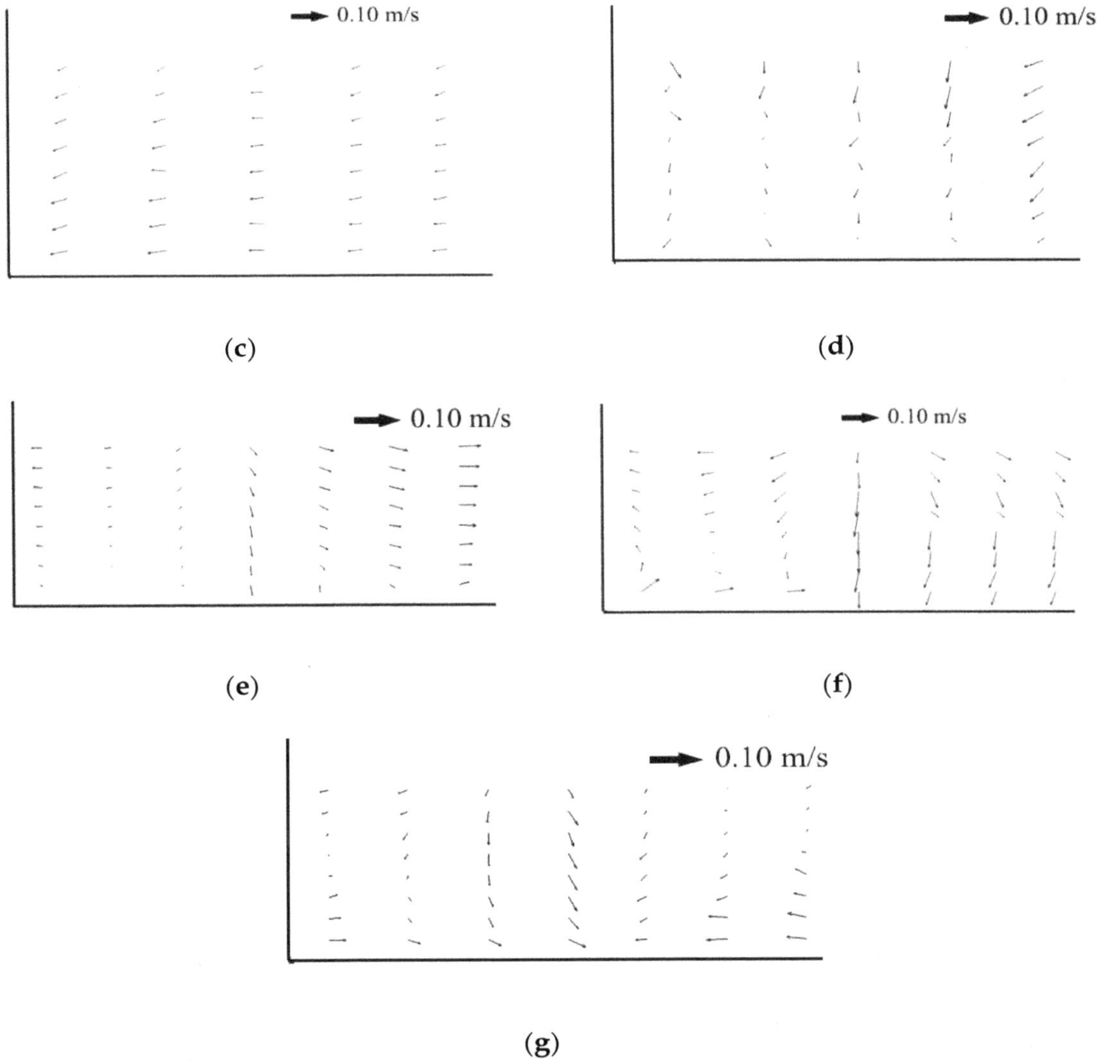

Figure 6. Comparison of secondary flow structures without vegetation. (**a**) Cross Section a; (**b**) Cross Section b; (**c**) Cross Section c; (**d**) Cross Section d; (**e**) Cross Section e; (**f**) Cross Section f; (**g**) Cross Section g.

4.3. Turbulent Kinetic Energy

The relational graphs of the change of turbulent kinetic energy with depth were shown in Figures 7 and 8, respectively. According to Figures 7 and 8, the results showed that the turbulent kinetic energy in the tributaries was much larger with vegetation than without it. The maximum value with vegetation was greater than 20 Pa (Figure 7a–d), compared to a maximum of less than 15 Pa (Figure 8a–d) without vegetation. The turbulent kinetic energy of the vegetated area was less than that of the non-vegetated area, and the turbulent kinetic energy showed a remarkable increase near the junction of the vegetated and non-vegetated areas. Although the numerical values of turbulent kinetic energy for each section at different depths were different, the distribution trends in the tributaries were consistent. After convergence, as shown in Figure 7e–g, the turbulent kinetic energy of the non-vegetated area was smaller than that of the vegetated area.

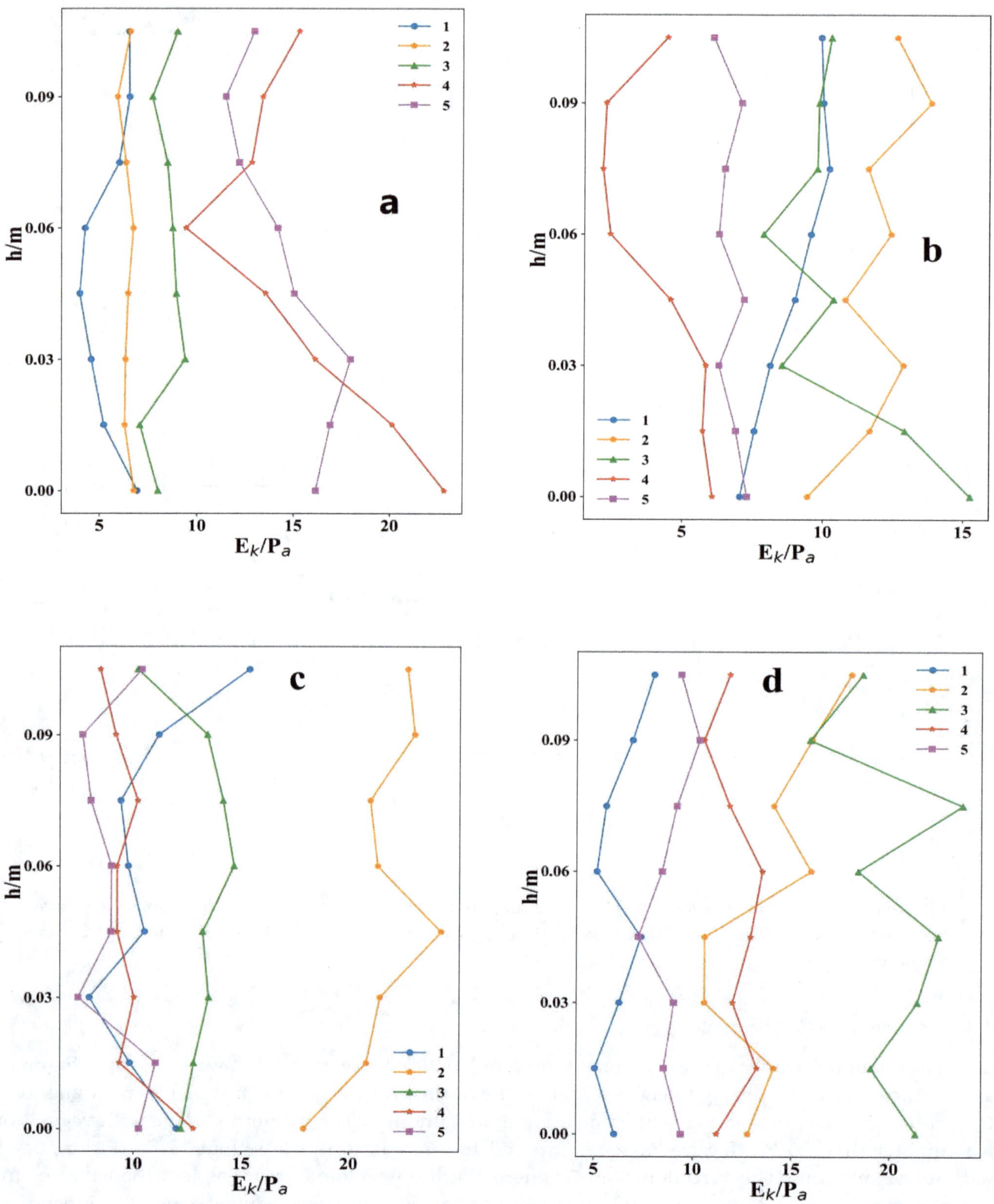

Figure 7. *Cont.*

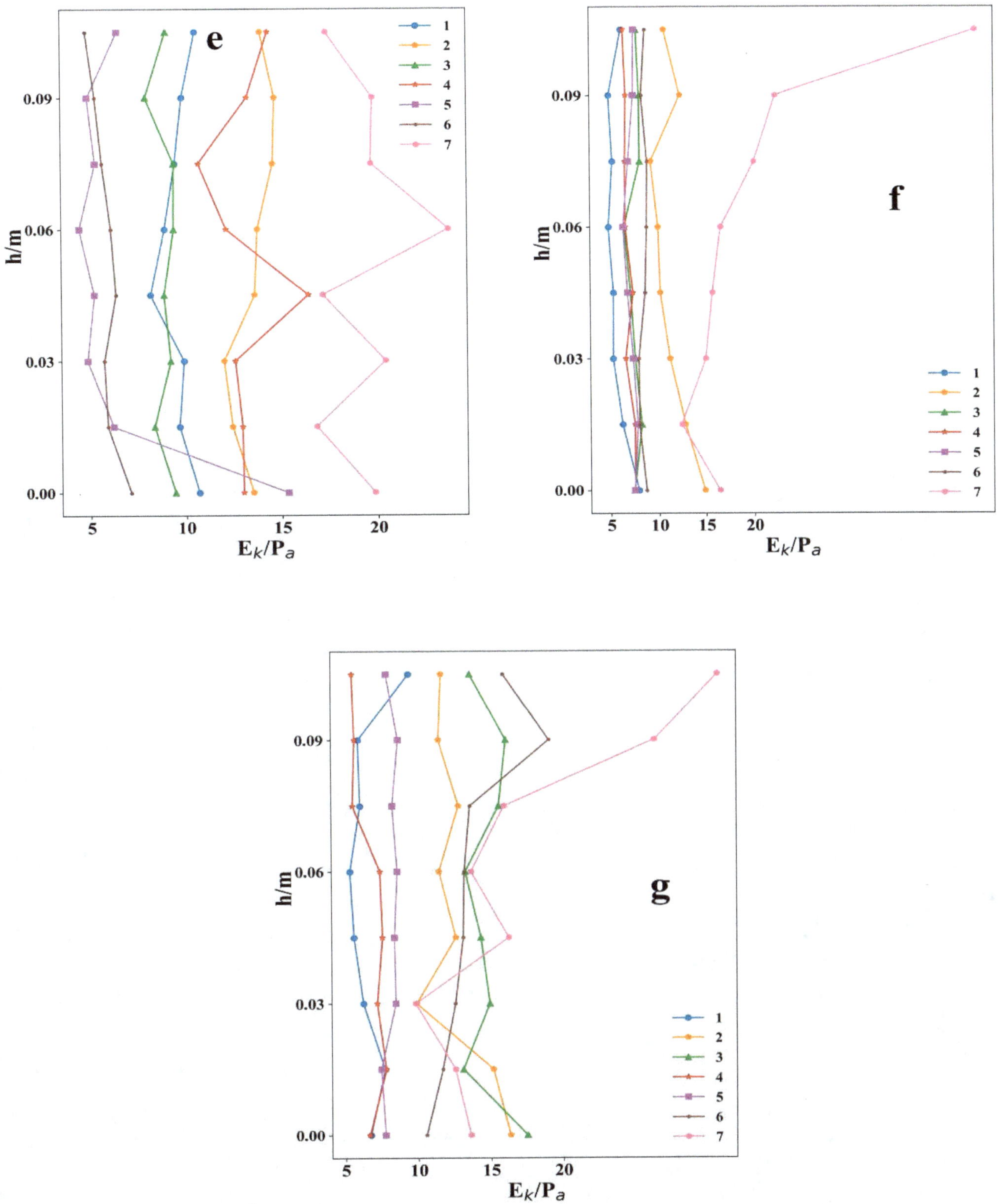

Figure 7. Turbulent kinetic energy (E_k/Pa); (**a**) Cross Section a at vegetation cases; (**b**) Cross Section b at vegetation cases; (**c**) Cross Section c at vegetation cases; (**d**) Cross Section d at vegetation cases; (**e**) Cross section e at vegetation cases; (**f**) Cross section f at vegetation cases; (**g**) Cross section g at vegetation cases.

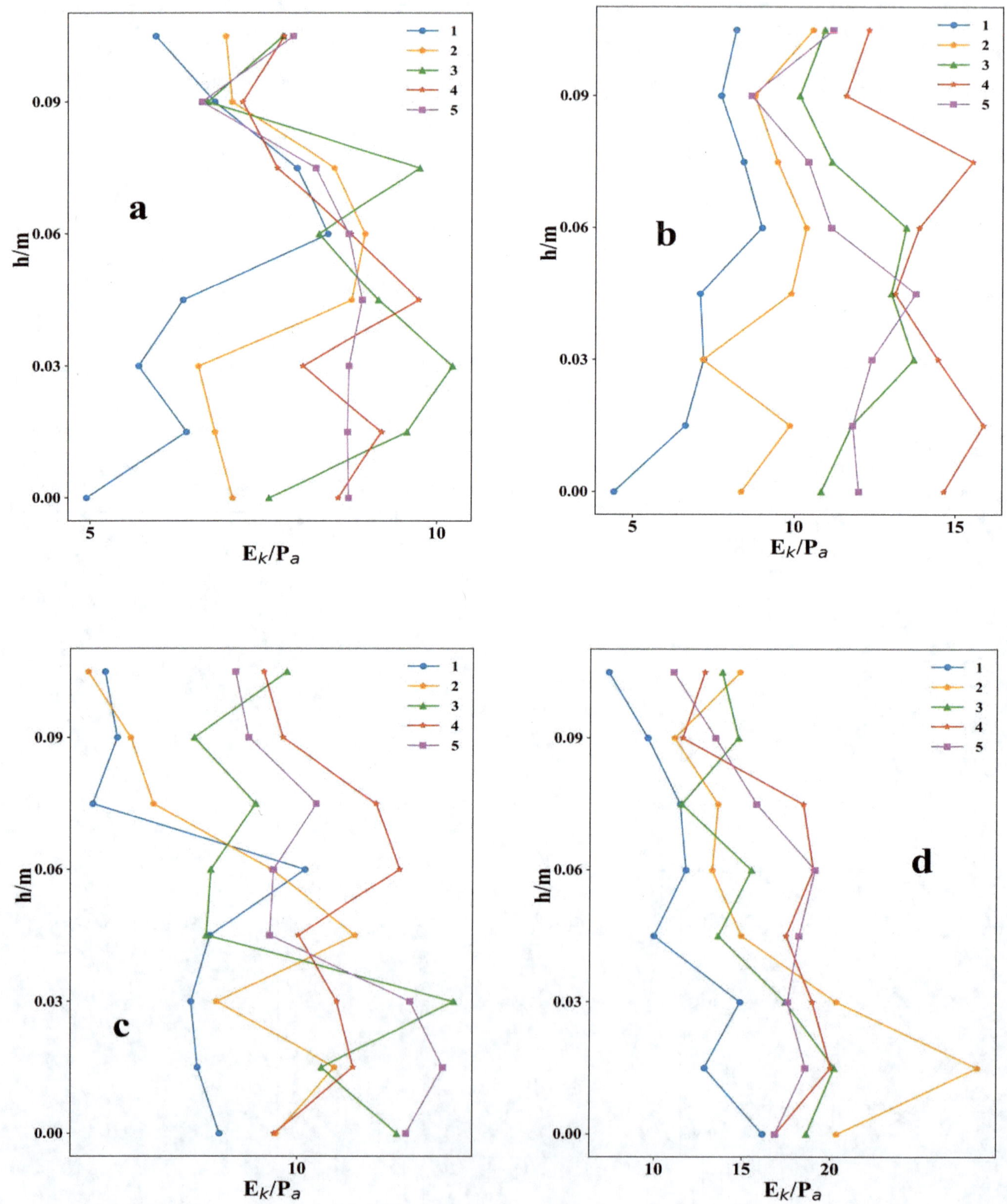

Figure 8. *Cont.*

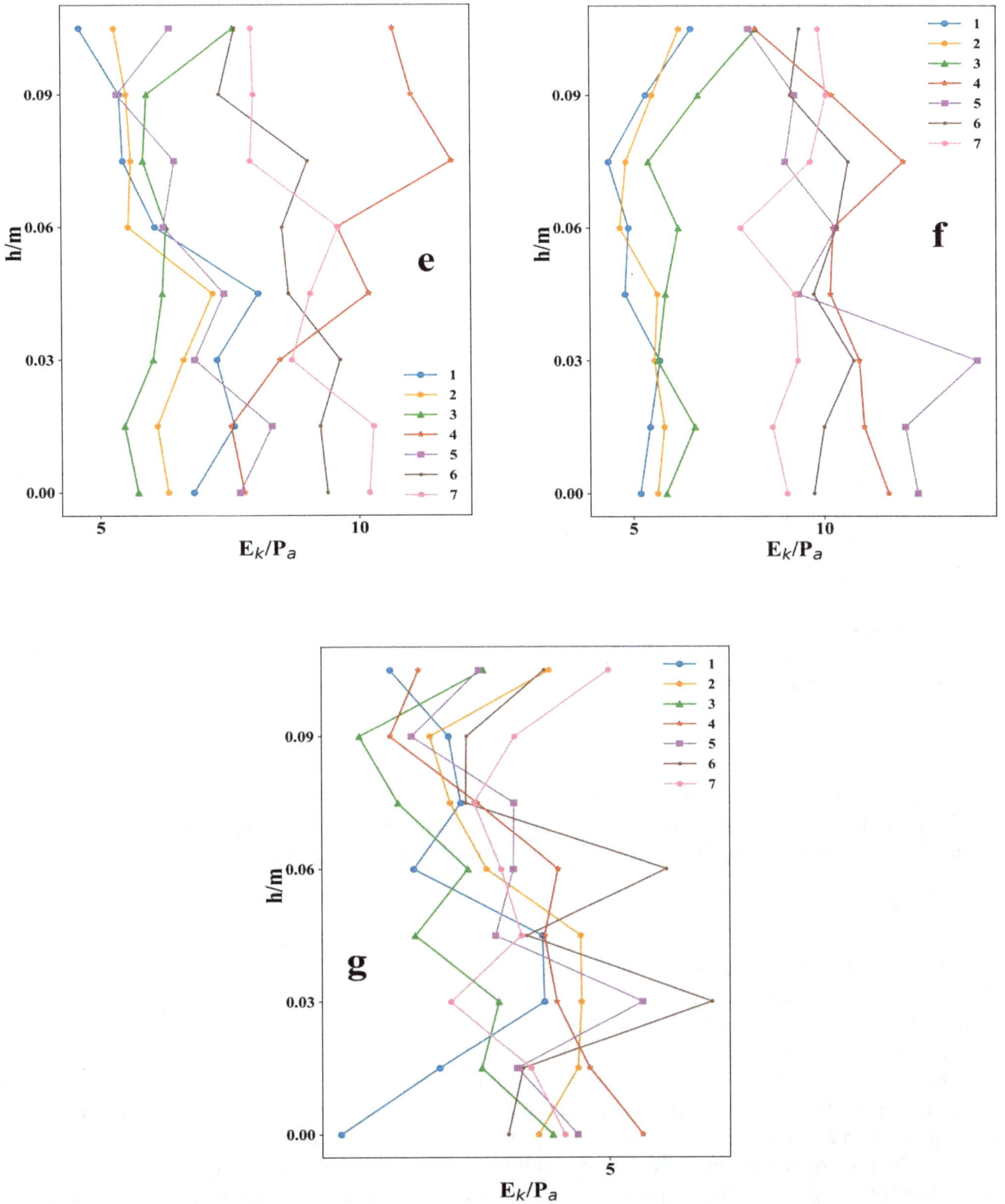

Figure 8. Turbulent kinetic energy (E_k/Pa); versus relative depth (h/m) with vegetation, versus relative depth (h/m) without vegetation. (**a**) Cross Section a at non-veg cases; (**b**) Cross Section b at non-veg cases; (**c**) Cross Section c at no-veg cases; (**d**) Cross Section d at no-veg cases; (**e**) Cross Section e at no-veg cases; (**f**) Cross Section f at no-veg cases; (**g**) Cross Section g at no-veg cases.

5. Discussion

In this study, with the effect of vegetation, the streamwise velocities in the vegetated area were much lower than in the non-vegetated area (Figures 3 and 4), which is consistent with the field study conducted by Chambers et al. [44]. They showed that in natural rivers with slower currents and therefore in the vegetated region, the water current was expected to be lower than in the main stream. The velocities in the vegetated case have a higher gradient than in the non-vegetated case, which can be ascribed to the fact that the retardation caused by vegetation makes the transverse distribution slightly more non-uniform, consistent with the conclusion drawn in the study by Mossa et al. [45]. Along the streamwise direction, the velocities in the vegetated area decreased and those in the non-vegetated area increased when heading downstream (Figures 3 and 4). The reason for this may have been that the main circulation in the Y-shaped confluence changed the internal flow structures, and the velocity distribution varied accordingly, which is consistent with the results from Wang et al. [42]. The vegetation prevented vertical mixing of fluid, and because of resistance, the velocity near the vegetated side was significantly lowered.

A large velocity gradient was apparent near the junction of the vegetated and non-vegetated areas. We can see that all the figures show that a large velocity gradient was apparent near the junction of the veg and non-veg areas, indicating there are must be remarkable mass and momentum exchanges at the junction of these two areas. The secondary flow is occurred for the momentum exchange [46]. The experimental results for the confluence of the Y-shaped channels with vegetation (Figure 5) and without vegetation (Figure 6) were compared. In the vegetation area, no clear circulation is found in the whole area.

The presence of vegetation caused a great change in the internal flow structure and made the flow in non-vegetated areas more intense. When water flows through vegetation, the flow resistance increases and the corresponding velocity decreases. At the same time, the flow will generate turbulence, consuming the flow energy. The turbulent kinetic energy of the non-vegetated area was smaller than that of the vegetated area. This is similar to Liu et al.'s results [47].

Hydraulic properties of riverine vegetation have been widely researched. Many laboratory, numerical, analytical and field studies have been conducted to account for the complex dynamic interactions between vegetation and moving fluids. The channel characteristics of natural rivers are seen to constitute an interdependent system which can be described by a series of graphs having simple geometric forms [48]. There are many Y-shaped confluence channels in natural river networks. This study can be helpful in providing some physiographic implications for practical use.

6. Conclusion and Future Works

Due to the presence of vegetation, the flow velocities in the Y-shaped confluence channel were redistributed. Velocities in the vegetation area were much smaller than those in non-vegetation area due to the presence of vegetation. Compared with the flow without vegetation, the streamwise velocity in the vegetated area was much lower than in the non-vegetated area because of the effect of vegetation. A large velocity gradient is generated between the vegetated and non-vegetated areas, indicating a remarkable mass and momentum exchange at the junction of these areas. Under the combined effect of vegetation and the Y-shaped confluence, the streamwise velocities in vegetated areas were much lower than those in non-vegetated areas, and the streamwise velocity of non-vegetated areas in the mainstream in the case with vegetation increased significantly compared with the case without vegetation.

In the tributaries, due to the presence of vegetation, the high-velocity area moved rapidly to the middle of the channel in non-vegetated areas, and the secondary flow phenomenon disappeared. In the mainstream, when vegetation was introduced, circulation disappeared, and the degree of lateral mixing decreased.

The presence of vegetation brought about great changes in internal flow structure. It caused the flow in non-vegetated areas to become more intense, and the turbulent kinetic energy of the tributaries in the cases with vegetation was significantly lower than in the cases without vegetation.

At the junction of vegetated and non-vegetated areas, the turbulent energy increased significantly. The turbulent energy of the non-vegetated areas was significantly greater than in the same position in the absence of vegetation.

This work focused on a single type of aquatic plant in the Y-shaped confluence channel. However, rivers actually include several types of plants. Further studies could investigate different vegetation types and different discharges in the Y-shaped channel. This study focused on flume experiments, but further investigations could combine this approach with numerical modeling.

Author Contributions: Conceptualization, X.T. and X.L.; Methodology, T.Y.; Software, R.L.; Validation, J.L., Z.W. and X.T.; Formal Analysis, T.Y.; Investigation, Z.W.; Resources, Z.W.; Data Curation, T.Y.; Writing-Original Draft Preparation, X.T.; Writing-Review & Editing, X.L.; Visualization, X.L.; Supervision, Z.H. and X.L.; Project Administration, X.L.; Funding Acquisition, X.L.

Acknowledgments: We are grateful to the Hydraulic Laboratory, Hohai University, for the use of laboratory facilities and to Xu Jingzhao and Mei Shengchen for their thoughtful comments on measurements during the laboratory research.

References

1. Ashmore, P.; Parker, G. Confluence scour in coarse braided streams. *Water Resour. Res.* **1983**, *19*, 392–402. [CrossRef]
2. Best, J.L. Sediment transport and bed morphology at river channel confluences. *Sedimentology* **1988**, *35*, 481–498. [CrossRef]
3. Best, J.L.; Reid, I. Separation zone at open-channel junctions. *J. Hydraul. Eng.* **1984**, *110*, 1588–1594. [CrossRef]
4. Best, J.L.; Roy, A.G. Mixing-layer distortion at the confluence of channels of different depth. *Nature* **1991**, *350*, 411–413. [CrossRef]
5. Biron, P.M.; Lane, S.N. Modelling hydraulics and sediment transport at river confluences. In *River Confluences, Tributaries and the Fluvial Network*; Rice, S.P., Roy, A.G., Rhoads, B.L., Eds.; John Wiley: Hoboken, NJ, USA, 2008; pp. 17–43.
6. Biron, P.; De Serres, B.; Roy, A.G.; Best, J.L. Shear layer turbulence at an unequal depth channel confluence. In *Turbulence: Perspectives on Flow and Sediment Transfer*; Clifford, N., Ed.; Wiley: New York, NY, USA, 1993.
7. Bo, L.A.N. The comprehensive analysis of the special property at the tributary junction of Mountain river. *J. Chongqing Jiaotong Inst.* **1998**, *17*, 91–96. (In Chinese)
8. Best, J.L. Flow dynamics at river channel confluences: Implications for sediment transport and bed morphology. In *Recent Developments in Fluvial Sedimentology*; Ethridge, F.G., Flores, R.M., Harvey, M.D., Eds.; SEPM Society for Sedimentary Geology: Tulsa, OK, USA, 1987; pp. 27–35.
9. Taylor, E.H. Flow characteristics at rectangular open-channel junctions. *J. Trans.* **1944**, *109*, 893–902.
10. Rhoads, B.L.; Kenworthy, S.T. Flow structure at an asymmetrical stream confluence. *Geomorphology* **1995**, *11*, 273–293. [CrossRef]
11. Rhoads, B.L.; Sukhodolov, A.N. Field investigation of three-dimensional flow structure at stream confluences: 1. Thermal mixing and time-averaged velocities. *Water Resour. Res.* **2001**, *37*, 2393–2410.
12. Riley, J.D.; Rhoads, B.L.; Parsons, D.R.; Johnson, K.K. Influence of junction angle on three-dimensional flow structure and bed morphology at confluent meander bends during different hydrological conditions. *Earth Surf. Process. Landf.* **2015**, *40*, 252–271. [CrossRef]
13. Gualtieri, C.; Filizola, N.; de Oliveira, M.; Santos, A.M.; Ianniruberto, M. A field study of the confluence between Negro and Solimões Rivers. Part 1: Hydrodynamics and sediment transport. *C. R. Geosci.* **2018**, *350*, 31–42.
14. Penna, N.; De Marchis, M.; Canelas, O.B.; Napoli, E.; Cardoso, A.H.; Gaudio, R. Cardoso and Roberto Gaudio. Effect of the Junction Angle on Turbulent Flow at a Hydraulic Confluence. *Water* **2018**, *10*, 469. [CrossRef]
15. Schnauder, I.; Sukhodolov, A.N. Flow in a tightly curving meander bend: Effects of seasonal changes in aquatic macrophyte cover. *Earth Surf. Process. Landf.* **2012**, *37*, 1142–1157. [CrossRef]
16. Weber, A.; Zhang, J.; Nardin, A.; Sukhodolov, A.; Wolter, C. Modelling the Influence of Aquatic Vegetation

on the Hydrodynamics of an Alternative Bank Protection Measure in a Navigable Waterway. *River Res. Appl.* **2016**, *32*, 2071–2080. [CrossRef]
17. Łoboda, A.M.; Bialik, R.J.; Karpiński, M.; Przyborowski, Ł. Two simultaneously occurring Potamogeton species: Similarities and differences in seasonal changes of biomechanical properties. *Pol. J. Environ. Stud.* **2019**, *28*, 1–17. [CrossRef]
18. Aberle, J.; Järvelä, J. Flow resistance of emergent rigid and flexible floodplain vegetation. *J. Hydraul. Res.* **2013**, *51*, 33–45. [CrossRef]
19. Järvelä, J.; Aberle, J.; Dittrich, A.; Schnauder, I.; Rauch, H.P. Flow–Vegetation–Sediment Interaction: Research Challenges. In Proceedings of the International Conference River Flow, London, UK, 6–8 September 2006; Ferreira, R.M.L., Alves, E.C.T.L., Leal, J.G.A.B., Cardoso, A.H., Eds.; Taylor & Francis: London, UK, 2006; Volume 2, pp. 2017–2026.
20. Nepf, H.M. Drag, turbulence, and diffusion in flow through emergent vegetation. *Water Resour. Res.* **1999**, *35*, 479–489. [CrossRef]
21. Ghisalberti, M.; Nepf, H. Mass Transport in Vegetated Shear Flows. *Environ. Fluid Mech.* **2005**, *5*, 527–551. [CrossRef]
22. Nikora, V. Hydrodynamics of aquatic ecosystems: An interface between ecology, biomechanics and environmental fluid mechanics. *River Res. Appl.* **2010**, *26*, 367–384. [CrossRef]
23. Nikora, V.; Lamed, S.; Nikora, N.; Debnath, K.; Cooper, G.; Reid, M. Hydraulic resistance due to aquatic vegetation in small streams: field study. *J. Hydraul.* **2008**, *134*, 1326–1332. [CrossRef]
24. Termini, D. Effect of vegetation on fluvial erosion processes: Experimental analysis in a laboratory flume. *Procedia Environ. Sci.* **2013**, *19*, 904–911. [CrossRef]
25. Termini, D. Flexible Vegetation Behavior and Effects on Flow Conveyance: Experimental Observations. *Int. J. River Basin Manag.* **2015**, *13*, 401–411. [CrossRef]
26. Nezu, I.; Sanjou, M. Turbulence structure and coherent motion in vegetated canopy open-channel flows. *J. Hydro-Environ. Res.* **2008**, *2*, 62–90. [CrossRef]
27. Zhao, K.; Cheng, N.-S.; Wang, X.; Tan, S.K. Measurements of Fluctuation in Drag Acting on Rigid Cylinder Array in Open Channel Flow. *J. Hydraul. Eng.* **2014**, *140*, 48–55. [CrossRef]
28. Łoboda, A.M.; Przyborowski, Ł.; Karpiński, M.; Bialik, R.J.; Nikora, V.I. Biomechanical properties of aquatic plants: The effect of test conditions. *Limnol. Oceanogr. Methods* **2018**, *16*, 222–236. [CrossRef]
29. Niklas, K.J. Plant Biomechanics. In *An Engineering Approach to Plant Form and Function*; University of Chicago Press: Chicago, IL, USA, 1992; ISBN 0-226-58641-6.
30. American Society for Testing and Materials (ASTM) D790-03. *Standard Test Methods for Flexural Properties of Unreinforced and Reinforced Plastics and Electrical Insulating Materials*; ASTM International: West Conshohocken, PA, USA, 2003.
31. Carollo, F.G.; Ferro, V.; Termini, D. Flow velocity measurements in vegetated channels. *J. Hydraul. Eng.* **2002**, *128*, 664–673. [CrossRef]
32. Li, C.G.; Xue, W.Y.; Huai, W.X. Effect of vegetation on flow structure and dispersion in strongly curved channels. *J. Hydrodyn.* **2015**, *27*, 286–291. [CrossRef]
33. Nezu, I.; Nakagawa, H.; Tominaga, A. Secondary Currents in a Straight Channel Flow and the Relation to Its Aspect Ratio. In *Turbulent Shear Flows 4*; Springer: Berlin/Heidelberg, Germany, 1985.
34. Karcz, I. Secondary currents and the configuration of a natural stream bed. *J. Geophys. Res.* **1966**, *71*, 3109–3112. [CrossRef]
35. Wang, Z.Q.; Cheng, N.S. Time-mean structure of secondary flows in open channel with longitudinal bedforms. *Adv. Water Resour.* **2006**, *29*, 1634–1649. [CrossRef]
36. Przyborowski, Ł.; Łoboda, A.; Bialik, R. Experimental Investigations of Interactions between Sand Wave Movements, Flow Structure, and Individual Aquatic Plants in Natural Rivers: A Case Study of *Potamogeton Pectinatus* L. *Water* **2018**, *10*, 1166. [CrossRef]
37. Blanckaert, K.; Vriend, H.J. Turbulence characteristics in sharp open-channel bends. *Phys. Fluids* **2005**, *17*, 717–731. [CrossRef]
38. Li, K.-F.; Wang, D.; Yang, K.; Zhu, M. Distribution of Turbulent Energy in Open Channel with Vegetation Patch. *Water Resour. Power.* **2017**, *2*, 122–124. (In Chinese)
39. Huai, W.; Li, C.; Zeng, Y. Curved open channel flow on vegetation roughened inner bank. *J. Hydrodyn.* **2012**,

24, 124–129. [CrossRef]

40. Nikora, N.; Nikora, V.; O'Donoghue, T. Erratum for "Velocity Profiles in Vegetated Open-Channel Flows: Combined Effects of Multiple Mechanisms" by Nina Nikora, Vladimir Nikora, and Tom O'Donoghue. *J. Hydraul. Eng.* **2014**, *140*, 08014003. [CrossRef]
41. Caroppi, G.; Västilä, K.; Järvelä, J.; Rowin´ski, P.M. Experimental investigation of turbulent flow structure in a partially vegetated channel with natural-like flexible plant stands. In *New Challenges in Hydraulic Reseearch and Engineering, Proceedings of the 5th IAHR Europe Congress, Trento, Italy, 12–14 June 2018*; Armanini, A., Nucci, E., Eds.; Taylor & Francis: New York, NY, USA, 2018; pp. 615–616.
42. Wang, X.; Yan, Z.; Guo, W. Three-dimensional simulation for effects of bed discordance on flow dynamics at Y-shaped open channel confluences. *J. Hydrodyn.* **2007**, *19*, 587–593. [CrossRef]
43. Vaghefi, M.; Akbari, M.; Fiouz, A.R. An experimental study of mean and turbulent flow in a 180 degree sharp open channel bend: Secondary flow and bed shear stress. *KSCE J. Civ. Eng.* **2016**, *20*, 1582–1593. [CrossRef]
44. Chambers, P.A.; Prepas, E.E.; Hamilton, H.R.; Bothwell, M.L. Current velocity and its effect on aquatic macrophytes in flowing waters. *Ecol. Appl.* **1991**, *1*, 249–257. [CrossRef]
45. Mossa, M.; Ben Meftah, M.; De Serio, F.; Nepf, H.M. How vegetation in flows modifies the turbulent mixing and spreading of jets. *Sci. Rep.* **2017**, *7*, 6587. [CrossRef] [PubMed]
46. Tominaga, A.; Nezu, L.; Ezaki, K.; Nakagawa, H. Three-dimensional turbulent structure in straight open channel flows. *J. Hydraul. Res.* **1989**, *27*, 149–173. [CrossRef]
47. Liu, C.; Luo, X.; Liu, X.; Yang, K. Modeling depth-averaged velocity and bed shear stress in compound channels with emergent and submerged vegetation. *Adv. Water Resour.* **2013**, *60*, 148–159. [CrossRef]
48. Leopold, L.B.; Maddock, T. *The Hydraulic Geometry of Stream Channel and Some Physiographic Implications*; US Government Printing Office: Washington, DC, USA, 1953.

13

The Evaluation of Regional Water-Saving Irrigation Development Level in Humid Regions

Lu Zhao [1,2], Lili Zhang [3], Ningbo Cui [1,4,*], Chuan Liang [1] and Yi Feng [1]

1 State Key Laboratory of Hydraulics and Mountain River Engineering & College of Water Resource and Hydropower, Sichuan University, Chengdu 610025, China; luya1121@163.com (L.Z.); lchester@sohu.com (C.L.); fengyscu@163.com (Y.F.)

2 Provincial Key Laboratory of Water-Saving Agriculture in Hill Areas of Southern China, Chengdu 610066, China

3 International School of Technical Education, Sichuan College of Architectural Technology, Deyang 618000, China; lindaz616@163.com

4 Key Laboratory of Agricultural Soil and Water Engineering in Arid and Semiarid Areas, Ministry of Education, Northwest A & F University, Yangling 712100, China

* Correspondence: cuiningbo@126.com.

Abstract: Water-saving irrigation development level (WIDL) refers to reasonably and accurately judging a water-saving area based on the analysis of all factors affecting the water-saving irrigation development. The evaluation of regional WIDL is the premise of scientific planning guidance to irrigation work. How to select reasonable evaluation indexes and build a scientific and comprehensive model to evaluate WIDL is of great significance. In this study, the comprehensive evaluation index system of WIDL in 21 cities (states) of the Sichuan province in China (a typical humid region in southern China) was constructed, and the TOPSIS (Technique for Order Preference by Similarity to an Ideal Solution) method was improved to evaluate WIDL. Results showed that the overall development level of water-saving irrigation was "poor" in Sichuan province. The water-saving irrigation level turned out to be "good" in three regions with advantageous geographical conditions and developed economies, "general" in four regions with good economic levels where agronomy water saving has been popularized, and "poor" in fourteen regions of mountainous and hilly areas, especially Ganzi, Aba, and Liangshan, located in the Northwest plateau of Sichuan province, with poor natural resources and insufficient economies. The evaluation results were in good agreement with the actual situation, and in this area, there is enormous potential for the development of water-saving irrigation strategies. This study provides an important technical approach for the evaluation of water-saving irrigation development in humid regions of Southern China.

Keywords: water-saving irrigation development level; Sichuan province; TOPSIS method

1. Introduction

Water scarcity has become a major restraint factor for social and economic development among many regions, especially in semiarid and arid regions [1,2]. Agriculture consumes the largest amount of water resources among human activities: Irrigation water withdrawals represent 70% of the total human use of renewable water resources [3–6]. About 18% of croplands worldwide, or about 2% of the total land surface, are irrigated and meet 40% of the global food demand [7].

There are various available measures to solve global water scarcity, such as water re-pricing, water re-use, desalination, water diversion and distribution, improvements in water delivery systems, alternative plants, and water conservation through efficient irrigation, among which water-saving irrigation is one of the most feasible and effective measures attributing to significant water-savings [8–11]. The factors affecting water-saving irrigation mainly feature engineering (drip irrigation, micro-irrigation, etc.), agronomy (regulated deficit irrigation, water-saving varieties, etc.) and management (water price, government police, etc.) measures [12–18].

The natural conditions, such as hydrogeology, channel soil, irrigation soil, crop species, the management and maintenance level of irrigation districts, farmer's habits of water use, and water price policy vary in different regions, which will lead to different engineering technologies, economic benefits, environmental benefits, and development level of water-saving irrigation in various regions. Although many studies have focused on the evaluation of water-saving irrigation technology or comprehensive benefits [19–23], less research has been conducted on the comprehensive evaluation of the regional water-saving irrigation development level (WIDL). WIDL refers to reasonably and accurately judging a water-saving or high-efficiency water-saving area based on the analysis of all factors affecting water-saving irrigation development. It is an objective summary of past achievements and also a judgment of the current starting point. In most cases, we used the agricultural integrated gross irrigation quota to express water-saving irrigation development level. However, the regional WIDL is the organic embodiment of engineering, agronomy, and management the water-saving level under the conditions of optimal allocation of water resources. WIDL has never been reported in the humid regions of Southern China. Although water resources in these regions are relatively abundant, with annual precipitation greater than 1000 mm, seasonal drought often occurs, combined with serious engineering water shortages, leading to more serious droughts. Therefore, it is necessary to develop water-saving irrigation in the humid regions of Southern China. With the continuous enrichment of multi-index comprehensive evaluation methods such as the fuzzy comprehensive evaluation method [24], analytic hierarchy process method [25,26], set pair analysis [27], TOPSIS method [28] and so on, the comprehensive evaluation of WIDL is worthy of study and improvement.

Attaining the weights of indexes is a prerequisite for a comprehensive evaluation since the weight of the index represents the relative importance of the index in the evaluation system, and the accuracy of the weight directly affects the final evaluation results [29]. There are two kinds of methods to judge weight. One is the subjective method (such as analytic hierarchy process). Weights of indexes are obtained by the subjective judgment of experts. This method reflects the knowledge and experience of experts but easily leads to deviations due to personal subjective elements. The other is the objective weighting method (such as entropy weight method). Weights are judged according to the relationship among the original data with a strong mathematical theoretical basis. In order to make the evaluation results more convincing, in this study weights were obtained by combining the subjective and objective methods.

Sichuan province is a typical humid region of Southern China, with annual precipitation of 1000–1200 mm. The objective of this paper is to evaluate WIDL in 21 cities (states) of Sichuan province. How could we evaluate the development level of water-saving irrigation strategies? First, we established a comprehensive evaluation index system of WIDL from three aspects of engineering, agronomy, and management. Second, in order to simplify the evaluation indexes and enhance the accuracy of index weight, the principal component analysis (PCA) was used to extract the principal indexes, and the combined weight method was used to judge the relative importance of each index. Third, the TOPSIS method was improved to evaluate WIDL in 21 cities (states) of Sichuan province.

2. Materials and Methods

2.1. Study Area

Sichuan province, covering 21 cities (states), is the main producing area of grain in China (Figure 1). In this area, 500,000 small and medium-sized irrigation districts were built up until 2013. The effective irrigation area and water-saving irrigation area are 2647 thousand hm^2 and 1125 thousand hm^2, respectively. High-efficiency water-saving irrigation areas, with sprinkler irrigation, micro-irrigation, and pipe conveying irrigation of 37.9, 9.5, and 50 thousand hm^2 respectively, account for only 8% of the water-saving irrigation area. The demand for irrigation water in Sichuan province is 5595 m^3 per hm^2, the per capita income of the agriculture population is 5239 CNY, and the generalization area of agriculture water-saving technology is 667 thousand hm^2. A total of 3540 water user associations have been built to control 714.1 thousand hm^2 of irrigation croplands. A property rights system reform has been implemented in 380 thousand small water conservation projects. The current agriculture water price standard is 50% of the average water supply. The agricultural irrigation water fee in Sichuan province is supposed to be 320 million CNY, but the actual yield is less than 80%.

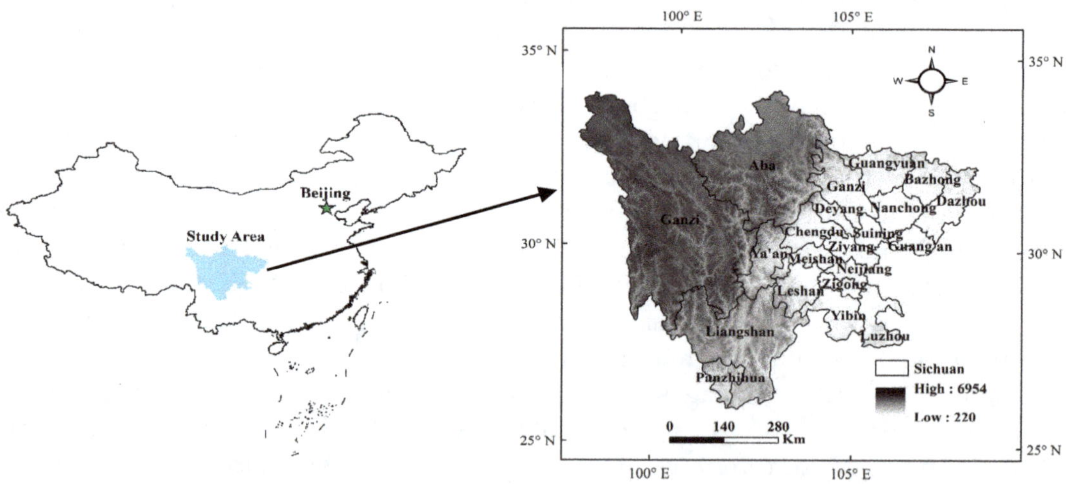

Figure 1. Twenty-one cities (states) of Sichuan province.

2.2. Methods

2.2.1. Principal Components Extracting of Evaluation Indexes

The principal component analysis (PCA) is one of the most widely applied tools which allows researchers to manipulate more variables [30–34]. The aim of the PCA algorithm is to reduce the dimensionalities of variables and meanwhile keep much information about the variables. When m principal components whose cumulative contribution rate $\geq$85% are selected in the real issue, this indicates that the first m principal components contain the total information of all indexes. In the paper, the indexes with the highest correlation with m principal components were chosen, which reduced the number of evaluation indexes and provided convenience for practical problems analysis.

2.2.2. Weight Assignments for the Indexes

The analytic hierarchy process (AHP) is a non-structure decision theory built by operational research experts Saaty [35]. It is applied perfectly to complex problems which are difficult to fully address with quantitative indexes for analysis. The basic idea is that the decision maker decomposes the complex problem into several levels and elements. A simple comparison, judgment, and calculation were carried out for the elements in order to get the weights of the different elements and the pending program. The entropy weight method (EWM) is found to be very useful in multi-attribute problems [36]. The smaller the value of the information entropy of the given parameter is, the larger the amount of contribution it will provide for the comprehensive evaluation.

We multiply the weights of the j-th index calculated by the above two methods, and normalize the product to obtain the combination weight ω_j:

$$\omega_j = \frac{\alpha_j \times \beta_j}{\sum\limits_{j=1}^{n} (\alpha_j \times \beta_j)}, 1 \leq j \leq n \tag{1}$$

where α_j, β_j are the weights of the j-th index determined by AHP and EWM, separately.

2.2.3. The Improved TOPSIS Method

TOPSIS (Technique for Order Preference by Similarity to an Ideal Solution), first developed by Hwang and Yoon (1981), is a simple ranking method which attempts to choose alternatives in the shortest distance from the positive ideal solution and the farthest distance from the negative ideal solution simultaneously [37,38]. The TOPSIS method can be summarized in a series of steps.

Step 1 involves the construction of the original performance rating matrix. A set of cases ($M = (M_1, M_2, \ldots, M_m)$) are compared with respect to a set of attributes ($C = (C_1, C_2, \ldots, C_n)$). The performance matrix can be obtained as follows:

$$Z = \begin{bmatrix} & C_1 & C_2 & \cdots & C_n \\ M_1 & Z_{11} & Z_{12} & \cdots & Z_{1n} \\ M_2 & Z_{21} & Z_{12} & \cdots & Z_{2n} \\ \vdots & \vdots & \vdots & \vdots & \vdots \\ M_m & Z_{m1} & Z_{m2} & Z_{m3} & Z_{mn} \end{bmatrix} \quad (i = 1, 2, \cdots, m; \quad j = 1, 2, \cdots, n) \tag{2}$$

Step 2 involves the construction of the normalized performance rating matrix.

$$v_{ij} = \frac{Z_{ij}}{\sqrt{\sum_{i=1}^{m} Z_{ij}^2}} \quad (i = 1, 2, \cdots m; \quad j = 1, 2, \cdots, n) \tag{3}$$

Step 3 is the construction of the weighted normalized decision matrix. Attribute weights (w_j) have to be determined to indicate their relative importance and to calculate the weighted normalized values (r_{ij}) through:

$$r_{ij} = w_{ij} v_{ij} \quad (i = 1, 2, \cdots, m; \quad j = 1, 2, \cdots, n) \tag{4}$$

where w_j is the index weight determined by both AHP and EWM.

Step 4 indicates the determination of the positive ideal solution (PIS) and negative ideal solution (NIS).

$$\begin{aligned} PIS = A^+ &= \left\{ r_1^+(x), r_2^+(x), \cdots, r_j^+(x), \cdots, r_n^+(x) \right\} \\ &= \left\{ \max_i (r_{ij}(x) | j \in J_1), \min_i (r_{ij}(x) | j \in J_2), | i = 1, 2, \cdots, m \right\} \end{aligned} \tag{5}$$

$$\begin{aligned} NIS = A^- &= \left\{ r_1^-(x), r_2^-(x), \cdots, r_j^-(x), \cdots, r_n^-(x) \right\} \\ &= \left\{ \min_i (r_{ij}(x) | j \in J_1), \max_i (r_{ij}(x) | j \in J_2), | i = 1, 2, \cdots, m \right\} \end{aligned} \tag{6}$$

where J_1 and J_2 are the benefit and the cost attributes, and $r_j{}^+(x)$ and $r_j{}^-(x)$ are the maximum and minimum values for the j-th attribute.

Step 5 involves the calculation of the separation from the PIS and the NIS between alternatives. The determining distances $S_i{}^+$, $S_i{}^-$ of each scheme away from the PIS and the NIS are as given as:

$$S_i^+ = \sqrt{\sum_{j=1}^{n} \left[v_{ij}(x) - v_j^+(x) \right]^2}, \quad i = 1, \cdots, m \tag{7}$$

$$S_i^- = \sqrt{\sum_{j=1}^{n} \left[v_{ij}(x) - v_j^-(x) \right]^2}, \quad i = 1, \cdots, m \tag{8}$$

Step 6 involves the calculation of the similarities to the ideal solution. The scheme decision and relative closeness between kinds of schemes and ideal solution are as follows:

$$\varepsilon_i = \frac{S_i^-}{S_i^+ + S_i^-} \qquad i = 1, 2, \cdots, m \tag{9}$$

Scheme M_i is sequenced according to ε_i. The larger the value of ε_i, the closer scheme M_i is to the ideal solution, which is better. In the opposite case, the scheme is worse.

The above TOPSIS method (traditional method) is only a ranking method. It cannot judge the level of the scheme M_i according to ε_i. In the study, the corresponding classification thresholds (the number of classification standard is k) of each index were seen as k schemes, which were evaluated together with the original program using the TOPSIS method. ε'_j ($j \leq k$) is the relative closeness between the k schemes and ideal solution. Then, the relative closeness was graded as (ε'_1~(ε'_1 + ε'_2)/2), ... , (ε'_{k-2} + ε'_{k-1})/2)~(ε'_{k-1} + ε'_k)/2, (ε'_{k-1} + ε'_k)/2), by which we can determine which level the scheme M_i is in.

3. Results

3.1. Construction of Evaluation System

The regional water-saving strategies were divided into three categories: Engineering, agronomy, and management water-saving. The indexes selected and the relationships among them were used as the foundation to establish the comprehensive evaluation index system of WIDL.

3.1.1. Engineering Water-Saving Evaluation Indexes

Engineering construction is the foundation of water-saving irrigation, and the contents are selected from the perspective of advanced and economic technology and reducing losses caused in the course of water transportation. Considering the present operation level and factors influenced by the engineering in the irrigation area, this research selected 12 indexes (Table 1) based on technical specifications for water-saving irrigation engineering (Chinese National Standard GB/T 50363-2006), and a detailed description of the following indexes was presented:

L_1 represents the ratio of the water-saving irrigation area to the effective irrigation area. This ratio indirectly reflects the water-saving level. The effective irrigation area refers to those areas of farmland, equipped with water sources and irrigation engineering subsidiaries, and their ability to conduct irrigation in normal years.

L_2 represents the ratio of the established high-efficiency water-saving irrigation area to the total water-saving irrigation area. High-efficiency water-saving engineering is also named as pressure irrigation, normally including some advanced irrigation methods such as sprinkler irrigation, micro-irrigation, drip irrigation, etc. Applying high-efficiency water-saving irrigation technology greatly increases water utilization efficiency.

L_3 represents the water efficiency of irrigation. This refers to the ratio of irrigated water available to crops in the field to the volume of water transported from the canal head. Since 2005, calculation of the water efficiency of irrigation has been performed all over China.

L_4 represents the economic benefits per hm^2 of water-saving irrigation. This is a comprehensive index used to evaluate the effects of water-saving irrigation. A complicated calculation method is required to figure out the value of it, and a lot of factors should be taken into consideration. On one hand, the comprehensive benefits of water-saving irrigation for economic crops vary a lot from food crops; on the other hand, the costs of economic crops and food crops vary greatly because of the use of different indexes such as operation costs, water-saving costs, and energy-saving costs, etc., used in the different engineering projects.

L_5 represents the water usage amount per hm^2 for irrigation. Water usage amount refers to the water consumed for the growth of crops, including loss amount in water delivery. It varies a great deal in different areas for different crops and in different weather conditions. The water usage amount equals the sum of irrigation water consumption in each growth periods of crops while considering the water loss caused by the irrigation system.

L_6 represents the ratio of water-saving irrigation area to cultivated area. Cultivated area refers to the field for growing crops and includes irrigation paddy fields, upland fields, and irrigated land. The cultivated area is larger than the irrigation area. The ratio of water-saving irrigation area to cultivated area, on the one hand, objectively reveals the local agricultural irrigation development level, and on the other hand, it reflects the present development level of water-saving agriculture.

L_7 represents the main crops yield per unit. This index significantly reflects the production benefit as well as the scientific development of the planting industry. The usual method used to increase the per unit yield of the main crops is the rational close planting method which provides relatively large space for sunlight and helps control organic consumption in respiration within a relatively small amount at the same time; an optimum density is selected by considering those two factors to increase the organic accumulation.

L_8 represents per capita GDP. Per capita GDP significantly reflects economic development and people's living standards. Water-saving irrigation commences with a rather late start, a large scale of engineering projects is under construction, and heavy investment for water-saving engineering construction is eagerly demanded. Investing capacity is closely related to regional economic development, so per capita GDP is considered as one index in engineering the water-saving evaluation.

L_9 represents the percentage of canal lining. This refers to the ratio of the calculated area of canal seepage proof to the maximum flow section area.

L_{10} represents the percentage of working lining channels Restrained by a depressed economy and poor technology during the period of construction, most channels were built below the construction standard in the irrigation area. Furthermore, after working for years, the channels have suffered some degree of damage. Therefore, maintenance for the channel system is needed. The percentage of working lining channels is significant to reflect the maintenance of damaged channels.

L_{11} represents the water usage amount for agriculture. This covers the water used for field irrigation, fishery, and forestry and fruit industries. Water used for forestry and the fruit industry belongs to the classification of agricultural irrigation water. Water used for agriculture is less likely to increase because of water shortage. Sichuan province measures the water usage amount for agriculture, and thus the amount can be directly obtained.

L_{12} represents agriculture investment. Agriculture investment reflects local economic development level as well as the degree of attention paid to agriculture. Funds invested in water-saving irrigation construction and agricultural investment come from various sources such as national finance, local finance, credit funds, collective economy, individual investment, or even foreign investment. With rapid economic development, the amount of funds invested in water-saving irrigation construction from local finance, collective agricultural economic organizations, and individual investments tends to increase year by year.

3.1.2. Agronomy Water-Saving Evaluation Indexes

Agronomy water-saving refers to a certain comprehensive agricultural technology which integrates the resources of water, soil, and crops by optimizing the cropping and farming system to effectively reduce soil evaporation and luxury transpiration of crops, and further improves water efficiency with regards to water-saving and productivity. The main features of the agronomy water-saving strategy are the rational layout of farm crops and the improvement of planting methods. Five indexes were selected from the contents to establish the evaluation system and a detailed description of the following indexes was presented as follows.

M_1 refers to the ratio of the rice dry nursery seedling area to the rice planting area. The rice dry nursery seedling ratio is distinctive with a short seedling stage and convenience of management, with the advantages including saving field planting resources, water conservation, high economic benefit, disease resistance, cold resistance, and so on. Compared with the water-raised seedling, the rice dry nursery seedling saves roughly 50–90% of water use.

M_2 represents the ratio of the area with the "Thin-Shallow-Wet-Dry" technique to the rice cultivated area. The "Thin-Shallow-Wet-Dry" irrigation technology is a scientific water-saving method used to plant rice. "Thin" means that a thin layer of water is needed for seedlings to take root when planting. "Shallow" means that rice seedlings turn green in shallow water. "Wet" refers to the practice of keeping the water-holding seedlings at an early tilling stage. "Dry" means stopping irrigation and adapting the seedlings to an external environment to train the seedlings at a late tilling stage. This technology is consistent with the water demand of the rice growth at different stages, which helps to save irrigation water as well as to invigorate its physiological activities to save water and increase production.

M_3 refers to the per capita income of the agriculture population. The development and popularization of water-saving irrigation technology has much to do with the economic development level, as applying a series of measures and technologies is concerned with the agronomy water-saving demand investment. So far, government finance-oriented investment has not been established in the application of agronomy technology. Therefore, the application of this technology is closely related to the income of the local population.

M_4 refers to the ratio of the drought-tolerant crops cultivated area to the cultivated area of all crops. The outstanding advantage of drought-tolerant crops is the water-saving advantage, hence one of the effective measures to solve water shortages in dry farming areas is to grow drought-tolerant crops. This practice has been popularized in Northwest of China; however, more efforts should be made to implement this in Sichuan province.

M_5 refers to water use efficiency. Water use efficiency (WUE) refers to the quantity of economic product yielded by water consumption and it is significant as a reflection of the water-saving irrigation efficiency. At present, WUE is the major economic index used to evaluate water-saving irrigation benefits.

3.1.3. Management Water-Saving Evaluation Indexes

Management water-saving includes all management and maintenance work after the completion of a water-saving construction project. Sixteen indexes, which were divided into 6 qualitative indexes and 10 quantitative indexes, were included in this section. Six indexes, i.e., the establishment of subsidiary policies and regulations (N_1), the degree of support from the government (N_4), propaganda and education level (N_5), water-saving engineering design level (N_7), the rationality of the irrigation system (N_{10}), and the sound level of the water-saving incentive mechanism (N_{14}), which are difficult to evaluate with data-measurement, were obtained using the investigation method. We described the 16 indexes as follows:

N_1 refers to the establishment of subsidiary policies and regulations. The perfection of legal systems and regulations is effective for the practice of irrigation water management. Thus, for the local government, promoting water-saving technology is a top priority in compiling an agricultural development plan, and much attention should be paid to develop, demonstrate, and extend key water-saving technologies.

N_2 refers to the reform execution situation of the property system. The reform execution situation of the property system refers to the proportion of the number of identified property rights and identified management and maintenance of main body projects to the number of small-scale agricultural water conservancy projects.

N_3 refers to the construction level of the technology popularizing system. This refers to the proportion of existing technical service institutions to the technical service institutions which are supposed to be established.

N_4 refers to the degree of support from the government. The attention paid to water saving strategies from governments is of benefit for the earlier execution of related water-saving regulations, measures, and funding.

N_5 refers to the propaganda and education level. Propaganda promoting water saving targets both government officials and citizens, especially farmers and other irrigation water users. Improving the public's water-saving consciousness is a long-term task. Water saving habits and consciousness should be cultivated among the public, and water saving knowledge should be popularized as well.

N_6 refers to the user participation level. This indicates the proportion of user-based management irrigation areas to total irrigation areas.

N_7 refers to the water-saving engineering design level. It is key to the promotion of water-saving irrigation to guarantee engineering quality, as well as ensure good operation of the system across long periods of time.

N_8 refers to the degree of perfection of engineering subsidiaries. This indicates the proportion of existing canal-attached facilities to attached facilities which are supposed to be constructed according to statistical data from the irrigation region.

N_9 refers to the management and maintain level of water-saving engineering. This indicates the proportion of the maintenance expense of irrigation engineering to the required maintenance cost.

N_{10} refers to the rationality of irrigation system. With the enforcement of the rationality of the irrigation system, the optimum design of water-saving irrigation systems has become increasingly important. The irrigation system design should be developed on the basis of comprehensive technologies of agricultural water-saving irrigation and experimental data, which underlies the basic research foundation of this method.

N_{11} refers to monitoring the coverage rate of soil moisture. The monitoring coverage rate of soil moisture refers to the proportion of existing monitoring stations to the total number of planned soil moisture monitoring stations.

N_{12} refers to the degree to which water measurement is popularized, indicating the proportion of practicing water measurement areas to total irrigation areas.

N_{13} refers to the ratio of planned water use irrigation areas to total irrigation areas.

N_{14} refers to the sound level of water-saving incentive mechanisms. Establishing water rights-based optimal water dispatching and allocation is the core of the water-saving incentive mechanism. It is urgent to accelerate the confirmation of agricultural irrigation water rights, to develop the water right trade market, and to establish agricultural water saving incentive mechanisms.

N_{15} refers to the popularity rate of the measurement charges of water, indicating the proportion of practicing water measurement charges areas to total irrigation areas.

N_{16} refers to the rationality of the water price. The rationality of the water price refers to the proportion of the collected price for agricultural irrigation water to the cost.

3.1.4. The Comprehensive Evaluation Index System of WIDL

The hierarchy of the comprehensive evaluation index system was classified into three layers: The target layer, system layer, and index layer. According to the actual situation in Sichuan province and available data, 12, 5, and 16 indexes were adopted for engineering, agronomy, and management WIDL, respectively. The target layer (first-class index) was the evaluation result of WIDL, the system layer (second-class index) contained engineering, agronomy and management WIDL, and the index layer (third-class index) included the 33 specific evaluation indexes (Table 1). The 33 indexes were divided into 2 categories according to the method of data acquisition. One included the 6 qualitative indexes using the questionnaire method, and the other included the 27 quantitative indexes from the 2014 Sichuan province Provincial Water Conservancy Statistical Yearbook, the 2014 Sichuan province Water Management Yearbook, and the 2014 Sichuan province Agricultural Yearbook.

Table 1. The index system of the comprehensive evaluation model for the water-saving irrigation development level (WIDL).

Target Layer	System Layer	Index Layer	Source
Water-saving irrigation developing level	Engineering water-saving	Ratio of water-saving irrigation area to effective irrigation area (%) (L_1)	B,D
		Ratio of established high-efficient water-saving irrigation area to the total water-saving irrigation area (%) (L_2)	C
		Water efficiency of irrigation (L_3)	C
		Economic benefits per hm^2 of water-saving irrigation (CNY) (L_4)	D
		Water usage amount per hm^2 for irrigation (m^3) (L_5)	B,D
		Ratio of water-saving irrigation area to cultivated area (%) (L_6)	B,C
		Main crops yield per unit (kg/hm^2) (L_7)	B
		Per capita GDP (CNY) (L_8)	D
		Percentage of canal lining (%) (L_9)	C
		Percentage of working lining channels (%) (L_{10})	C
		Water usage amount for agriculture (m^3/hm^2) (L_{11})	C,D
		Agriculture investment (CNY/hm^2) (L_{12})	B,C
	Agronomy water-saving	Ratio of rice dry nursery seedling area to rice planting area (%) (M_1)	B,D
		Ratio of area with the technique of "Thin-Shallow-Wet-Dry" to rice cultivated area (%) (M_2)	D
		Per capita income of agriculture population(CNY) (M_3)	B
		Ratio of drought tolerant crops cultivated area to all crops cultivated area (%) (M_4)	B,D
		Water use efficiency (Kg/m^3) (M_5)	C,D
	Management water-saving	Establishment of subsidiary policies and regulations (N_1)	A
		Reform execution situation of property system (%) (N_2)	B,C
		Construction level of technology popularizing system (N_3)	B,C
		Degree of support from the government (N_4)	A
		Propaganda and education level (N_5)	A
		User participation level (N_6)	B,C
		Water-saving engineering design level (N_7)	A
		Degree of engineering subsidiaries perfection(N_8)	B
		Management and maintained level of water-saving engineering (N_9)	C
		Rationality of irrigation system (N_{10})	A
		Monitoring coverage rate of soil moisture (%) (N_{11})	C,D
		Degree of popularizing water measurement (%) (N_{12})	B,C
		Ratio of planned water use (N_{13})	C
		Sound level of water-saving incentive mechanisms (N_{14})	A
		Popularity rate of measurement charges of water (%) (N_{15})	B,C
		Rationality of water price (N_{16})	C

Note: A, B, C, and D respectively indicate the qualitative indexes, the 2014 Sichuan province Provincial Water Conservancy Statistical Yearbook, the 2014 Sichuan province Water Management Yearbook, and the 2014 Sichuan province Agricultural Yearbook.

3.2. Index Quantification

We adopted a questionnaire method among the staff in the water administration department in Sichuan province and 21 cities (states) to quantify the 6 qualitative indexes. The staff had ample knowledge of water-saving irrigation management and were very familiar with the conditions of Sichuan province. Therefore, they could give a proper judgment or assessment for each index. Their work provided reliable information for our research. The current survey data of 6 qualitative indexes were quantified as follows:

$$Q = (f_1 \times M_1 + f_2 \times M_2 + f_3 \times M_3 + f_4 \times M_4 + f_5 \times M_5)/N \tag{10}$$

where M_1~M_5 ($M_1 = 10$, $M_2 = 8$, $M_3 = 6$, $M_4 = 4$, $M_5 = 2$) were the quantitative index values corresponding to "better, good, general, poor, poorer". f_1~f_5 were the sample numbers in the corresponding class and N is the total sample numbers.

Taking Sichuan province as an example, 20 qualified staff from the water administration department of Sichuan province were invited to the survey to quantify the 6 indexes. The staff had ample knowledge of water-saving irrigation and were very familiar with the conditions of Sichuan province. Therefore, they could give a proper judgment or assessment for each index. Their work provided reliable information for our research. The quantitative values of the 6 indexes in Sichuan province were obtained by Equation (10) as seen in Table 2.

Table 2. The quantification of the management WIDL indexes in Sichuan province.

No.	Qualitative Indexes	Quantitative Value
1	N_1	6.8
2	N_4	6.7
3	N_5	6.1
4	N_7	7.2
5	N_{10}	6.2
6	N_{14}	4.7

Similarly, 6 qualitative evaluation indexes of WIDL in 21 cities (states) of the Sichuan province were quantified, and 33 indexes of WIDL in the Sichuan province and 21 cities (states) were shown in Table 3.

3.3. Principal Component Extracting

Although the 33 evaluation indexes contained more comprehensive information on the development level of water-saving irrigation, they will increase the complexity of the evaluation process. We therefore chose PCA to simplify the engineering, agronomy, and management evaluation indexes, respectively. The principal components of engineering WIDL in Sichuan province and 21 cities (states) were analyzed using PCA in the DPS program. The eigenvalues and contribution rates of the principal components were shown in Table 4. From Table 4, the accumulative contribution rate of the first 5 principal components was up to 89.64% and above 85%. Therefore, the first 5 principal components can represent all the information of the 12 engineering WIDL.

The weights of coefficients between each index and the five principal components (F_1–F_5) were then shown in Table 5. The indexes with maximum weight coefficients were selected as the simplified principal indexes. Thus, the engineering WIDL evaluation indexes were simplified to 6 indexes as the water efficiency of irrigation (L_3), the economic benefits per hm^2 of water-saving irrigation (L_4), the ratio of water-saving irrigation area to cultivated area (L_6), the percentage of canal lining (L_9), and the agriculture investment (L_{12}).

Similarly, three indexes, including per capita income of agriculture population (M_3), water use efficiency (M_5), and the ratio of the area with the technique of "Thin-Shallow-Wet-Dry" to the rice cultivated area (M_2) were extracted from five indexes of agronomy WIDL using PCA with the accumulative contribution rate of 88.6%. The evaluation indexes of management WIDL could be simplified to six items as follows: the degree of support from the government (N_4), the degree of popularizing water measurement (N_{12}), the sound level of water-saving incentive mechanisms (N_{14}), the construction level of technology popularizing systems (N_3), the monitoring coverage rate of soil moisture (N_{11}), and the water-saving engineering design level (N_7). So, 33 evaluation indexes were simplified to 14 indexes, which greatly facilitated our evaluations (Table 6).

The grading standard of each index is the basis for determining the WIDL grade. Five grades of evaluation standards were established for each index of the regional WIDL according to the mean and standard deviation grading method and the actual situation of Sichuan province. These standards, (better, good, general, poor, and poorer) correspond to the I–V levels shown in Table 6 [39]. Although this standard will be further improved or updated in the future, at this stage we established a relatively objective evaluation system to ensure the accuracy of the evaluation results.

Table 3. The thirty-three indexes of WIDL in Sichuan province and 21 regions.

Region	L_1	L_2	L_3	L_4	L_5	L_6	L_7	L_8	L_9	L_{10}	L_{11}	L_{12}	M_1	M_2	M_3	M_4	M_5	N_1	N_2	N_3	N_4	N_5	N_6	N_7	N_8	N_9	N_{10}	N_{11}	N_{12}	N_{13}	N_{14}	N_{15}	N_{16}
Sichuan Province	42.00	1.55	0.40	975	5595	28	5254	21,369	45.16	36.80	3121	515	7.50	14.70	2802	22	1.20	6.80	0.67	0.67	6.70	6.10	0.31	7.20	0.73	0.45	6.20	0.43	0.10	0.66	4.70	0.08	0.33
Chengdu	67.50	3.00	0.43	1286	7365	63	6134	41,253	72.00	32.40	7550	746	25.60	17.30	4485	31	1.12	7.00	0.98	0.90	9.20	6.40	0.50	7.30	0.83	0.80	7.20	0.55	0.15	0.98	6.00	0.13	0.44
Zigong	29.10	0.04	0.38	630	5280	19	5856	23,613	41.15	38.20	1843	705	2.70	0.01	3188	23	1.33	6.00	0.8	0.80	5.80	5.70	0.37	5.50	0.74	0.40	5.80	0.34	0.08	0.50	3.00	0.07	0.40
Panzhihua	66.16	0.72	0.43	1388	9690	51	5459	43,959	33.29	46.60	4580	510	6.50	55.60	3463	19	1.28	7.60	0.68	0.65	8.00	6.80	0.39	5.80	0.96	0.36	7.20	0.60	0.08	0.72	5.80	0.08	0.60
Luzhou	35.04	0.23	0.40	837	3345	21	5441	16,698	45.41	31.00	1694	659	2.00	1.70	3165	25	1.30	6.70	0.77	0.90	6.90	6.00	0.49	8.00	0.78	0.16	6.90	0.28	0.05	0.53	3.90	0.05	0.35
Deyang	53.49	0.07	0.41	1082	9090	45	6435	25,335	58.53	46.00	6944	524	50.00	41.00	3585	33	1.13	7.70	0.98	0.80	7.50	7.00	0.40	7.70	0.9	0.59	7.70	0.50	0.15	0.96	4.70	0.15	0.62
Mianyang	41.53	0.80	0.40	1074	6150	33	5332	20,053	61.00	36.80	4430	478	20.20	0.80	3179	34	1.36	7.20	0.66	0.70	7.20	6.20	0.39	6.80	0.6	0.72	5.60	0.51	0.13	0.82	4.40	0.13	0.58
Guangyuan	36.95	1.27	0.38	788	4215	20	5306	12,313	40.45	48.00	1696	585	2.40	54.80	2000	11	1.31	5.60	0.38	0.60	6.60	5.90	0.13	6.90	0.35	0.32	6.60	0.30	0.10	0.45	4.60	0.05	0.38
Suining	28.40	0.51	0.39	767	3495	25	5257	14,498	46.05	48.80	1992	566	9.70	22.00	2828	18	1.11	6.30	0.72	0.75	5.00	5.80	0.25	5.80	0.93	0.13	5.80	0.51	0.10	0.50	4.00	0.05	0.46
Neijiang	39.81	4.14	0.39	844	2700	29	4923	18,022	43.85	37.60	2007	579	4.70	17.10	2986	25	1.21	7.20	0.79	0.70	6.60	5.40	0.24	6.00	0.78	0.32	4.60	0.40	0.02	0.49	3.40	0.02	0.50
Leshan	18.91	5.73	0.39	870	9825	13	4720	22,490	47.26	36.00	4269	464	0.60	8.90	3241	26	1.06	5.40	0.69	0.75	6.10	6.30	0.50	6.10	0.89	0.55	6.00	0.45	0.15	0.62	5.00	0.12	0.55
Nanchong	37.59	0.05	0.37	806	3060	28	5589	13,212	52.59	52.60	1738	479	1.40	0.80	2646	29	1.41	7.30	0.91	0.50	6.60	6.60	0.30	8.40	0.74	0.88	6.90	0.38	0.13	0.78	5.10	0.11	0.70
Meishan	39.14	6.24	0.39	900	7530	41	5617	18,586	50.83	30.00	2987	454	10.10	32.50	3284	20	1.23	7.70	0.65	0.60	7.30	6.70	0.15	2.80	0.72	0.45	5.50	0.38	0.13	0.72	2.80	0.10	0.48
Yibin	51.01	0.46	0.39	854	2520	26	5585	19,499	55.32	32.80	2036	690	6.20	6.70	3068	31	1.08	5.60	0.71	0.75	3.40	5.30	0.26	3.40	0.62	0.68	5.10	0.39	0.10	0.55	4.90	0.06	0.45
Guang'an	28.28	0.22	0.39	780	2490	18	5571	15,588	34.13	46.00	1790	509	2.70	7.90	2915	19	1.32	6.40	0.73	1.00	6.80	6.40	0.50	8.00	0.78	0.55	4.20	0.30	0.09	0.45	2.80	0.07	0.38
Dazhou	34.19	0.01	0.38	770	1710	19	5221	14,623	52.04	54.00	1073	812	10.30	27.60	2943	29	1.43	7.70	0.74	0.70	6.20	5.50	0.24	6.80	0.79	0.50	6.00	0.31	0.06	0.46	6.00	0.06	0.33
Ya'an	38.23	0.00	0.38	714	10,200	32	4243	18,881	32.03	44.40	2847	431	2.10	0.01	2829	10	1.14	6.10	0.8	0.80	6.30	5.60	0.25	7.80	0.77	0.42	5.50	0.35	0.06	0.45	5.30	0.06	0.30
Bazhong	27.44	0.18	0.37	819	1560	14	5372	8717	34.54	52.00	4479	800	3.00	19.10	2031	10	1.33	6.50	0.73	0.50	8.00	7.50	0.38	7.50	0.73	0.32	7.50	0.30	0.07	0.40	2.80	0.07	0.45
Ziyang	39.99	0.82	0.38	966	4065	26	4318	16,644	41.79	20.00	1957	500	16.10	0.01	2988	23	1.29	5.80	0.56	0.50	5.80	5.30	0.37	7.50	0.87	0.33	5.00	0.40	0.05	0.55	3.50	0.05	0.50
Aba	9.65	2.49	0.36	630	3540	3	3100	14,662	28.43	23.60	1421	439	0.01	0.01	1881	5	1.01	3.10	0.001	0.001	3.00	3.20	0.001	3.20	0.35	0.001	2.80	0.001	0.001	0.001	5.30	0.001	0.001
Ganzi	16.39	0.70	0.36	602	1905	6	2824	11,659	27.80	46.00	590	327	0.01	0.01	1309	4	1.01	7.00	0.001	0.001	6.50	7.80	0.001	6.20	0.45	0.001	6.90	0.001	0.001	0.001	5.10	0.001	0.001
Liangshan	58.16	0.05	0.37	713	9600	23	4358	17,560	49.82	31.00	4120	302	0.01	0.01	2438	28	1.25	6.10	0.41	0.30	7.80	7.20	0.12	7.20	0.94	0.13	7.00	0.001	0.02	0.24	2.90	0.001	0.25

Table 4. The eigenvalues and contribution rates of the principal components.

No.	Eigenvalues	Contribution Rate (%)	Cumulative Contribution Rate (%)
F_1	6.0462	50.3852	50.3852
F_2	1.7518	14.5984	64.9836
F_3	1.424	11.8667	76.8504
F_4	0.7881	6.5679	83.4182
F_5	0.7467	6.2224	89.6406

Table 5. The loading matrix of the principal components.

Evaluation Indexes	F_1	F_2	F_3	F_4	F_5
L_1	0.35	0.15	−0.04	0.15	−0.34
L_2	0.08	−0.55	−0.11	−0.13	0.41
L_3	0.57	0.09	−0.01	0.23	0.18
L_4	0.37	0.05	0.52	0.27	0.16
L_5	0.27	−0.33	0.29	−0.41	−0.33
L_6	0.58	0.05	−0.07	−0.01	−0.04
L_7	0.25	0.36	−0.32	−0.21	0.24
L_8	0.35	−0.07	0.14	0.28	0.01
L_9	−0.03	0.57	0.24	−0.52	0.27
L_{10}	0.25	0.06	−0.56	−0.18	−0.03
L_{11}	0.17	0.14	0.63	0.13	0.27
L_{12}	−0.32	0.26	−0.12	0.47	0.65

Table 6. The classification standard of the WIDL evaluation indexes.

Indexes	I	II	III	IV	V
L_3	>0.6	0.5~0.6	0.45~0.5	0.40~0.45	≤0.40
L_4	>2250	1650~2250	1050~1650	450~1050	≤450
L_6	>60	60~45	45~30	30~15	≤15
L_9	>75	60~75	50~60	35~50	≤35
L_{12}	≥1500	1200~1500	600~1200	300~600	<300
M_2	>50	35~50	15~35	5~15	≤5
M_3	>8000	6000~8000	4000~6000	2000~4000	≤2000
M_5	>1.75	1.35~1.75	1.0~1.35	0.8~1.0	≤0.8
N_3	>0.86	0.71~0.86	0.56~0.71	0.40~0.56	≤0.40
N_4	9~10	7~9	5~7	3~5	0~3
N_7	>7.94	7.02~7.94	6.10~7.02	5.18~6.10	≤5.18
N_{11}	>0.50	0.40~0.50	0.30~0.40	0.20~0.30	≤0.20
N_{12}	>0.12	0.10~0.12	0.07~0.10	0.04~0.07	≤0.04
N_{14}	>5.41	4.78~5.41	4.15~4.78	3.52~4.15	≤3.52

3.4. Weight Determination

Assigning weights to the 14 indexes was a key step in our evaluation. The weight vector of the second-class indexes determined by AHP was $\alpha = (0.395, 0.262, 0.344)$, which passed the consistency test. The weights of the third-class indexes were determined considering both subjective weights vector α by AHP and objective weights vector β by EWM, and the combined weight ω was acquired by Equation (9). The final weights of each index were shown in Table 7, and the distributions of several indexes with high weights in 21 regions were shown in Figure 2. From Table 7, the ratio of water-saving irrigation areas to cultivated land area (L_6), which was the direct reflection of the development level of water-saving irrigation, had the highest weight of 0.132, followed by the per capita income of the agriculture population (M_3), the water use efficiency (M_5), the degree of support from government (N_4), and the water-saving engineering design level (N_7) with higher weights of about 0.1. The construction level of technology popularizing system (N_3), the monitoring coverage rate of soil moisture (N_{11}), the degree of popularizing water measurement (N_{12}), and the sound level of water-saving incentive mechanisms (N_{14}) had the smallest weights (<0.05).

Table 7. The weights of each evaluation index.

Target Layer	System Layer	Second-Class Weights	Indexes Layer	Third-Class Weights	Final Weights
Regional water-saving irrigation developing level	Engineering water-saving	0.395	L_3	0.167	0.066
			L_4	0.163	0.064
			L_6	0.334	0.132
			L_9	0.167	0.066
			L_{12}	0.169	0.067
	Agronomy water-saving	0.262	M_2	0.246	0.064
			M_3	0.378	0.099
			M_5	0.376	0.098
	Management water-saving	0.344	N_3	0.119	0.041
			N_4	0.274	0.094
			N_7	0.266	0.092
			N_{11}	0.116	0.040
			N_{12}	0.114	0.039
			N_{14}	0.111	0.038

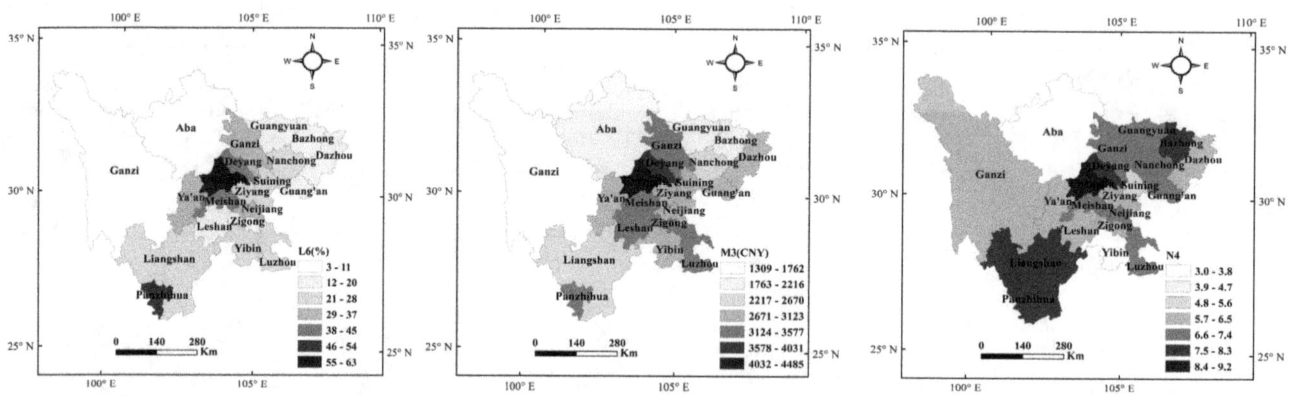

(a) (b) (c)

Figure 2. The key evaluation indexes between 21 regions of the Sichuan province. (**a**) The ratio of water-saving irrigation area to cultivated area (%) (L_6); (**b**) the per capita income of the agriculture population (CNY) (M_3); (**c**) the degree of support from the government (N_4).

3.5. The Calculation of Relative Closeness

In the study, we made some improvements when adopting TOPSIS method. Firstly, the corresponding classification thresholds of 14 indexes were seen as 5 schemes. Based on the classification thresholds in Table 6 and the principal indexes in the 21 cities (states) and in Sichuan province, decision matrix Z was constructed as follows. In matrix Z, the first 5 lines were the thresholds of the 14 indexes corresponding to the five categories (better, good, general, poor, and poorer), and the next 22 lines were the values of the 14 indexes in Sichuan province and 21 cities (states).

$$Z = \begin{vmatrix} 0.6 & 150 & 60 & 75 & 1500 & 1.75 & 50 & 8000 & 9 & 0.12 & 5.41 & 0.86 & 0.5 & 7.94 \\ 0.5 & 110 & 45 & 60 & 1200 & 1.35 & 35 & 6000 & 7 & 0.1 & 4.78 & 0.71 & 0.4 & 7.02 \\ 0.45 & 70 & 30 & 50 & 600 & 1 & 15 & 4000 & 5 & 0.07 & 4.15 & 0.56 & 0.3 & 6.1 \\ 0.4 & 30 & 15 & 35 & 300 & 0.8 & 5 & 2000 & 3 & 0.04 & 3.52 & 0.4 & 0.2 & 5.18 \\ 0 & 0 & 0 & 0 & 0 & 0 & 0 & 0 & 0 & 0 & 0 & 0 & 0 & 0 \\ 0.4 & 65 & 28 & 45.16 & 315.3 & 1.2 & 14.7 & 2802 & 6.7 & 0.1 & 4.7 & 0.67 & 0.43 & 7.2 \\ \vdots & \vdots & \vdots & \vdots & \vdots & \vdots & \vdots & \vdots & \vdots & \vdots & \vdots & \vdots & \vdots & \vdots \\ 0.37 & 47.5 & 23 & 49.82 & 302 & 1.25 & 0.01 & 2438 & 7.8 & 0.02 & 2.9 & 0.3 & 0.001 & 7.2 \end{vmatrix} \quad (11)$$

The standard decision matrix V was then obtained using the dimensionless method in Equation (3). The weight decision matrix R was established by multiplying the standard decision matrix V and weights w (Table 7) according to Equation (4). Finally, according to the 14 benefit indexes based on principal component analysis, the positive ideal solution (PIS) was obtained using Equation (5), and the negative ideal solution (NIS) was obtained using Equation (6). We calculated the distance S^+, S^- of 27 schemes to PIS and NIS using Equations (7) and (8), and further obtained the relative closeness ε between the 27 schemes and the ideal solution.

S^+ = (0.0055, 0.0267, 0.0523, 0.0737, 0.1023, 0.0571, 0.0432, 0.0627, 0.0437, 0.0601, 0.046, 0.0562, 0.0582, 0.057, 0.0554, 0.0661, 0.0593, 0.0523, 0.0598, 0.0622, 0.0561, 0.0626, 0.063, 0.0602, 0.0791, 0.0771, 0.0663);

S^- = (0.0991, 0.0763, 0.0525, 0.0334, 0, 0.0517, 0.0774, 0.0472, 0.0706, 0.0518, 0.0662, 0.0567, 0.0549, 0.0488, 0.0509, 0.0461, 0.0539, 0.0572, 0.0467, 0.0496, 0.053, 0.0506, 0.0493, 0.0499, 0.0339, 0.0385, 0.0474);

ε = (0.9125, 0.6301, 0.5498, 0.4124, 0, 0.4754, 0.6418, 0.4294, 0.6174, 0.4627, 0.5900, 0.5022, 0.4855, 0.4610, 0.4789, 0.4110, 0.4762, 0.5224, 0.4388, 0.4437,0.4858, 0.4470, 0.4389, 0.4535, 0.2996, 0.3328, 0.4168);

The first 5 relative closeness (ε) values were 0.9125, 0.6301, 0.5498, 0.4124, and 0, respectively. So, the grading standard was (>(0.9125 + 0.6301)/2, (0.6301 + 0.5498)/2~(0.9125 + 0.6301)/2, (0.5498 + 0.4124)/2~(0.6301 + 0.5498)/2, (0.4124 + 0)/2~(0.5498 + 0.4124)/2, 0~(0.382 + 0)/2), that (>0.7713, 0.5899~0.7713, 0.4811~0.5899, 0.2062~0.4811, 0~0.2062), corresponding to the 5 categories (better, good, general, poor, and poorer). The WIDL in each region was judged using the grading standard, and the evaluation results are shown in Table 8 and Figure 3.

Table 8. The relative closeness (ε) of the water-saving irrigation development level in 21 regions of Sichuan province.

Region	ε	Level	Region	ε	Level
Sichuan province	0.4754	Poor	Nanchong	0.4762	Poor
Chengdu	0.6418	Good	Meishan	0.5224	General
Zigong	0.4294	Poor	Yibin	0.4388	Poor
Panzhihua	0.6174	Good	Guang'an	0.4437	Poor
Luzhou	0.4627	Poor	Dazhou	0.4858	General
Deyang	0.5900	Good	Ya'an	0.4470	Poor
Mianyang	0.5022	General	Bazhong	0.4389	Poor
Guangyuan	0.4855	General	Ziyang	0.4535	Poor
Suining	0.4610	Poor	Aba	0.2996	Poor
Neijiang	0.4689	Poor	Ganzi	0.3328	Poor
Leshan	0.4110	Poor	Liangshan	0.4168	Poor

From Table 8, the overall development level of water-saving irrigation was "poor" in Sichuan province with a ε value of 0.4754. Only 3 regions were in "good", 4 regions were "general", and the other 14 regions were "poor". By consulting experts or relevant managers from the Sichuan province

Water Conservancy Bureau, we determined that this result was consistent with the general situation of Sichuan province and was consistent with the previous reports of Lou [39].

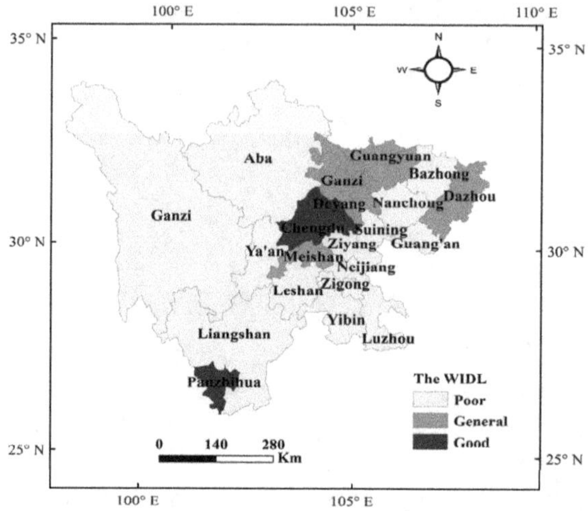

Figure 3. The water-saving irrigation development level (WIDL) in 21 regions of Sichuan province.

Only Chengdu, Panzhihua, and Deyang were at the "good" level, with ε values of about 0.6. The economic development levels in these regions were higher; the per capita GDP of Chengdu and Panzhihua was in the top 3 of Sichuan province, more than 60 thousand CNY higher than the average of 32.5 thousand CNY of Sichuan province. The per capita income of the agriculture population was relatively higher, so local people had a strong awareness of water conservation. The ratio of water-saving irrigation area to total cultivated land in those 3 regions ranks in the top 3. With better economic development level than other regions in Sichuan province and more funds invested to water-saving irrigation engineering maintenance in these regions, the existing lining canals were in better conditions than in the other 18 regions. The same was true with information management in irrigation areas. In 2004, Panzhihua was listed as the "national agricultural water-saving demonstration city", and it made a lot of exploration in engineering and management construction. The government attached great importance to water-saving irrigation development, leading to the improvement of water-saving incentive mechanisms.

The four regions of Meishan, Guangyuan, Dazhou, and Mianyang were at the "general" level, with a ε value of about 0.5. In these regions, the engineering and agronomy water-saving development level was higher than the average level of Sichuan province. The per capita GDP of these regions were up to or exceeding the average level of Sichuan province. With strong support by local government, the indexes of water use efficiency, the percentage of working lining canals, and the ratio of water-saving irrigation area to the total cultivated area were better in these 4 regions than in the other regions in Sichuan province. In GuangYuan, the agriculture planting structure was adjusted according to the differences of the water demand and drought resisting characteristics in different kinds of crops. The "Thin-Shallow-Wet-Dry" area of the rice was relatively high.

Fourteen regions were in "poor" condition, such as Neijiang, Bazhong, Ganzhi, and Aba, with ε values of less than 0.48. Most of these regions are located in mountainous and hilly areas of Sichuan province, especially Ganzi, Aba, and Liangshan, which are located on the Northwest plateau of Sichuan province, with poor natural resources and insufficient economies. This led to a lack of investment in country public utilities and a funds shortage for the construction and maintenance of water-saving engineering. With the water efficiency of irrigation below 0.4 in these regions, the ratio of the water-saving irrigation area to the total cultivated area is at the inadequate–average level in Sichuan province, only reaching around 5% in Ganzi and Aba. Agronomy water-saving and management water-saving levels were also relatively poor in these regions. Moreover, information management for irrigation (soil moisture content supervision and measurement technology) was hardly conducted in Ganzhi, Aba, and Liangshan.

4. Discussion

The shortage of water resources has become an important factor restricting the sustainable development of regional economies. The development of water-saving irrigation methods to improve the utilization efficiency of water resources is an important way to deal with the engineering water shortage and seasonal drought conditions in the humid regions of Southern China. A reasonable and accurate assessment of WIDL is not only a basis for further scientific planning and construction to promote the development of regional water-saving irrigation, but also a basis to promote and guarantee the sustainable development of agricultural production. Therefore, the objective evaluation of regional WIDL has great significance to promote the development of water-saving irrigation along a scientific, healthy, and positive track, and to ensure sustainable development of the national economy.

According to the evaluation results, great efforts should be made to increase water-saving irrigation levels in Sichuan province. Firstly, more focus should be placed on increasing the investment in agricultural water-saving irrigation and promoting water-saving engineering. The governmental authority should implement the policy that 10% of land transaction fees are extracted to invest into farmland water conservancy construction in order to solve problems such as the inadequate service of necessary facilities for key projects in the irrigation region, the low qualified canal (channel) number, the aging canal system structures, and so on. On the other hand, investment and financial reform should increase, and diversified investment should be established to play the role of the financial fund, and to fully utilize preferential policy, incentive measures, and credit aid to attract new major participants in agricultural operations such as leading enterprises, major planting farmers, and cooperatives to invest in water-saving efficiency.

Secondly, more focus should be placed on adjusting agricultural industry distribution and extending water-saving technologies. It is necessary to establish the agricultural industry layout so that it is appropriate for local water resources, land, and ecosystems, and to adjust the agricultural water utilization structure. Highly efficient water-saving irrigation strategies in accordance with water resources, planting structure, and geographic conditions, land management should be adopted. In hilly regions with water scarcity and poor water conservancy facilities, such as Guangyuan, Bazhong, and Nanchong, etc., the principle of exploring water sources combined with saving water should be conducted to plan water conservancy engineering in a scientific way. In plains and large-scale agricultural management areas, such as Chengdu and Deyang, high water-saving irrigation efficiency should be practiced with the assistance of a traceability system in agricultural production chains and a brand marketing system for agricultural products. Internet-accelerated speed must be introduced to advance the development of high-efficiency water-saving technologies. Efforts should be made to practice information management in agricultural water saving engineering, extend efficient water-saving technology in irrigation, and enhance the transformation of the achievements made in agricultural water-saving technology. Advanced water-saving technology from both at home and abroad should be assimilated and utilized.

Thirdly, more focus should be placed on amplifying the water price system and creating incentives for water-saving irrigation. Legislative, economic, and administrative incentives should be executed to reform the total agricultural water consumption and water quota. Firstly, an index system of water allocation and permits of water-withdrawal volume control in the administrative districts at the provincial, municipal, and county level ought to be established respectively, which would be convenient for decomposing the index throughout the three levels from the top-down. Secondly, water pricing for agricultural use should be reformed to encourage volumetric pricing, improve terminal water pricing, and enhance the ratio of agricultural water use. Meanwhile, compensation and incentive mechanism for saving water should be established to encourage local officials and individual farmers to actively practice water-saving irrigation.

Finally, more focus should be placed on enhancing management and maintenance capabilities. It is important to perfect the management and maintenance system by clearly defining the responsibilities of project property owners and water administrative supervisors. Meanwhile, technological guidance should be provided to constitute regulations and standards for highly-efficient water-saving irrigation and to compile standardized management manuals. Furthermore, integrated institutions of water conservancy services, professional teams, water users associations, and village-level water management should be established to guarantee long-term water use efficiency.

5. Conclusions

Regional WIDL evaluation is a task which involves multi-target and multi-attribution methods. In the paper, we choose 3 second-class indexes of engineering, agronomy, and management water-saving along with 33 third-class indexes to establish a comprehensive evaluation index system for WIDL. The principal components analysis method was adopted for simplifying the water-saving indexes of engineering, agronomy, and management, and the number of third-class indexes was simplified into 14. The combined weight which considered both objective weight and subjective weight was applied to judge the relative importance of each index. The combination weight offset the deficiency of the evaluation index weights determined only by expert subjective experience or indexes of sample data. Moreover, the TOPSIS method was improved to evaluate WIDL in 21 cities (states) of Sichuan province.

Results showed the overall development level of water-saving irrigation was "poor" in Sichuan province with a ε value of 0.4754. The water-saving irrigation level turned out to be "good" in the following three regions with advantageous geographical conditions and developed economies: Chengdu, Panzhihua and Deyang. The water-saving irrigation level was "general" in the following four regions with good economic levels where agronomic water saving has been popularized: Meishan, Guangyuan, Dazhou, and Mianyang. Water-saving irrigation level was and "poor" in fourteen regions of mountainous and hilly areas in Sichuan province, especially in Ganzi, Aba, and Liangshan. Located on the Northwest plateau of Sichuan province, these areas feature poor natural resources and insufficient economies. According to the evaluation results, efforts should be made to increase the water-saving irrigation level in Sichuan province. We should increase the investment in agricultural water-saving irrigation and enhance management and maintenance, especially in "poor" regions. Besides, we should adjust the agricultural industry distribution and extend water-saving technology to be more appropriate for local water resource, land, and ecosystem conditions. Furthermore, we should amplify the water price system and create new incentives for the promotion of water-saving irrigation strategies.

This paper also has limitations and shortcomings. The evaluation of the data collection needs to be broader to fully represent the actual conditions of the studied areas. In the meantime, the evaluation standard can only be adapted in Sichuan province, and more evaluation methods should be used to verify the results. Although water-saving irrigation policies, technologies, agronomy measures, and evaluation criterion vary in other areas, the evaluation of WIDL in Sichuan province established a demonstration of the potential effect of these strategies in other humid regions of southern China.

Author Contributions: All authors were involved in designing and discussing the study. L.Z. (Lu Zhao) performed the calculation, result analysis, and wrote the article; C.L. and Y.F. contributed data collection and result calculation and analysis; N.C. and L.Z. (Lili Zhang) contributed to the article's discussion and edition.

Acknowledgments: We would like to thank the National Climatic Centre of the China Meteorological Administration for providing the climate database used in this study. Thanks are extended to the editors and anonymous reviewers for their valuable comments.

References

1. Salinas, J.G.; Pinto, M.; Pinto, M. Water security as a challenge for the sustainability of La Serena-Coquimbo conurbation in northern Chile: Global perspectives and adaptation. *Mitig. Adapt. Strateg. Glob. Chang.* **2016**, *21*, 235–1246. [CrossRef]
2. Niu, G.; Li, Y.P.; Huang, G.H.; Liu, J. Interactive Fuzzy-Boundary Interval Programming for Water Resources Management of the Hetao Basin, China. *J. Irrig. Drain. Eng.* **2016**, *142*, 04016056. [CrossRef]
3. Piedad, M.; Lin, H.; David, D.; Chen, C.D. The impact of on-farm water saving irrigation techniques on rice productivity and profitability in Zhanghe Irrigation System, Hubei, China. *Paddy Water Environ.* **2004**, *2*, 207–215.
4. Fisher, G.; Tubiello, F.N.; van Velthuizen, H. Climate change impacts on irrigation water requirements: Effects of mitigation, 1990–2080. *Technol. Forecast. Soc.* **2007**, *74*, 1083–1107. [CrossRef]
5. Sun, S.K.; Wang, Y.B.; Liu, J.; Wu, P.T.; Cai, H.J. Sustainability Assessment of Regional Water Resources Under the DPSIR Framework. *J. Hydrol.* **2016**, *532*, 140–148. [CrossRef]
6. Lee, S.H.; Yoo, S.H.; Choi, J.Y.; Engel, B.A. Effects of climate change on paddy water use efficiency with temporal change in the transplanting and growing season in South Korea. *Irrig. Sci.* **2016**, *34*, 443–463. [CrossRef]
7. UNCSD. Comprehensive Assessment of the Freshwater Resources of the World. United Nations Commission on Sustainable Development. 1997. Available online: http://www.un.org/esa/documents/ecosoc/cn17/1997/ecn171997-9.htm/ (accessed on 25 February 2016).
8. Ket, P.; Garré, S.; Oeurng, C.; Hok, L.; Degré, A. Simulation of Crop Growth and Water-Saving Irrigation Scenarios for Lettuce: A Monsoon-Climate Case Study in Kampong Chhnang, Cambodia. *Water* **2018**, *10*, 666. [CrossRef]
9. Cai, X.M.; Rosegrant, M.W. Irrigation technology choices under hydrologic uncertainty: A case study from Maipo River Basin. *Water Resour. Res.* **2004**, *4*, 1–10. [CrossRef]
10. Tal, A. Rethinking the sustainability of Israel's irrigation practices in the Drylands. *Water Res.* **2016**, *90*, 387–394. [CrossRef]
11. Tal, A. Seeking sustainability: Israel's evolving water management strategy. *Science* **2006**, *313*, 1081–1084. [CrossRef]
12. Horst, M.G.; Shamutalov, S.S.; Pereira, L.S.; Gonçalves, J.M. Field assessment of the water saving potential with furrow irrigation in Fergana, Aral Sea basin. *Agric. Water Manag.* **2005**, *77*, 210–231. [CrossRef]
13. Mushtaq, S.; Dawe, D.; Lin, H.; Moya, P. An assessment of the role of ponds in the adoption of water-saving irrigation practices in the Zhanghe Irrigation System, China. *Agric. Water Manag.* **2006**, *83*, 100–110. [CrossRef]
14. Ørum, J.E.; Boesen, M.V.; Jovanovic, Z.; Pedersen, S.M. Farmers' incentives to save water with new irrigation systems and water taxation-A case study of Serbian potato production. *Agric. Water Manag.* **2010**, *98*, 465–471. [CrossRef]
15. Nyakudya, I.W.; Stroosnijder, L. Effect of rooting depth, plant density and planting date on maize (*Zea mays* L.) yield and water use efficiency in semi-arid Zimbabwe: Modelling with Aqua Crop. *Agric. Water Manag.* **2014**, *146*, 280–296. [CrossRef]
16. Li, X.L.; Zhang, X.T.; Niu, J.; Tong, L.; Kang, S.Z.; Du, T.S.; Li, S. Irrigation water productivity is more influenced by agronomic practice factors than by climatic factors in Hexi Corridor, Northwest China. *Sci. Rep.* **2016**, *6*, 37971. [CrossRef]
17. Pérez-Sarmiento, F.; Mirás-Avalos, J.M.; Alcobendas, R.; Alarcón, J.J.; Mounzer, O.; Nicolas, E. Effects of regulated deficit irrigation on physiology, yield and fruit quality in apricot trees under Mediterranean conditions. *Span. J. Agric. Res.* **2016**, *14*, 1–12. [CrossRef]
18. Peng, Y.; Zhang, J.M.; Meng, J.P. Second order force scheme for lattice Boltzmann model of shallow water flows. *J. Hydraul. Res.* **2017**, *55*, 592–597. [CrossRef]
19. Jerry, R.W.; Orlan, H.B.; Gary, J.D. A microcomputer model for irrigation system evaluation. *South. J. Agric. Econ.* **1988**, *7*, 56–69.
20. Anagnostopoulosa, K.P.; Petalasb, C. A fuzzy multicriteria benefit-cost approach for irrigation projects evaluation. *Agric. Water Manag.* **2011**, *98*, 1409–1416. [CrossRef]

21. Adusumilli, N.; Davis, S.; Fromme, D. Economic evaluation of using surge valves in furrow irrigation of row crops in Louisiana: A net present value approach. *Agric. Water Manag.* **2016**, *174*, 61–65. [CrossRef]
22. Xue, J.; Ren, L. Evaluation of crop water productivity under sprinkler irrigation regime using a distributed agro-hydrological model in an irrigation district of China. *Agric. Water Manag.* **2016**, *178*, 350–365. [CrossRef]
23. Zhang, Q.W.; Cui, N.B.; Feng, Y.; Gong, D.Z.; Hu, X.T. Improvement of Makkink model for reference evapotranspiration estimation using temperature data in Northwest China. *J. Hydrol.* **2018**, *566*, 264–273. [CrossRef]
24. Luo, Y.F.; Fu, H.L.; Xiong, Y.J.; Xiang, Z.; Wang, F.; Bugingo, Y.C.; Khan, S.; Cui, Y.L. Effects of water-saving irrigation on weed infestation and diversity in paddy fields in East China. *Paddy Water Environ.* **2017**, *15*, 593–604. [CrossRef]
25. Okada, H.; Styles, S.W.; Grismer, M.E. Application of the analytic hierarchy process to irrigation project improvement Part I. Impacts of irrigation project internal processes on crop yields. *Agric. Water Manag.* **2008**, *95*, 199–204. [CrossRef]
26. Okada, H.; Styles, S.W.; Grismer, M.E. Application of the analytic hierarchy process to irrigation project improvement Part II. How professionals evaluate an irrigation project for its improvement. *Agric. Water Manag.* **2008**, *95*, 205–210. [CrossRef]
27. Tan, C.; Song, Y.; Che, H. Application of set pair analysis method on occupational hazard of coal mining. *Saf. Sci.* **2017**, *92*, 10–16.
28. Mao, N.; Song, M.J.; Deng, S.M. Application of TOPSIS method in evaluating the effects of supply vaneangle of a task/ambient air conditioning system on energy utilization and thermal comfort. *Appl. Energy* **2016**, *180*, 536–545. [CrossRef]
29. Sun, H.Y.; Wang, S.F.; Hao, X.M. An Improved analytic hierarchy process method for the evaluation of agricultural water management in irrigation districts of north China. *Agric. Water Manag.* **2017**, *179*, 324–337. [CrossRef]
30. Abazar, S.; Amir, P.; Ramin, B.; Heidar, Z. Pre-processing data using wavelet transform and PCA based on support vector regression and gene expression programming for river flow simulation. *J. Earth Syst. Sci.* **2017**, *126*, 65. [CrossRef]
31. Loucif, B.; Larbi, H. The effect of simple imputations based on four variants of PCA methods on the quantiles of annual rainfall data. *Environ. Monit. Assess.* **2018**, *190*, 569. [CrossRef]
32. Shieh, M.Y.; Chiou, J.S.; Hu, Y.C.; Wang, K.Y. Applications of PCA and SVM-PSO based real-time face recognition system. *Math. Probl. Eng.* **2014**, *2014*, 1–12. [CrossRef]
33. Zhang, J.; Song, W.; Jiang, B.; Li, M. Measurement of lumber moisture content based on PCA and GS-SVM. *J. For. Res.* **2018**, *29*, 557–564. [CrossRef]
34. Lin, T.K. PCA/SVM-based method for pattern detection in a multisensor system. *Math. Probl. Eng.* **2018**, *2018*, 1–11. [CrossRef]
35. Saaty, T.L. How to make a decision: The analytic hierarchy process. *Eur. J. Oper. Res.* **1990**, *48*, 9–26. [CrossRef]
36. Zhang, H.; Polytechnic, W. Application on the entropy method for determination of weight of evaluating index in fuzzy mathematics for wine quality assessment. *Adv. J. Food Sci. Technol.* **2015**, *7*, 195–198. [CrossRef]
37. Hwang, C.L.; Yoon, K. *Multiple Attribute Decision Making-Method and Applications, A State-of-the-Art Survey*; Springer: New York, NY, USA, 1981.
38. Aghajani, M.; Mostafazadeh-Fard, B.; Navabian, M. Assessing Criteria Affecting Performance of the Sefidroud Irrigation and Drainage Network Using TOPSIS-Entropy Theory. *Irrig. Drain.* **2017**, *66*, 626–635. [CrossRef]
39. Lou, Y.H.; Kang, S.Z.; Cui, N.B. Application of set pair analysis in the comprehensive water-saving irrigation development level evaluation. *J. Sichuan Prov. Univ. (Eng. Sci. Ed.)* **2014**, *46*, 20–27.

PERMISSIONS

All chapters in this book were first published in MDPI; hereby published with permission under the Creative Commons Attribution License or equivalent. Every chapter published in this book has been scrutinized by our experts. Their significance has been extensively debated. The topics covered herein carry significant findings which will fuel the growth of the discipline. They may even be implemented as practical applications or may be referred to as a beginning point for another development.

The contributors of this book come from diverse backgrounds, making this book a truly international effort. This book will bring forth new frontiers with its revolutionizing research information and detailed analysis of the nascent developments around the world.

We would like to thank all the contributing authors for lending their expertise to make the book truly unique. They have played a crucial role in the development of this book. Without their invaluable contributions this book wouldn't have been possible. They have made vital efforts to compile up to date information on the varied aspects of this subject to make this book a valuable addition to the collection of many professionals and students.

This book was conceptualized with the vision of imparting up-to-date information and advanced data in this field. To ensure the same, a matchless editorial board was set up. Every individual on the board went through rigorous rounds of assessment to prove their worth. After which they invested a large part of their time researching and compiling the most relevant data for our readers.

The editorial board has been involved in producing this book since its inception. They have spent rigorous hours researching and exploring the diverse topics which have resulted in the successful publishing of this book. They have passed on their knowledge of decades through this book. To expedite this challenging task, the publisher supported the team at every step. A small team of assistant editors was also appointed to further simplify the editing procedure and attain best results for the readers.

Apart from the editorial board, the designing team has also invested a significant amount of their time in understanding the subject and creating the most relevant covers. They scrutinized every image to scout for the most suitable representation of the subject and create an appropriate cover for the book.

The publishing team has been an ardent support to the editorial, designing and production team. Their endless efforts to recruit the best for this project, has resulted in the accomplishment of this book. They are a veteran in the field of academics and their pool of knowledge is as vast as their experience in printing. Their expertise and guidance has proved useful at every step. Their uncompromising quality standards have made this book an exceptional effort. Their encouragement from time to time has been an inspiration for everyone.

The publisher and the editorial board hope that this book will prove to be a valuable piece of knowledge for researchers, students, practitioners and scholars across the globe.

LIST OF CONTRIBUTORS

Charles R. Ortloff
CFD Consultants International, 18310 Southview Avenue, Los Gatos, CA 95033, USA
Research Associate in Anthropology, University of Chicago, Chicago, IL 60637, USA

Arun Kamath and Hans Bihs
Department of Civil and Environmental Engineering, Norwegian University of Science of Technology, NTNU, 7491 Trondheim, Norway

Gábor Fleit
Department of Hydraulic and Water Resources Engineering, Budapest University of Technology and Economics, Moegyetem rkp. 3, 1111 Budapest, Hungary

Shuai Li, Xiaoqing Chen and Jiangang Chen
Key Laboratory of Mountain Hazards and Earth Surface Processes, Chinese Academy of Sciences, Chengdu 610041, China
Institute of Mountain Hazards and Environment, Chinese Academy of Sciences and Ministry of Water Conservancy, Chengdu 610041, China

Jianmin Zhang
State Key Laboratory of Hydraulics and Mountain River Engineering, Sichuan University, Chengdu 610065, China

Jie Ren, Xiuping Wang, Bo Chen and Lili Men
State Key Laboratory of Eco–hydraulics in Northwest Arid Region of China, Xi'an University of Technology, Xi'an 710048, China

Yinjun Zhou
Changjiang River Scientific Research Institute, Wuhan 430010, China

Jing Zhang, Lingxin Zhang and Jian Deng
State Key Laboratory of Fluid Power and Mechatronic Systems, Zhejiang University, Hangzhou 310027, China
Key Laboratory of Soft Machines and Smart Devices of Zhejiang Province, Zhejiang University, Hangzhou 310027, China
Department of Mechanics, Zhejiang University, Hangzhou 310027, China

Kiyosi Kawanisi and Mohamad Basel Al Sawaf
Department of Civil and Environmental Engineering, Graduate School of Engineering, Hiroshima University, 1-4-1 Kagamiyama, Higashi Hiroshima 739-8527, Japan

Mochammad Meddy Danial
Department of Civil and Environmental Engineering, Graduate School of Engineering, Hiroshima University, 1-4-1 Kagamiyama, Higashi Hiroshima 739-8527, Japan
Faculty of Engineering, Tanjungpura University, Pontianak 78124, Indonesia

Yong Peng
State Key Laboratory of Hydraulics and Mountain River Engineering, Sichuan University, Chengdu 610065, China

Matteo Rubinato
Civil and Structural Engineering Department, University of Sheffield, Sheffield S1 3JD, UK

Fangfang Wang, Ang Gao, Shiqiang Wu, Senlin Zhu and Jiangyu Dai
State Key Laboratory of Hydrology-Water Resources and Hydraulic Engineering, Nanjing Hydraulic Research Institute, Nanjing 210029, China

Qian Liao
Department of Civil Engineering and Mechanics, University of Wisconsin-Milwaukee, Milwaukee, WI 53201, USA

Ruidi Bai, Chang Liu, Bingyang Feng, Shanjun Liu and Faxing Zhang
State Key Laboratory of Hydraulics and Mountain River Engineering, Sichuan University, Chengdu 610065, China

Xin Li, Maolin Zhou, Jianmin Zhang and Weilin Xu
State Key Laboratory of Hydraulics and Mountain River Engineering, Sichuan University, Chengdu 610065, China

Zhaojia Tang, Liming Yu, Fenghua Wang and Na Li
Faculty of Modern Agriculture Engineering, Kunming University of Science and Technology, Kunming 650500, China

Liuhong Chang
School of Hydraulic Engineering, Changsha University of Science and Technology, Changsha 410114, China

Xuneng Tong, Xiaodong Liu, Ting Yang, Zulin Hua, Zian Wang, Jingjing Liu and Ruoshui Li
Key Laboratory of Integrated Regulation and Resource Development on Shallow Lake of Ministry of Education, 1 Xikang Road, Nanjing 210098, China

Chuan Liang and Yi Feng
State Key Laboratory of Hydraulics and Mountain River Engineering & College of Water Resource and Hydropower, Sichuan University, Chengdu 610025, China

Lu Zhao
State Key Laboratory of Hydraulics and Mountain River Engineering & College of Water Resource and Hydropower, Sichuan University, Chengdu 610025, China
Provincial Key Laboratory of Water-Saving Agriculture in Hill Areas of Southern China, Chengdu 610066, China

Lili Zhang
International School of Technical Education, Sichuan College of Architectural Technology, Deyang 618000, China

Ningbo Cui
State Key Laboratory of Hydraulics and Mountain River Engineering & College of Water Resource and Hydropower, Sichuan University, Chengdu 610025, China
Key Laboratory of Agricultural Soil and Water Engineering in Arid and Semiarid Areas, Ministry of Education, Northwest A & F University, Yangling 712100, China

Index

S

T

V

W

www.ingramcontent.com/pod-product-compliance
Lightning Source LLC
LaVergne TN
LVHW080949210926
848485LV00004B/807
9781647404260